ALLE ZEIT WACH 1842

Evgenii A. Romankevich

Geochemistry of Organic Matter in the Ocean

With 68 Figures and 72 Tables

Springer-Verlag
Berlin Heidelberg NewYork Tokyo 1984

Dr. Evgenii A. Romankevich
P. P. Shirshov Institute of Oceanology
Academy of Science
23, Krasikov
Moscow, 117218, USSR

Title of the original Russian edition:
Geokhimiia organicheskogo veshchestva v okeane

ISBN 978-3-642-49966-1 ISBN 978-3-642-49964-7 (eBook)
DOI 10.1007/978-3-642-49964-7

Library of Congress Cataloging in Publication Data. Romankevich, Evgeniĭ Aleksandrovich. Geochemistry of organic matter in the ocean. Translation of: Geokhimiia organicheskogo veshchestva v okeane. 1. Chemical oceanography. 2. Organic geochemistry. 3. Biogeochemistry. I. Title. GC116.R6613. 1984. 551.46. 84-5596

Softcover reprint of the hardcover 1st edition 1984

Typesetting: K + V Fotosatz GmbH, Beerfelden

2132/3130-543210

Contents

Author's Preface to the English Edition

My work *Geochemistry of organic matter in the ocean* first appeared in Russian in 1978. Since then much progress has been made in the exploration of various forms of organic matter in the ocean: dissolved, colloidal, organic matter suspended in particles and that contained in bottom sediments and in interstitial waters. The appropriate evidence is found in hundreds of articles and several review works, such as Andersen (1977), *Biogéochimie de la matière organique à l'interface eau-sédiment marine* (1980), Duursma and Dawson (1981).

A great amount of new information has been obtained in the Soviet Union's scientific institutions on the composition and distribution in natural waters and bottom sediments of organic matter and its separate components playing a crucial role in the formation of the chemical and biological structure of the ocean and its productivity, in the biogeochemistry of the elements and geochemistry of organic matter in the Earth's sedimentary cover.

The areas of exploration have expanded over the past four-and-a-half years to embrace many new, little-known regions, including the Arctic seas. In contrast to the three preceeding decades, the research has been focused on investigating the existing forms, the distribution and accumulation of organic matter in near-continental oceanic zones between land and sea, and in river estuaries.

This research was based, as a rule, on the use of advanced working procedures and sampling techniques. Great care was attached to preventing the pollution of samples in the process of sample taking and to obtaining samples that would give a more objective characteristic of the matter under study (500-litre water bottles, hermetic samples of water and bottom sediments, specific filtration regimes and the conditions of separation of interstitial water, etc.).

The new data have compelled research workers in some cases to revise the earlier information obtained by certain marine expeditions and to discard it as unreliable.

The author has attempted to re-work the monograph to include the new information available to him. I am glad to be able to use this opportunity to express my sincere gratitude to those Soviet and foreign colleagues, G. Arrhenius, E. T. Degens, T. V. Drozdova, K. O. Emery, E. M. Galimov, R. Ishiwatary, T. Koyama, A. Lerman, F. T. Manheim, S. Nishizawa, A. Nissenbaum, T. R. Parsons, J. P. Riley, R. W. Stewart, N. M. Strakhov, N. B. Vassoyevich, V. V. Weber and P. M. Williams, who have sent their comments and proposals in connection with the appearance of the Russian edition of the book and have thus contributed to updating its English version, as well as to V. B. Kouznetsov who was responsible for the preparation of the translation for the revised English edition. I also owe a great deal to my assistants L. N. Krayushkina and S. L. Veshner, who did the graphic work.

Author's Preface to the Russian Edition

Today works on various problems connected with the distribution, composition, dynamics and balance of different forms of organic matter (OM) in the seas and oceans total a few thousand. Neither a comprehensive treatment nor a critical analysis of these results or of the methods employed to obtain them are the purpose of this book. Its aim is different. It sums up certain results of investigations on the distribution, composition, transformation, balance and biogeochemical cycling of organic matter in the seas and oceans at the stages of sedimentogenesis and early diagenesis. In treating these problems, the author has considered it necessary to concentrate his efforts on: estimating the degree of general balance between organic matter production, its supply to the seabed and its burial; throwing light on the patterns of distribution of different forms of organic matter (dissolved, suspended, contained in bottom sediments and in interstitial water), or absolute masses and total organic matter in the ocean as a whole and in the various structural-geomorphologic zones, such as the submarine margin, marginal and central ocean-floor areas; describing the interrelationships between organic matter and certain environmental factors and, finally, revealing the nature of the transformation of organic matter in the system plankton – suspended material – bottom sediments. The motives for writing this book were the synthesis of the results of the author's research on the problem over the last 20 years, as well as the absence of a similar review work in either domestic or foreign literature.

In the world ocean the quantity of dissolved organic matter greatly exceeds that of suspended or particulate organic matter. Suspended material is, however, practically the only form in which organic matter is transferred into the bottom sediments. Dissolved organic matter, the supply and loss of which in the ocean are generally balanced, is converted into a suspended state when utilized by the organisms as a structural material, as well as through sorption on particles and by colloidal coagulation. This work therefore devotes much attention to organic matter contained in suspended material and bottom sediments. One part of this, and usually a small one, forms the most dynamic and short-living portion of all the organic matter – namely the living material. Very many published results on the distribution and composition of producers and consumers of organic matter in the ocean are cited because they are relevant to understanding the processes of sedimentogenesis and the general cycle of organic matter and because they permit comparisons between the composition of suspended material and bottom sediments.

The book reviews in much detail the results obtained over the past twenty five years at the P. P. Shirshov Institute of Oceanology, the USSR Academy of Sciences, relating to quantitative determinations of the content in sea and ocean waters, in suspended material, bottom sediments and interstitial waters of carbon and nitrogen, or organic matter, phosphorus, carbohydrates and of their group and monosaccharide composition, of lipids and their composition, and of amino acids. The suspended material and bottom sediments were also analyzed to determine the content of proteins, humic acids, humins and ATP. Optical methods proved useful in studying the morphology of both particulate organic matter and bottom sediments. Use was made of the results obtained in studying the origin of the components of the sediments, absolute datings and biostratigraphy.

The data on organic matter were obtained during more than forty cruises of the research vessels "Vityaz", "Ob", "Academik Kurchatov", "Dmitrii Mendeleyev". Observations were made in different areas in the Pacific, Indian and Atlantic oceans within the depth range from 30 m to 10710 m. Since 1955 the author has been involved in eleven cruises of the "Vityaz", "Academik Kurchatov", "Dmitrii Mendeleyev" studying organic matter composition of suspended material, sea water, bottom sediments, and the lithological composition of the latter.

Research into the geochemistry of organic matter was begun by the author in 1954 under the guidance of corresponding member of the USSR Academy of Sciences P. L. Bezrukov and Professor V. N. Florovskaya. For his interest in the problem and the development of relevant ideas the author is greatly indebted to advice, personal contacts and the works of V. G. Bogorov, S. V. Bruyevich, N. B. Vassoyevich, V. V. Veber, L. A. Zenkevich, A. P. Lisitsin, N. M. Strakhov. In his research activities the author has also enjoyed consultation with and advice from A. N. Guseva, T. V. Drozdova, A. P. Zhuze, A. A. Ilyina, S. M. Manskaya, I. N. Mitskevich, E. A. Ostroumov, K. F. Rodionova, A. D. Semeonov, M. N. Sokolova, and Z. A. Filatova.

The bulk of the work in sampling and studying the organic matter composition in both sea water and suspended material, as well as developing analytical methods was done at the Institute of Oceanology by the following scientific workers: V. E. Artemiev, A. N. Beliaeva, L. I. Korzhikova, L. N. Krayushkina, S. V. Liutsarev, I. A. Melnikov, N. D. Starikova, I. M. Urbanovich, O. G. Yablokova, M. V. Baturina, with all of whom the author has both worked closely and discussed the results.

Most analyses of bottom sediments for C_{org} content were made at the USSR Academy of Sciences' Institute of Oceanology using a standard procedure and under the direction of E. A. Ostroumov and M. P. Nesterova; at the Atlantic and Southern Affiliates of the Institute this work was directed by E. M. Emelyanov and V. S. Sokolov. Contributors to the study of the organic matter composition of bottom sediments have been V. V. Baranova and M. N. Yestafyeva, members of the Chair of Geology and Geochemistry of Combustible Minerals in the Geological Faculty of the Moscow State University; S. G. Batrakov from the Lipid Chemistry Laboratory, USSR Academy of Sciences' Institute of the Chemistry of Natural Compounds; S. S. Shanin from the Gamaleya Institute of Epidemiology and Microbiology, USSR Academy of Medical Sciences; T. N. Dyuzhikova and others from the All-Union Geological Prospecting Petroleum Institute.

Radiochemical researches were made under L. A. Khristianova's leadership at the USSR Academy of Sciences' Vernadsky Institute of Geochemistry and Analytical Chemistry; biostratigraphic studies of the sediments were supervised by A. P. Zhuze and Kh. M. Saïdova at the Institute of Oceanology. On the basis of these and other investigations, we calculated the age of sediments and the absolute masses of organic matter. Other constituents making up the composition of the sediments were determined under the guidance of V. P. Petelin, E. A. Ostroumov and M. P. Nesterova at the geological and chemical departments and the analytical laboratory of the Institute of Oceanology. Some results obtained in studying the organic matter in suspended material and bottom sediments and in investigating phytoplankton and phytobenthos productivity were kindly supplied by Yu. A. Bogdanov, V. B. Vozzhinskaya, A. I. Gorskaya, E. M. Emelyanov, O. I. Koblents-Mishke, O. B. Maksimov, Yu. I. Sorokin, and Z. A. Filatova.

The author would like to express sincere thanks to all these persons for their generous assistance with materials, invaluable help during the research stage and useful advice in preparing the manuscript for the press.

E. A. ROMANKEVICH

Editor's Preface

A remarkable trend in modern times is the mounting interest in the world ocean. Its investigation and comprehensive reclamation hold great promise for mankind. Studies of the ocean from different angles gather in momentum from year to year. The ocean is increasingly a focus of attention on account of its vast biological resources. Hydrobiologists are examining these resources and are now drawing up plans to secure maximum and rational utilization of marine flora and fauna, the ultimate aim being to ensure a change-over from collecting and hunting practices to *"mariculture"* (just as on land the transition to *agriculture* has been practically universal). The geologists are out to put the immense mineral wealth hidden in oceanic waters, in oceanic sediments, as well as beneath the ocean floor to economic use. The significance of the world ocean, both economic and political, is today enormous.

Oceanology, a complex science formed by the fusion of certain departments of physics, chemistry, geology, biology and geography, is making rapid progress at present. Having sprung up as an integral science and on account of the exceptional complexity and multifaceted nature of its subject matter, oceanology has split up into separate disciplines. This is not unusual, the rapid progress of science inevitably involving "cross-pollination" of different sciences, their "intraspecific" and remote "hybridization" (as an illustration of the latter we would like to point to molecular paleontology). Integration and differentiation go hand in hand, as they always do. Second- and third-generation sciences are no exceptions to this rule. One such integrational science is biogeochemistry, encompassing those departments of biology, geology and chemistry that contribute to the investigation of the various functions of the biosphere, particularly the hydrobiosphere.

The principal role in the biogeochemical processes occurring both in the ocean and in oceanic sediments is played by organisms and by carbonic (D. I. Mendeleyev favoured the term 'hydrocarbonic') organic matter (OM). A. P. Vinogradov (1964) rightly noted the difficulty of "separating the geochemical role played by living matter from that of organic substances formed after the death of the organisms". In fact, the enquiry into the role of both these aspects is a highly important task of biogeochemistry in its broadest interpretation.

This monograph is essentially devoted to *ocean biogeochemistry*. It is, in effect, a new landmark in the development of this science.

Some people consider ocean biogeochemistry to be an off-spring of oceanology and organic geochemistry, or it can be regarded as an important department of the theory of the biosphere elaborated by V. I. Vernadski (1967, 1980), or as part of ocean geochemistry initiated by A. P. Vinogradov (1964), or, finally, as a department of lithology which is greatly indebted for its progress to the Soviet researchers N. M. Strakhov, P. L. Bezrukov, A. P. Lisitsin, et al.

It appears to be more correct, however, to regard ocean biogeochemistry, which is indisputably of great scientific and practical significance, as an independent field of knowledge with its own objects and methods or research, its problems and goals.

E. A. Romankevich's very comprehensive study is devoted exclusively to oceanic organic matter. Whatever is known today about it to oceanologists has been included in the book. On the whole the author has been successful in surmounting the many difficulties stemming from inadequately elaborated biochemical and organochemical classifications (the former pertaining to organisms and the latter to their decay products and newly formed chemical compounds) and from the complexity and confusion in the respective terminology.

The ocean, in fact, represents a colossal system of dynamic equilibrium between the saline solution, the liquid and solid inputs from land, the atmosphere, the ocean-inhabiting organisms, dissolved and suspended matter and bottom sediments. The relationships between these masses of organic matter in the ocean are highly significant although not universally recognized. In comparison to the total C_{org} contained in the ocean living matter (2.8×10^9 tons), the quantity of C_{org} contained in suspended organic matter is ten times as great and that contained in dissolved organic matter 650 times as great (180×10^{10} tons). Such a relationship proved quite unexpected to a great many people.

In the exceptionally complex system which the ocean represents, second- and third-order systems occur – the dynamic equilibrium between the water and the sediments, between the water-dissolved gases and the atmosphere, etc.

In this monograph the author deals with all such systems – those in which either the living or dead carboniferous organic matter plays a more or less prominent part. The cycle of organic matter in the ocean is discussed at length and biological, physico-geographical and lithologic-geochemical factors are assessed in their effect on this many-sided process. A detailed, step-by-step consideration covers all stages in the complex cycle of organic matter beginning with those living organisms – the producers that are at its base, through the trophic chains including consumers and reducers which ensure the decay of dead plant and animal remains and continues as far as the burial and diagenetic transformation of organic matter in the bottom sediments.

Before the author embarked on his research it had been generally assumed that the influence of the land on the input of organic matter to the ocean was rather insignificant. E. A. Romankevich was the first to show that the most important allochthonous sources of organic matter supply to the ocean are not only rivers but also eolian transfer. An adequate quantitative basis has been laid for the concept of circumcontinental distribution of living, suspended and precipitated organic matter in the ocean and for the notion of the inherited character of the accumulation of absolute masses of organic matter in the bottom sediments. Wherever possible, measurable and numerical quantities have been introduced. The aspects of balance in the processes of organic matter transformation are always in the focus of the author's attention.

The latest references were consulted in assessing phytoplankton and phytobenthos production, the supply of allochthonous organic matter to the ocean with river inflow and eolian transfer including cosmogenic matter and technogenous pollutions.

The author has made it his task to trace the fate of all the main classes of biochemical compounds and of the fragments of new organic compounds originating from them, beginning with living matter (plankton) and ending with organic matter compounds in the mass of sediments at a depth of several decimetres and metres. He has coped successfully with the laborious task of processing extensive data including both the results of his own research and those from the abundant literature. An important aspect of this diachronic study is that wherever possible the analysis deals with narrow fractions and monomers.

The established quantitative patterns in the variability of organic matter composition today permit us to conjecture much more fully than was formerly possible, the highly complex transformations of original biopolymers, their partial decay and the conversion of monomers into geopolymers adapted for existence under the conditions of the geochemical sedimentary facies.

A typical process of the organic matter compositional variability at the sedimentogenesis and diagenesis stage is shown to be humification, involving a decline in the content of all classes of biochemical compounds accompanied by an increase in specific heteropolycondensed structures (humic acids, fulvic acids and humins). They are formed after the melanoidin (amykagine) reaction and are classed under so-called sapropellic acids.

The two zones of most active biogeochemical transformation or organic matter in the ocean, the photic layer and the surficial sedimentary layer, have been adequately characterized by the author.

From the lithogenetic (sedimentologic) viewpoint we are bound to give much credit to the author's biogeochemical research in the ocean. His results in fact constitute a new contribution to the theory of marine and ocean sedimentation and to lithology generally, a very important aspect of the geological sciences, providing researchers with methods of obtaining adequate and allround information on recent and fossil sediments, and thus facilitating an insight into the geological history of those regions of the world which are made up of sedimentary rocks.

The characterization of E. A. Romankevich's work would be incomplete if we disregard the significance of his biogeochemical researches for the evaluation of the oil- and gas-forming potentialities in different parts of the immense area covered by the world ocean.

Off-shore oil production has lately accounted for more than 20% of the world total. We have every reason to believe that in the near future the share of offshore oil from beneath the ocean floor will increase still further. The scientifically valid, efficient explorations of new oil and hydrocarbon gas deposits are based at present on the sediment-migrational theory of hydrocarbon formation. The historical-genetic geological-geochemical exploration technique ensuing from this theory necessitates adequate information on organic matter contained in the sediments and being parental to oil and hydrocarbon gases. The new systematized data are essential contributions to our knowledge of sedimentary organic matter, such data being indispensible for correct, though often tentative estimates of the oil- and gas-bearing prospects of different areas in the world ocean.

As is known, there are different views on the oil- and gas-bearing prospects of deep oceanic areas. The data contained in Romankevich's monograph show that up to 90% of total organic matter is being deposited in submarine continental margins, including sediments of continental slopes. The sediments deposited in the abyssal depths of the immense ocean floor are, however, very deficient in organic matter and we have no reasons to expect major oil finds associated with them.

Oil and hydrocarbon gases are not, however, the only mineral wealth associated with the ocean. P. L. Bezrukov rightly referred to the ocean as the colossal and practically untapped storage place of different minerals (Bezrukov 1971, Bezrukov et al. 1970). The comprehensive work of E. A. Romankevich will to a certain extent facilitate the search for some of them. The book contains immense, meticulously verified, properly systematized and up-to-date data on all the aspects of the ocean as the principal element of the biohydrosphere with its biogeochemical functions of which the most important are associated with living matter and with organic matter formed after the death of the organisms.

The monograph *Geochemistry of Organic Matter in the Ocean* will certainly be helpful to a wide circle of readers interested in the hydrobiology, physicochemistry and organic chemistry of the ocean as well as in the sedimentogenesis and early processes of diagenesis occurring there. Numerous oceanologists, sedimentologists, broadly specialized lithologists, geologists, petroleum geochemists and many other specialists will find in this book up-to-date information on many important aspects of ocean biogeochemistry.

N. B. VASSOYEVICH

Introduction

A specific feature of current scientific progress is a sharp increase of our knowledge about the world ocean, associated in many respects with a desire to put to rational use its nutritive, mineral and power resources. The avalanche-like growth of information about various processes in the ocean was the result of new opportunities for its study, using up-to-date research ships, earth satellites and submersible vessels, and of the progress achieved in research techniques and in quantitative methods of studying the oceanosphere and the hydrosphere as a whole.

Of great importance for understanding the potential of and limits to development of the hydrosphere are the combining of different research methods and the integration of information obtained with their aid. The large amount of information which can be gained from studying the components of organic matter at the molecular and atomic levels should in particular be noted.

The fact that organic matter is a highly potent factor in the evolution of the biosphere has been universally recognized. The eminent natural scientist Vernadsky was among the first to realize the outstanding role of living matter and of the products of the vital activities of organisms, above all organic matter, in the biosphere, particularly in the seas and oceans, and made the relevant quantitative calculations (1921, 1934, 1965). Following his research and the work of his collaborators and followers, both in this and in other countries, it became clear that the study of the factors governing distribution, transformation and cycling of organic matter and its accumulation in sediments is of considerable practical and theoretical value, particularly for the progress of lithology and for understanding the formation patterns of deposits of numerous useful minerals, above all oil.

Today, off-shore oil accounts for over 25% of the total oil output. By 1985 this share will in all probability have increased to 40%. This trend has been one of the factors which have sparked off interest in the fate of organic matter in the oceans and seas. Geochemical studies of organic matter are usually included in complex oceanological expeditions which are regularly sponsored in different countries.

Research into the geochemistry of different forms of organic compounds has lately become even more important due to increasing contamination of the ocean with industrial wastes. Owing to annual discharges into the oceans and seas of several million tons of technogenic organic compounds, the cycle of carbon and its derivatives has deteriorated compared with what it used to be, with man himself increasingly suffering from the disturbed natural equilibria. Thus it has now become crucial to determine the exact concentrations in the seas and oceans of the different classes of organic matter, their residence time and the manner in which they can be extracted from the natural cycle (Vinogradov 1973); The Global Carbon Cycle (1977).

The main reason for the continuously rising interest of a wide section of the scientific community in the transformation, cycling, and fate of the various forms of organic matter in the oceans, which is the subject matter of the rapidly developing sciences – organic geochemistry and biogeochemistry of the ocean – is thus due to the fact that they are interrelated with the pressing problems of today's world: a threat of depleting oil and gas resources on dry land, explorations for oil and gas in the ocean, protection against industrial pollution, rational utilization of biological and mineral resources of the ocean.

The studies of organic matter contained in sea water and bottom sediments were initiated by the works of Verigo (1881 – 1888) and Natterer (1892 – 1894, cit. Duursma 1965). In subsequent years the study of organic matter in the seas and oceans was conducted along various lines in accordance with the tasks facing lithology, ecology, biochemistry of marine organisms, and hydrochemistry.

In the lithologic-geochemical area of research, starting with the work of Arkhangelsky (1927a, b) and Trask (1932), the main attention was given to the identification of the composition and fate of organic matter in bottom sediments, of its role in diagenesis and in the accumulation of the material parental to oil and gas. Organic matter of sea water (dissolved and particulate) was usually exempt from investigation.

Biohydrochemical research, starting with the works of Pütter (1909), Krogh (1931) and Skopintsev (1950), concentrated largely on dissolved organic matter and its effect on the trophic (rarely sensory) functions of the organisms and/or on the hydrochemical system of the sea water. At first the particulate matter was little studied, but later, when research workers became fully aware of the exceptional importance of particulate organic matter, it was given greater attention in the investigations related to diversified aspects of lithology, geochemistry and trophodynamics.

In the present work, in considering the geochemistry and overall cycle of organic matter in the ocean, we take account of the summarized flows of substances: organisms ⇄ dissolved organic matter ⇄ particulate organic matter ⇄ precipitated organic matter. All these forms of organic matter are interrelated and make up a single system maintaining a dynamic equilibrium. Its evolution is proceeding jointly with the mineral carriers of organic matter (suspended material, bottom sediments, water). Thus a study of the balance and cycling of organic matter cannot be undertaken unless we consider jointly all the components of the system.

Abbreviations

OM	– organic matter
C_{org}	– carbon of organic matter
DOM	– dissolved organic matter
DOC or C-DOM	– dissolved organic carbon
POM	– particulate organic matter
POC or C-POM	– particulate organic carbon
N-DOM	– nitrogen of dissolved organic matter
N-POM	– nitrogen of particulate organic matter
N_{org}	– nitrogen of organic matter
P_{org}	– phosphorus of organic matter
P-DOM	– phosphorus of dissolved organic matter
P-POM	– phosphorus of particulate organic matter
ATP	– adenosinetriphosphate
ASS	– alcohol-soluble saccharides
WSP	– water-soluble polysaccharides
WISP	– water-insoluble polysaccharides
DL	– lipids, dissolved and colloidal
PL	– particulate lipids
HA	– humic acids
FA	– fulvic acids
ROM	– residual organic matter

1. Sources of Organic Matter in the Ocean

Organic compounds, which are present in every drop of ocean water, in every particle of suspended material, have either been first created in the ocean by photosynthesizing plants (autochthonous organic matter) or reached the ocean from land (allochthonous organic matter). Undoubtedly the allochthonous component had also once been initially created in the process of photosynthesis, but thereafter its progress was tortuous and often long, involving an incomplete utilization in trophic chains, deposition, consolidation and "maturing" until it arrived in the ocean. Genetically it is related to the land phytomass, the soil humus and organic matter in the old rocks. Living organic matter being supplied as part of some or other source into the ocean from dry land, for instance with river waters, constitutes an insignificant portion of total allochthonous organic matter. It can only serve as an indicator of fresh-water conditions (Rubenchik 1972).

The larger part of allochthonous organic matter exhibits a much greater resistance to decay than does the bulk of labile autochthonous organic matter. The latter as a result of strenuous nutritive relations is almost completely utilized and undergoes restructuring into the animal and bacterial organic matter which is rich in unstable proteins. Therefore, the estimates of annual contributions of the various sources (not phytoplankton alone) to the total amount of organic matter in the world ocean (Table 1) can obviously be very interesting.

1.1 Primary Production of Organic Matter

The chief producer of organic matter in the ocean is the phytoplankton. The production due to phytobenthos is much less. The contribution of bacterial primary chemo- and photosynthesis has been little studied, but on the whole it appears to be small. The rest of the organisms – zooplankton, zoobenthos, protozoans, saprophitic bacteria, as well as chemoorganotrophic microorganisms – are not primary producers, since they are carrying on synthesis on the basis of organic matter formed earlier. The only criterion for classifying organic matter as either primary or secondary is the source of the energy spent on its formation (Vinberg 1960, Kuznetsov 1970). The carbon source and the oxidizing substrate are of no importance in this case. Primary oceanic production is, therefore, assumed to include all kinds of organic matter formed by phototrophic (photosynthesizing) organisms (plankton and benthonic algae, phanerogams, mangrove vegetation, phototrophic bacteria – both photolithotrophic and photoorganotrophic), as well as some part of the organic matter formed by chemotrophic (chemolithotrophic) bacteria. Thus, by far the greater part of organic matter formed by auto-

Table 1. Annual generation and supply of organic carbon of the world ocean[a]

Source of C_{org}	10^6 t C_{org} yr^{-1}	g C_{org} m^{-2} yr^{-1}	C_{org} of total sum (%)	C_{org} in original matter (%)	Supply of original matter
Phytoplankton, net production	25,000	69.19	93.94	31[b]	$110 \cdot 10^9$t[b]
Phytobenthos	600	1.66	2.25	37.5[b]	$1.6 \cdot 10^9$t[b]
River water discharge	212	0.587	0.80	5 mg $\cdot l^{-1}$	42,390 km^3
Glacial runoff (the Antarctic alone)	2	0.006	<0.01	0.5 – 1.0 mg $\cdot l^{-1}$	2310 km^3
Subterranean runoff	66.4	0.184	0.25	27 mg $\cdot l^{-1}$	2200 km^3
Solid river discharge	392	1.085	1.47	2.5[b]	$15.7 \cdot 10^9$t[b]
Eolian transfer	320	0.886	1.20	20.0[b]	$1.6 \cdot 10^9$t[b]
Solid glacial runoff	1.5	0.004	<0.01	0.1[b]	$1.5 \cdot 10^9$t[b]
Wave abrasion material	2.0	0.006	<0.01	0.4[b]	$0.5 \cdot 10^9$t[b]
Material from submarine volcanoes (without dissolved substances)	10	0.028	0.04	0.2 – 0.3[b]	$(2-3) \cdot 10^9$t[b]
Anthropogenic pollution[c]	5	0.014	0.02	80	$6 \cdot 10^6$t
Cosmogenic material	<0.004	<0.001	<0.01	0.04	$<10 \cdot 10^6$t
Total	26,611	73.65	100	–	–
Including					
autochthonous	25,600	70.85	96.2	–	–
allochthonous	1,011	2.80	3.8	–	–

[a] This table was compiled from literary data and the author's materials. The area of the ocean is assumed to equal 361.3×10^6 m^2. (*The World Water Balance and the Earth's Water Resources*, 1974)
[b] Of dry mass
[c] Pollutions not included in other sources

trophic organisms (though not all of it) belongs to primary production. Secondarily produced organic matter comprises all organic compounds which were formed using the oxidation energy of organic substances regardless of the manner of the appearance of the latter. The organic matter formed by heterotrophs in the case of dark fixation of CO_2 is mostly secondary, having been obtained from the disintegration energy of organic compounds which were taken into account in calculating the primary production.

1.1.1 Phytoplankton

In the case of all photosynthesizing organisms the photosynthesis reactions induced by light and catalyzed by the pigment system are expressed by the following equation:

$$CO_2 + 2AH_2 \xrightarrow[\text{chlorophyll}]{\text{light}} 2A + (CH_2O) + H_2O,$$

where AH_2 is the hydrogen carrier.

Photosynthesis in sea water can proceed if radiation exceeds 0.18 cal $cm^{-2} h^{-1}$. At this energy level phytoplankton is capable of effecting the photolysis of water, of generating ATP from mineral phosphates and reducing the maximally oxidized carbon source (CO_2) to carbohydrates. Photosynthesis in the ocean therefore takes place in a relatively thin topmost layer of 2 – 250 m (average 40 m) thickness, within the limits of which the light intensity is not less than 400 lx (optimum 1000 – 4000 lx). Beneath the compensation depth the expenditure of energy in respiration exceeds the assimilation potentialities of the phytoplankton and it continues to exist for a certain period of time by utilizing previously accumulated organic matter and afterwards perishes (Moïseyev 1969, Parsons et al. 1977, Vinogradov and Lisitzin 1981). In the uppermost layers of the photic zone the optimum illumination for phytoplankton photosynthesis generally amounts to 1.8 – 9 cal $cm^{-2} h^{-1}$ (Koblents-Mishke and Vedernikov 1977).

The quantity of phytoplankton in the ocean is directly dependent on the concentration of biogenic elements in the trophogenic layer and at the lower boundary of the pycnocline (a layer of maximum density gradient). Therefore, wherever the upwelling waters, turbulent diffusion or river inflow secure the supply of nutrient salts to the photic zone the quantity of phytoplankton is observed to be high. Apart from these two main factors (light and biogens), the quantity of phytoplankton is affected by the depth of the main pycnocline (inverse correlation), the stratification stability of the superlying layers, and the extent of phytoplankton grazing by zooplankton. The combination of these factors determines the pattern of phytoplankton distribution in the ocean (Semina 1974, 1977, 1982).

However, direct cause-and-effect relationships between principal environmental factors and bioproductivity on the regional scale tend to become complicated by other influences and their combinations (for instance, a variation of baric fields, weather conditions, river runoff, ocean floor surface relief, anthropogenic effects, etc.) which may even produce an inverse pattern. These factors are the most influential in the highly active boundary oceanic zones: ocean – atmosphere, sea – land, water – bottom sediments (Aizatullin et al. 1979, Monin and Romankevich 1979, Gershanovich and Muromtsev 1982).

The main contribution to primary production of OM in the ocean is made by nannoplankton (less than 60 μm, Nienhuis 1981). Diatom algae synthesize more than 50% of the total organic matter in the ocean. Quite a significant role can be attributed to peridinians and coccolithines and, in some near-coastal areas and in the equatorial-tropical zone, to blue-green algae. Diatoms in the ocean are sharply predominant in the subarctic and antarctic areas, as well as in the upwelling zones at the western coasts of the continents. For instance, in the South West Africa region (5° – 23° S) diatoms, peridinians and blue-greens account, respectively, for 89.3%, 0.6%, and 9.7% of the total quantity of algae and for 91.7%, 7.2%, and 0.7% of summary biomass. In the East China Sea tropical plankton, diatoms are predominant at all stations but particularly in the Sea's autotrophically most productive areas. In central parts of the Pacific, Atlantic and Indian Oceans the areas of diatom predominance alternate with the zones where peridinians and coccolithines prevail (Semina 1974, 1977, 1982).

The magnitude and distribution of phytoplankton production are the cardinal characteristics of the production-destruction cycle of organic matter in the ocean.

Research workers give different estimates of phytoplankton production in the world ocean (in 10^9 tons C_{org} per year): Riley (1944) – 126, Steeman-Nielsen and Jensen (1957) – 20 – 25, Ryther (1969) – 20, Koblents-Mishke et al. (1970) – 23, Platt and Subba Rao (1975) – 31, Koblents-Mishke and Vedernikov (1977) – 25 – 30, Sorokin (1978) – 76, Woodwell et al. (1978) – 25, De Vooys (1977) – 44.

At present the most representative estimate of annual phytoplankton production in the world ocean is that obtained by Koblents-Mishke (1977): $25-30 \times 10^9$ tons C_{org} per year (gross production) and $20-25 \times 10^9$ tons C_{org} per year (net production). This estimate was made after a convincing analysis of the assumptions and errors inherent in different methods (Koblents-Mishke and Vedernikov 1977) and the summing up of more than eight thousand determinations of phytoplankton production. As a matter of fact, it is the net production (the difference between the total production of phytoplankton and expenditure of organic matter for its breathing) which is of greatest interest for our purposes. The net production characterizes the quantity of organic matter arriving at the first heterotrophic level and contributing to the formation of particulate and dissolved organic matter.

For the net phytoplankton production of 25×10^9 tons C_{org} per year, as assumed in the present work, in each square metre of the ocean's photic layer as much as 69 gC are incorporated into the cycle every year. As regards the average net productivity ($gC_{org}\,m^{-2}$ per year), the oceans can be arranged in the following ascending series: Pacific – about 56, Atlantic – about 83, Indian – about 97.

The distribution of phytoplankton production in the ocean (Fig. 1) on the whole follows that of the biomass and of the quantity of planktonic algae and is determined by the conditions of mineral nutrition and the amount of available luminous energy (Koblentz-Mishke and Vedernikov 1977). The latter factor exerts a limiting effect only in the near-polar regions where ice obstructs the penetration of light and the vegetation period is short.

In accordance with the zonal distribution of heat on the Earth and the character of circulation of the world ocean waters, primary production in the ocean is likewise subject to a zonal distribution. Its principal and most salient feature is the circumcontinental character of localization. This is manifested in a significant increase of phytoplankton production in transition from the oceanic waters to the neritic and coastal waters. In this respect the Antarctic is no exception, though the increase of production here is less significant compared to that near the other continents, apparently on account of low temperatures and the rapid submergence of the upper water layers.

The circumcontinental zonality of phytoplankton production in the ocean was not specifically noted. Sverdrup (1955) was the first to note, and later, on the basis of abundant factual material, specifically to underline the effect of the continents on the distribution of phytoplankton production in the ocean.

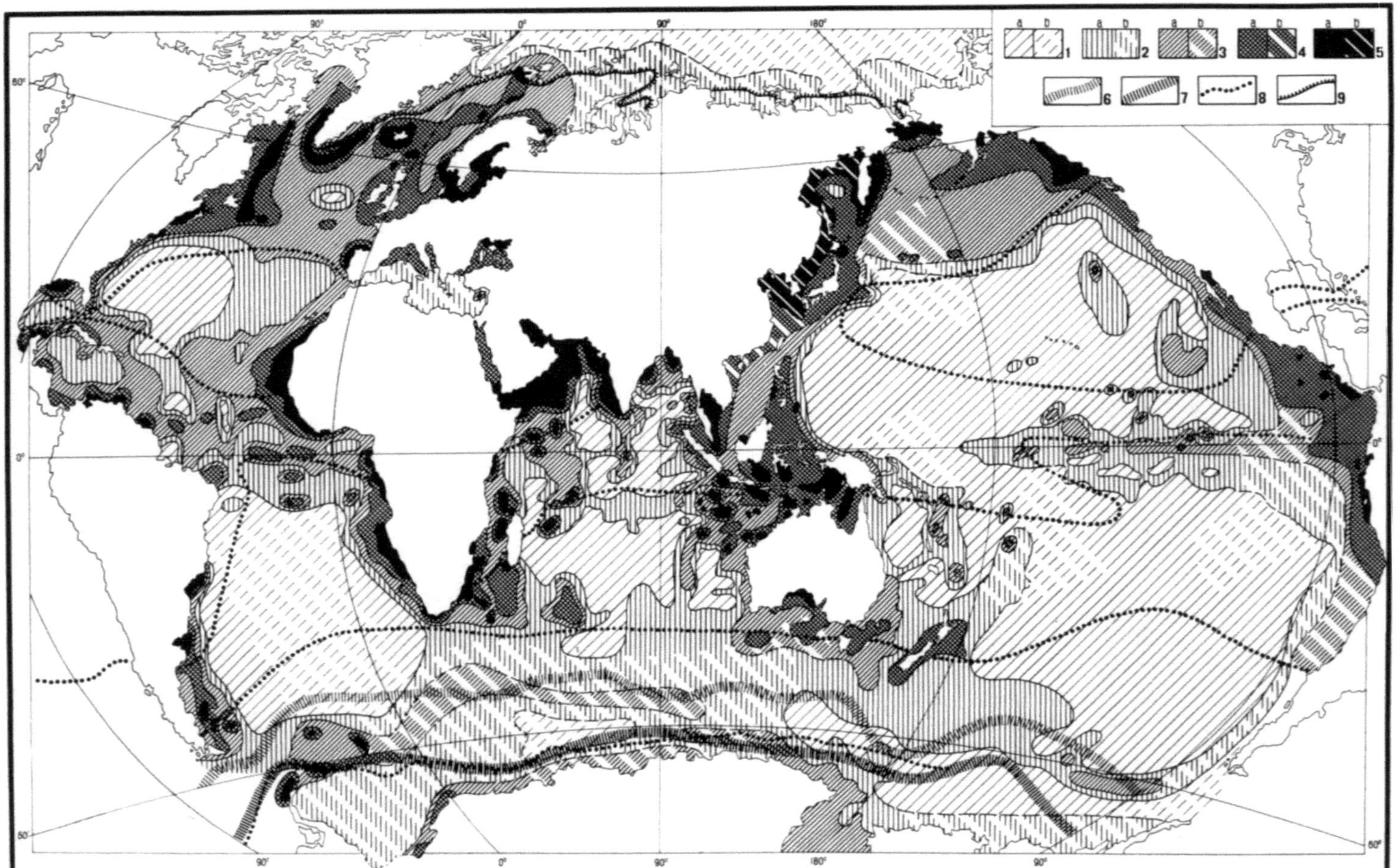

Fig. 1. Distribution of phytoplankton production in the ocean, mgC m^{-2} day^{-1}: *1* = 100; *2* = 100 – 150; *3* = 150 – 250; *4* = 250 – 500; *5* = >500; *a* direct measurements, *b* indirect data; *6* Antarctic convergence; *7* Antarctic divergence; *8* boundaries of climatic zones. *Toothed line:* the ice boundary in the Arctic (Koblents-Mishke 1977, with additions)

At present it has been established that in transition from coastal to oceanic waters, there is a manifold decrease in phytoplankton production, whereas in transition from high-latitudinal to tropical oceanic waters the decrease is less marked (Table 2). The productivity of coastal waters is superior to that of oceanic waters in the arctic by a factor of 6 – 15, in the temperate region by a factor of 5 – 10, in the tropical and subtropical zone by a factor of 7 – 20 [see, for instance, Radhakrishna (1969), Vinogradov and Lisitsin (1981)].

Table 3 clearly demonstrates an overall decrease of primary production in transition from the continental macroecosystem to the world ocean ecosystem,

Table 2. Average phytoplankton production values for various types of water of the oceans (Koblentz-Mishke et al. 1970, modified by Menzel 1974)

Type of water	Area occupied by the various types of water		Phytoplankton production	
	10^6 km^2	% of total	$g C_{org} m^{-2} yr^{-1}$	$10^9 t C_{org} yr^{-1}$
Oligotrophic waters of central parts of subtropical halistatic areas	148	40.3	25.6	3.79
Transitional waters between subtropical and polar zones; marginal zones of equatorial divergences	83	22.7	51.1	4.22
Waters of equatorial divergence and oceanic regions of subpolar zones	86	23.4	73.0	6.31
Coastal waters	39	10.6	124.1	4.80
Neritic waters	11	3.0	365.0	3.90
Total	367	100.0		23.0

Table 3. Global net primary production and standing crop (Woodwell et al. 1978, Nienhuis 1981)

Ecosystem type	Area, 10^6 km^2	Total net primary production		Total plant mass of carbon, $10^9 t C_{org}$
		$10^9 t C_{org} yr^{-1}$	$g C_{org} m^{-2} yr^{-1}$	
Continental ecosystems	149.0	52.8	354.4	826.5
Marine ecosystems including:	361.0	24.7	68.7	1.74
Algal bed and reef	0.6	0.7	1166.7	0.54
Estuaries	1.4	1.0	714.3	0.63
Upwelling zones	0.4	0.1	250.0	0.004
Continental shelf	26.6	4.3	161.6	0.12
Open ocean	332.0	18.7	56.3	0.45

Table 4. Phytoplankton primary production in some estuarine areas, $gC_{org}m^{-2}yr^{-1}$

Area	Production	Reference (cited by Wolff, 1980)	Method
Long Island Sound, USA	380	Riley (1956)	Oxygen
St. Margaret's Bay, Canada	190	Platt (1971)	^{14}C
Cochin Backwater, India	124	Qasim (1973)	^{14}C
Loch Etive, Scotland	70	Wood et al. (1973)	^{14}C
Wadden Sea, Netherlands	100–200	Cadée and Hegeman (1974)	^{14}C
Ems estuary, Netherlands	13–55	Cadée and Hegeman (1974)	^{14}C
Grevelingen estuary, Netherlands	146–200	Vegter (1977)	^{14}C

and within the latter's limits, from coastal areas to the open ocean (Woodwell et al. 1978).

The circumcontinental zonality of OM production in the ocean is associated with the existing system of water circulation. Wherever cold water flows approach the land, particularly where deep waters rich in biogenic elements rise up onto the shelf, primary production is observed to be significantly higher. Additional factors for productivity increases here are biogenic elements arriving with river runoff and from coastal abrasion, a rapid reutilization of OM in the shelf and continental slope areas and a continuous stirring of water down to the bottom of homogenization within the photic layer.

Favourable to phytoplankton producion are monsoon areas in the Arabian Sea, the Bay of Bengal and in the Indonesian seas, as well as the sea current divergence regions off the coasts of Kamchatka, Japan, Canada, Greenland, Brazil. Primary production in estuaries is on the whole rather high, though it is subject to wide temporal variations (Tables 3 and 4). However, particularly large masses of OM per unit area are being synthesized in off-shore upwellings, for example off the western coasts of the South and Central America, South West Africa, etc. In these zones the efficiency of solar energy utilization amounts to about 0.3%, i.e. as much as in the forests on land. The high level of OM reproduction by phytoplankton secures in these waters the abundance of heterotrophic life, the maximum content of POC for the ocean, as well as the maximum percentage concentrations and absolute masses of OM in the sedimentary mass (cf. Chap. 4). Thus, in accordance with the principle of actualism we can assert that the present circumcontinental zonality in the distribution of plankton production in the ocean has existed for millions of years in the past.

The latitudinal zonality in the distribution of phytoplankton production in the ocean is manifested in the existence of three zones of high bioproductivity: two moderate humid ones and an equatorial one separated in both hemispheres by tropical regions of general downwelling of waters and low bioproductivity.

The lowest primary production values are typical, on the one hand, of the high arctic glacial regions and, on the other, of tropical arid regions, which, as regards the efficiency of solar energy utilization (0.02%), are only slightly superior to the deserts on land (0.01%). The reasons for this are a short vegeta-

Table 5. Phytoplankton production in some seas and coastal oceanic areas

Basin	$g C_{org} m^{-2} yr^{-1}$	$10^6 t C_{org} yr^{-1}$	Source
White Sea	–	4.8	Maksimova (1973)
Baltic Sea	130	50	Ibid.
Black Sea (average)	112	93.1	Sorokin (1982)
Azov Sea	200 – 300	76 – 17	Kondratieva and Finenko (1971)
Caspian Sea	277	100	Datsko (1959)
Aral Sea	200	12	Khrustalev et al. (1977)
Mediterranean Sea	40 – 60	–	Kondratieva and Finenko (1971)
Red Sea	200	90	Ibid.
Peru shelf	1710 – 2220	–	Vedernikov and Starodubtsev (1971)
Peru shelf	~300 – 1600	–	Staresinic (1981)
Wallfish Bay	2700	–	Radhakrishna (1969)

tion period of phytoplankton, a lack of light (near-polar areas), a stable stratification of waters and an extreme poverty in biogenic elements (tropical areas). In high latitudes the OM production is adversely affected by the low rate of OM decay and of the reutilization of biogenic elements.

The productivity of the majority of inland, mediterranean, and marginal seas is, on average, much greater than that of the oceans (Table 5). In the seas the distribution of phytoplankton production is as irregular and is controlled by the same factors as in the oceans. Hydrodynamics continues to determine the regime of biogenic elements here and, along with light, the geographic distribution of phytoplankton. Furthermore, marine productivity is greatly dependent on local circulations, surge currents, hydrographic conditions, the amount of river runoff, coastal abrasion and water exchange with the adjoining marine basins. The same factors complicate normal patterns in the coastal oceanic areas. For this reason, the distribution of phytoplankton production there is of a more complex character, which duly affects all other biological and biogeochemical phenomena enhancing their diversity and making them still more intricate (Bruyevich 1953).

1.1.2 Phytobenthos

In the narrow coastal strip between the land and the sea (from the surf margin to a depth of 60 – 120 m, often 20 – 40 m) about 8000 species of unicellular and multicellular algae, about 100 species of phanerogams (sea weeds) and about 15 species of mangroves are found.

The algal belt surrounding dry land in a practically continuous strip is extremely irregularly inhabited by plants. At some rocky places their biomass attains 4.8 – 6.4 kg m^{-2} and productivity exceeds 6.8 g C_{org} m^{-2} per twelve hours (Towle and Pearse 1973). In the macrophytobenthal zone of the White Sea the output during the vegetation period is as high as 4.7 – 12.8 kg C_{org} m^{-2} (Vozzhinskaya 1974). On other, usually silty beds, algae may be absent altogether. The most widespread algal species in the ocean are Chlorophyta, Phaephyta and Rhodophyta, and they account for the bulk of the biomass and OM production. They are supplemented with smaller and more shade-phillic forms making up the lower tier, as well as with phanerogams (zostera, phyllospadix, potamogetonaceae, halophytes, as well as numerous epiphytes inhabiting the host thallus).

The quantitative and taxonometric diversity of aquatic plants is dependent on the combination of numerous factors characterizing the biotope: light, temperature, gaseous composition, salinity, coastal zone structure, nature of the seabed, currents, tides, surf, dissolved mineral and organic substances, etc. It correlates with natural geographic zones and the biological structure of the ocean (Zenkevich 1967, 1970, 1971). In the Earth's unified algal belt three main zones can be identified: arctic and antarctic, temperate and tropical (Vozzhinskaya 1977).

The arctic and antarctic zones are characterized by the absence of a continuous algal belt and by generally moderate algal biomass. The hard natural conditions account for the fact that mass growth of algae is observed only in the sublittoral area. On the littoral and upper sublittoral (phytal) they are either thinly distributed or completely absent. Brown seaweeds predominate among the algae, red and green seaweeds being less widespread.

The moderate zone is mostly characterized by the belt-like, rarely mosaic distribution of algae. Their density is very irregular and varies between some dozen grams to tens of kilograms, the average being several kilograms to a square metre. As regards the biomass of aquatic plants, the moderate zone is much richer than the arctic and antarctic and, apparently, the tropical zone. Brown algae (kelp and rockweeds) are predominant. Phanerogams are widespread in the seas (Black, Azov, Caspian, Sea of Japan). The eelgrass biomass in the Black Sea amounts to about 0.1×10^6 t C_{org}.

The tropical zone is not distinguished by abundant algal growth. Algal biomass here averages about 1 kg m^{-2}. The main producers of phytobenthogenic OM in tropical waters are sea weeds, mangrove trees and shrubs (Blinova and Vozzhinskaya 1971, De Vooys 1977). An additional source of OM are symbiotic algae of coral biogeocenoses (Wells et al. 1973). Their total production is unknown.

The mangrove vegetation is particularly well developed in the humid tropical regions of the Pacific and Indian Oceans. At particular sections of the coastal zone in humid tropics the mangroves form continuous thickets, supplying to the coastal zone, in the process of natural decay, large quantities of organic particles originating in the border area between land and sea. The mangrove forests happen to be an important local source of OM in the coastal areas of tropical seas. No definite figure is available for the total production of mangroves. It can be assumed that the productivity of mangrove thickets at the places of their continuous growth is as high as or even superior to that in humid tropical forests (2 – 10 kg C m^{-2}).

In contrast to the land, the abundant growth of phytobenthos in the coastal zone does not entail a large-scale development of herbivorous zooplanktonic organisms. This "... is due to the absence of a firm masticatory stomatic apparatus, similar to that of land animals, in the majority of marine zoobenthonic representatives, ... and to the presence in the seas and oceans of the enormous and rapidly produced phytoplankton mass which is completely absent on the land" (Zenkevich 1970, p. 19). Owing to a higher (compared to plankton) resistance of macrophyte OM to enzymatic decomposition the quantity of buried macrophyte-formed organic matter is liable to be out of proportion with the relation between phytoplankton and phytobenthos production. Unfortunately, we cannot as yet evaluate the scale of this phenomenon in natural conditions.

The ocean floor area occupied by phytobenthos amounts to approximately 1.8×10^6 km^2 (Uspensky 1956). De Vooys (1977) has calculated that the mondial standing crop of benthonic macrophytes makes up about 30×10^6 tons C_{org} yr^{-1}. Of the total, 20×10^6 tons C_{org} yr^{-1} is accounted for by giant brown algae (kelp) and 10×10^6 tons C_{org} yr^{-1} by rockweeds (brown algae) and red weeds.

Blinova and Vozzhinskaya (1971) and Vozzhinskaya (1977) have calculated that the annual phytobenthos biomass in the world ocean is about $(1.5-1.8) \times 10^9$ tons (green mass), which corresponds to $(300-360) \times 10^6$ tons dry mass, or about $(112-153) \times 10^6$ t C_{org}. The annual total production of phytobenthos in the ocean is 5–6 times as large as the biomass and amounts to $(560-680) \times 10^6$ t C_{org} (on the average ca. 600×10^6 t C_{org}) (Vozzhinskaya, pers. comm.). This value makes up 2.2% of the primary phytoplankton production. These estimates are many times greater than the evaluations of phytobenthos production in the world ocean made by Bogorov (1971), namely 30×10^6 tons of OM (about 15×10^6 tons C_{org} yr^{-1}), but still they are lower than those made in the mid-50's by Uspensky (1956): the biomass of phytobenthos in the world ocean of about 410×10^6 tons C_{org}, its production of 970×10^6 tons C_{org} yr^{-1}.

These data are chiefly based on the calculation of the biomass and production of macrophytes, brown, green and red algae. The production of submerged phanerogams and epiphytes was only partly taken into account. The OM formed by peridinian algae which are in symbiosis with corals, as well as that arriving in the ocean from marshes and mangroves, was not accounted for. Although no reliable direct data on these OM sources in the ocean are available, it appears useful, nontheless, to take into account the rough calculations of the global biomass and produce formed in the estuaries in which the main contribution is made by submerged phanerogams. As estimated by Woodwell et al. (1978), primary production in the estuaries amounts to about 1×10^9 tons C_{org} yr^{-1} or about 4% of the total global marine primary production.

In the seas the part of OM annually produced by littoral and sublittoral algae is usually higher than in the oceans and is believed to average 3–5% of the phytoplankton production. In some marine and oceanic coastal areas aquatic plants prove to be the principal OM producers. At some sections of the coastal zone they are known to be producing annually a hundred times as much OM as does phytoplankton (Blinova and Vozzhinskaya 1971, Vozzhinskaya 1977).

From the material on the distribution in the ocean of phytoplankton and phytobenthos production, kindly offered to the author by Koblentz-Mishke and

Table 6. Photosynthesis production over the main structural-geomorphological oceanic regions

Region	Area, 10^6 km^2	Annual production	
		$g C_{org} m^{-2}$	$10^9 t C_{org}$
Submarine continental margin	84	157	13.2
Ocean floor	276	45	12.4

Vozzhinskaya, it has been possible to calculate the quantity of organic matter formed by phototrophs over the main structural-geomorphological oceanic zones (Table 6).

The submarine margin comprises the following parts: the shelf, continental slope, continental rise and at some places the marginal parts of the abyssal plane. Its area calculated by the author is close to the value of 81.506×10^6 km^2 given by Leontiev et al. (1975).

As seen from Table 6, about 50% of primary OM production is formed over the submarine margin of the continents. The ocean floor areas account for the same quantity.

1.1.3 Bacterial Chemosynthesis

In the ocean numerous bacterial species are capable of forming organic matter, making use, as do green plants, of CO_2 as the unique source of carbon and of different reduced mineral compounds (H_2, H_2S, S^0, $S_2O_3^{2-}$, SO_2, NH_3, NO_2^-, Fe^{2+}, CO, etc.) as the source of energy. Chemolithotrophic organisms are much more abundant in the ocean than the photosynthesizing ones, though their overall mass is significantly less (Zobell 1972). They effect the same catalytic reaction as green plants, but obtain the energy for the reduction of CO_2 to C_{org} from the conjugate oxidation of relatively reduced compounds.

The assessment of the importance of chemolithotrophic bacteria as a primary source of OM in one or other water basin involves not only quantitative evaluation of their production but also determining whether the energy source for the formation of this organic matter is autochthonous or allochthonous. Chemosynthesis which makes use of the energy of reduced mineral compounds resulting from the disintegration of OM is a secondary process. In this case chemolithotrophs are only contributing to the conservation of energy at a higher level – in the form of chemical bonds of organic compounds, but the energy itself has already been taken into account when measuring the photosynthesis production and estimating the external supply of OM as part of one or other source. Chemosynthesis which uses the oxidation energy of allochthonous mineral compounds supplied with river and underground runoff or as part of volcanic gases and hydrotherms is, in fact, a primary process. The OM produced due to these energy sources should be taken into account in the total balance of OM in a basin. The same applies to the chemosynthesis based on the energy of methane and its homologues arriving in the ocean from the interior of the Earth.

The formation of OM in the water by chemosynthetical means has been discussed in a number of works (Rodina 1951, Vinberg 1960, Sorokin 1968, 1970, Kuznetsov 1970). Different viewpoints on the importance of chemosynthesis in primary production have been expressed (Rodina 1951, Vinberg 1960). As Vinberg (1960) put it " ... there are no grounds for an assumption, at least with respect to conventional water reservoirs, that the external supply of compounds, the oxidation of which may serve as the energy source for chemosynthetic bacteria, does play quantitatively a perceptible role in the general cycle of OM in a water reservoir". For the present ocean as a whole this statement is true, especially since the hydrotherms and gases escaping from the ocean floor are either completely or partly oxidized by oxygen, converting into compounds which are more stable under these conditions: H_2S-S converting into SO_2 and even into SO_3 and SO_4^{2-}; CO converting into CO_2. H_2 into H_2O; subject to partial oxidation is NH_3 and, apparently, methane whose content is usually about 3% (Vinogradov 1967). Furthermore, in deep-water oceanic regions numerous gases are present in the form of crystallohydrates, which impedes their utilization by bacteria.

The sum total of the supply of different reduced mineral compounds with hydrotherms and particularly that part of them which is utilized by bacteria, is unknown. In those water basins where the water mass or bottom sediments are periodically or continuously contaminated with hydrogen sulphide the role of deep-water sources of energy for chemosynthesis may be quite significant.

Sorokin (1970, 1982), using the radiocarbon method, has determined that in the Black Sea the chemosynthesis production amounts to 100 mg C m^{-2} day^{-1}. The formation of this quantity of OM calls for the oxidation of 0.3 g m^{-2} day^{-1} of methane or hydrogen sulphide (about 108 g m^{-2} $year^{-1}$, and for the entire hydrogen-sulphide zone of the Black Sea about 40×10^6 tons). The liberation of H_2S from the bottom sediments during OM oxidation and sulphate reduction, as well as the secondary process of sulphate reduction at the expense of OM in the mass of water at the boundary of the hydrogen-sulphide and the oxygen zones fails, apparently, to provide the required quantity of hydrogen sulphide. According to Sorokin (1970, p. 71), "... the role of the external energy source is played here by methane and, possibly, by other hydrocarbon gases arriving through the mass of sedimentary rocks into the Black Sea basin". Geodekyan's (1968) calculations revealed that in the South-Caspian trough as much as 36×10^9 tons of hydrocarbon gases have been liberated over about one million years. Other works also mention the liberation from the sedimentary rocks and in deep fracture zones of significant quantities of methane and other hydrocarbon gases (Sokolov 1971). Thus, the energy contained in these gases, regardless their origin, may avoid dissipation and once again be partly involved, in a chemosynthetic manner, in the synthesis of OM. Estimating the scale of this process in water basins is the task facing future workers.

Bacterial photosynthesis is principally effected by purple bacteria using light as the energy source, CO_2 as the carbon source, and mineral compounds (H_2S, H_2) as well as some organic substances as the oxidized substrate. These processes of OM synthesis may make progress wherever water contains H_2S and sufficient light is available for photosynthesis. In modern times such conditions are present only in isolated lakes where such organic matter production may be a few times as high as the photosynthesis of planktonic algae (Kuznetsov 1970). In the seas and oceans beyond coastal shallow-water areas, the occurrence of bacterial photosynthesis has not been established. It is interesting to note that in the bottom sediments of the Kuril-Kamchatka area the author together with A. A. Ilyina have discovered significant quantities of bacteriochlorophyll and bacterioviridine.

1.2 Input of Allochthonous Organic Matter

Allochthonous OM arrives in the seas and oceans with the river runoff, with the subterranean runoff bypassing rivers, with glacial runoff, with eolian material, as well as with the material of wave and glacial abrasion and of volcanic activity and, finally, as part of cosmogenic material. Most of these sources are supplying OM in the dissolved and particulate form. The greatest role in the supply of OM to the ocean belongs to rivers and eolian transport.

1.2.1 River and Subterranean Runoff

The quantity of moisture evaporating from the ocean surface (505×10^3 km^3) exceeds by about 47×10^3 km^3 the atmospheric precipitation falling into the ocean (458×10^3 km^3). (*The World Water Balance and the Earth's Water Resources*, 1974). Of the 47×10^3 km^3 about 42,390 km^3 is returned into the ocean with river waters and another 2310 km^3 predominantly in the form of ice (Table 7). The quantity of C_{org} contained in the moisture arriving from the ocean to dry land being less (or equal to zero as tacitly implied by the researches calculating the OM runoff to the ocean) than that supplied from the land, including C_{org} carried with

Table 7. Inflow of surface and subterranean waters into the world ocean (km^3)[a]

Ocean	Ocean area, 10^6 km^2	Inflow of surface waters from continents: Eurasia	Africa	North America	South America	Australia and Oceania	Antarctic	Total inflow of surface waters	Total inflow of surface and subterranean waters[b]
Pacific	178.1	7,450	–	2540	1,330	1800	975	14,100	14,800
Atlantic	91.7	2,300	3260	3260	10,370	–	570	19,860	20,760
Indian	76.2	3,560	748	–	–	568	765	5,640	6,150
Arctic	14.7	3,100	–	2040	–	–	–	5,140	5,220
World ocean	361.3	16,400	4110	7840	11,700	2370	2310	44,700	47,000

[a] *The World Water Balance and the Earth's Water Resources*, 1974

[b] The value of subterranean water inflow not drained by rivers was tentatively estimated for the individual oceans on the basis of the length of the coastline

the river and subterranean runoff, this additional quantity of organic matter must be taken into account in calculating the total balance of OM in the ocean. Apart from dissolved organic compounds, the rivers bring particulate organic matter to the ocean.

Dissolved Organic Matter Runoff. Clark gives data on the content of OM in the dry mineral residue (%) for eighteen rivers: Amazon 15.03, Delaware 16.00, Lough Neogh 16.40, Plata 49.59, Negro 53.89, Uruguay 59.90, Danube 3.25, James 4,14, Nile 10.36, Hudson 11.42, Rhein 11.93, Thames 12.10, Maumee 4.55, Cumberland 12.08, Genesee 12.80, Mohawk 15.34, Xingu 20.63, Tapajos 24.16 and assumes 10% as the average value (Clarke 1924). Proceeding from these data, Vernadsky calculted that organic matter makes up $1 \times 10^{-2} - 6.2 \times 10^{-5}$% of river water ($100 - 0.62$ mg l^{-1}) (Vernadsky 1960).

In 1955 Skopintsev and Krylova summarized the most reliable data on the content of total C_{org} (DOC + POC) in river waters and calculated that C_{org} accounts for 10 mg l^{-1}. Of this about 5 mg l^{-1} arrives in the ocean in the dissolved form (180×10^6 tons DOC). A similar value was obtained by Menzel (200×10^6 tons DOC) who extrapolated the values obtained for the Mississippi, Hudson and Amazon rivers to the total surface runoff into the ocean (Menzel 1974).

Alekin (in *The World Water Balance and the Earth's Water Resources*, 1974) employed the maps of permanganate oxidation and average runoff values for river waters and computed that as much as 34.2×10^6 tons DOC (1.74 t km^{-2}) are carried off into the ocean from the USSR territory and another 3.6×10^6 tons are deposited in the area which lacks discharge access to the ocean. Following an analysis of the factors determining the differences in the content of DOC in continental waters (compared to the USSR territory), he adopted the following values for the runoff of DOC (tons per km^2): Asia 2.0, Malayan Archipelago 2.0, Africa 1.5, North America 1.2, South America 2.5, Europe 1.2, Australia 1.8. Altogether, from all of the continents as much as 170×10^6 tons of DOC, i.e. 1.7 tons km^{-2}, are carried into the ocean from the territory which has the runoff access to the ocean.

Table 8 gives average perennial values of C_{org} runoff for 27 large rivers in the Soviet Union (Maltseva 1980). The river runoff calculations were made for different seasons (winter, spring high floods, summer-autumn) for two periods (1936 – 1965 and 1966 – 1975). To express the permanganate oxidation values in terms of C_{org}, use was made of the coefficients interconnecting these quantities under different physico-geographical conditions. An error inherent in calculating the transport of C_{org} by rivers is in the range of 30%. Comparison of the average perennial C_{org} transport by rivers for the two periods has shown that no significant variations on C_{org} transport for the majority of the investigated rivers of the USSR (54 rivers) had taken place. The amount of C_{org} discharge into the USSR's Arctic Seas is observed to decline from in the eastward direction: from 8.8 tons km^{-2} per year (Barents Sea) to 2 tons km^{-2} per year (Chukchee Sea). The bulk of C_{org} discharge into the Arctic Sea coincides with the high-flood period and increases eastwards from 45% to 80%.

The average value of C_{org} discharge from the USSR's territory into open seas and the Aral-Caspian basin amounts, according to average perennial data for 54

Table 8. Evacuation of organic substances by the USSR's major rivers for two periods (Maltseva 1980)

River and sampling site	1936–1965			1966–1975		
	Water runoff, km^3 $year^{-1}$	C_{org}, mg l^{-1}	Discharge OM, 10^3 tons	Water runoff, km^3 $year^{-1}$	C_{org}, mg l^{-1}	Discharge OM, 10^3 tons
North Dvina, Ust-Pinega vil.	107.7	16.8	3616	93.9	16.4	3075
Pechore, Ust-Tsilma vil.	106.0	13.6	2873	110.5	13.0	2883
Ob, Salekhard	391.5	10.9	8552	419.0	8.8	7372
Yenisei, Igarka	563.2	9.2	10375	562.0	7.4	8258
Lena, Kosyur vil.	518.2	10.5	10845	533.3	9.5	10123
Indigirka, near Vorontsovo	48.8	8.9	863.4	51.9	9.0	935.6
Kolyma, Srednekolymsk	71.1	6.7	957.0	62.6	8.3	1034
Amguema, mouth of river Shumnyi	8.9	4.0	71.4	8.1	6.7	108.2
Anadyr, Snezhnoe farm	32.4	6.8	441.4	28.4	7.2	408.1
Amur, Komsomolsk	328.9	9.5	6228	283.2	10.1	5703
Neva, Novosaratovka vil.	80.4	8.5	1389	71.7	8.6	1232
Luga, Kingisepp vil.	2.9	14.3	83.6	2.6	13.5	69.7
West Dvina, Daugavpils	14.8	11.8	349.8	11.5	13.3	306.7
Neman, Kaunas	15.1	6.8	209.0	12.6	7.1	178.6
South Bug, Aleksandrovka vil.	2.7	7.9	42.7	3.4	7.7	52.1
Dnieper, Kiev	43.1	8.3	712.6	46.5	9.9	917.5
Dnestr, Bendery	9.8	5.9	116.3	11.7	4.8	111.9
Don, Razdorskaya centre	21.9	9.3	405.8	17.6	9.0	317.7
Kuban, Tikhovsky vil.	11.9	7.2	172.0	10.4	6.8	141.8
Rioni, Sakochakidze vil.	12.6	3.0	76.1	12.4	2.1	53.0
Kura, Tbilisi	6.4	5.2	67.3	5.6	3.4	37.9
Terek, Mozdok	7.0	4.6	64.7	5.5	5.2	56.6
Volga, Verkhnelebyazhie vil.	227.0	8.0	3618	213.8	7.6	3257
Ural, Topoli vil.	8.8	10.4	181.9	8.1	7.1	115.0
Ili, Uchzharma vil.	14.8	6.1	181.4	13.4	5.4	114.6
Syr Darya, Kazalinsk	15.4	7.9	244.4	8.8	5.3	93.1
Amu Darya, Chatly kishlak	46.8	9.4	877.2	28.7	1.7	99.0

rivers, to 41×10^6 tons C_{org} per year. The distribution of C_{org} discharge seasonally corresponds to the water-runoff regime.

Table 9 gives information on the content of dissolved and particulate carbon in fluvial waters obtained by direct determinations of C_{org} from CO_{2org}, whilst Table 1 presents average values of the supply of DOC into the ocean with fluvial waters. At present we may assume that the lower limit of the content of DOC in the surface land waters coming into the ocean is 5 mg DOC l^{-1}. The annual surface water runoff being equal to 42,290 km^3 (without the Antarctic), this gives 212×10^6 tons DOC yr^{-1}.

With the antarctic ice (2310 km^3) containing, according to our data, 0.5 – 1 mg DOC l^{-1}, as much as about 2×10^6 tons DOC yr^{-1} is being brought into the ocean.

In calculating the total surface runoff of OM into the ocean it appears desirable that the cyclic DOC, i.e. that portion of DOC which is carried to the land as part of droplet-liquid water, is also taken into account. As revealed by our study, the rain waters over the central part of the Pacific Ocean[1] contain 0.4 – 0.7 mg DOC l^{-1}. Therefore the allochthonous component of OM in river runoff should be reduced by that amount. However, in reality things are far more complicated. The atmospheric precipitation on land due to moisture coming from the ocean amounts to 66,000 km^3. Thus, part of the precipitation of the oceanic origin (66,000 km^3 – 47,000 km^3 = 19,000 km^3) falls on the land, is once again evaporated and carried into the ocean. The content of DOC in the aqueous vapour over dry land is higher than over the oceans. In the atmosphere above the USSR's territory, for instance, this is 2.1 – 24.0, the average being 8.5 mg C_{org} l^{-1} (Semenov et al. 1967) and, apparently makes up fully for the loss of OM by the oceans. The fate of this C_{org}, which amounts to approx. $70 - 130 \times 10^6$ tons, is unknown. Further on we may assume the value of the fluvial discharge of dissolved carbon to be equal to 214×10^6 tons of DOC annually.

Discharge of Particulate Organic Matter. The annual input of particulate suspended material to the world ocean, as estimated by different authors, varies from 12.7×10^9 to 51.1×10^9 tons. The most part of average values cited in the literature are within the range of $12 - 19 \times 10^3$ tons × $year^{-1}$ (Lisitzin 1974, Gordeyev 1983). Alekseev and Lisitzyna summarized the materials for a period of many years on the discharge of suspended particles from the territory of the continents and obtained the following values (million tons of dry mass per year): world oceans 15,700, Pacific 5370, Indian 5960, Atlantic 4020, Arctic 342 (*The World Water Balance and the Earth's Water Resources,* 1974, p. 515).

The C_{org} content in the suspended material can averagely be assumed to be equal to 2.5% on the dry mass basis (see Table 9). The annual input of POC to the world ocean is about 392×10^6 tons, which is made up as follows: Pacific Ocean – 134, the Indian Ocean – 149, the Atlantic – 100, the Arctic Ocean – about 9. Thus, according to the author's data, the supply of dissolved and particulate organic matter to the world ocean by the rivers totals about 600×10^6

1 Water was filtered through glass fibre filters GF/C (pore size about 1 μm)

Table 9. The content of organic matter carbon in fluvial waters

River	Number of samples	Content of C_{org}				Source
		in solution[a], $mg\,l^{-1}$		in suspended material, % of dry mass		
		min – max	average	min – max	average	
1	2	3	4	5	6	7
Daugava	2	8.1 – 10.7	9.4	6.6 – 20.0	13.3	Artemiev et al. (1981)
North Dvina	2	12.9 – 15.3	14.1	2.4	~2.0	Skopintsev and Krylova (1955)
	12	8.8 – 18.2	12.4	6.8 – 52.8	20.2	
Kama	1	–	–	2.3	~2.0	Skopintsev and Krylova (1955)
Neva	2	6.4 – 6.9	6.6	–	~2.0	Skopintsev and Krylova (1955)
Dnieper	2	6.6 – 12.0	9.3	–	~2.0	Skopintsev and Krylova (1955)
Don	2	3.3 – 3.6	3.4	2.6	~2.0	Skopintsev and Krylova (1955)
	6	–	4.2	1.3 – 8.6	4.6	Artemiev (1981)
Kuban	2	1.6 – 3.3	2.4	1.5 – 1.7	~2.0	Skopintsev and Krylova (1955)
	4	1.9	–	1.4 – 3.0	2.1	Artemiev (1981)
Volga	2	6.7 – 10.7	8.7	1.2 – 2.3	~2.0	Skopintsev and Krylova (1982)
Kura	2	1.3 – 3.0	2.2	2.4 – 2.8	~2.0	Skopintsev and Krylova (1982)
Dnieper	2	12.3 – 12.6	12.4	–	–	Datsko (1959)
Don	6	5.7 – 15.6	10.9	–	–	Datsko (1959)
Kuban	4	4.2 – 9.2	7.2	–	–	Datsko (1959)
Dnestr	2	7.8 – 11.4	9.6	–	–	Datsko (1959)
Danube	2	6.4 – 7.2	6.8	–	–	Datsko (1959)
Rioni	2	5.1 – 8.1	6.6	–	–	Datsko (1959)
Dnieper	1	4.9	–	16.4	–	Artemiev (1981)
	2	7.0 – 9.0	8.0	6.67 – 8.73	7.70	Galgoleva (1959)
Don	2	4.0 – 7.0	5.5	2.14 – 3.78	2.96	Glagoleva (1959)
Kuban	2	1.3 – 2.5	1.9	1.35 – 1.54	1.45	Glagoleva (1959)
Danube	1	4.5	–	1.53	–	Glagoleva (1959)
	4	3.1 – 5.7	4.7	0.3 – 2.1	1.0	Artemiev (1981)
Rioni	1	0.9	–	0.74	–	Glagoleva (1959)
Chorokh	2	1.0	1.0	0.66 – 1.38	1.02	Galgoleva (1959)
Pripyat	1	16.0	–	–	–	Galgoleva (1959)
South Bug	5	2.1 – 3.5	2.8	3.2 – 11.8	7.7	Artemiev (1981)
Volga (mouth)	45	–	–	0.91 – 3.07	1.60	Pakhomova (1959)

Table 9 (continued)

River	Number of samples	Content of C_{org}				Source
		in solution[a], $mg\,l^{-1}$		in suspended material, % of dry mass		
		min – max	average	min – max	average	
1	2	3	4	5	6	7
Ob near Salekhard	3	0.84 – 5.68	–	1.19 – 2.00	2.00	Nesterova (1960)
Ket	1	1.78	–	4.07	–	Nesterova (1960)
Eastern Sikhote-Alin rivers (7 rivers)	20	–	–	2.92 – 10.1	6.57	Chudayeva (1980)
				1.71 – 10.24	5.78[b]	Chudayeva (1980)
Amur and Ussuri Gulfs drainage system (6 rivers)	6	–	–	2.11 – 5.49	3.63	Chudayeva (1980)
				2.48 – 3.78	3.17[b]	Chudayeva (1980)
Amazon, Mississipi, Hudson	–	–	4.5	–	–	Menzel (1974), Rickey et al. (1980)

[a] The larger part of C_{org} in volatile organic compounds was disregarded
[b] Entrained load

tons C_{org} annually. This value differs but little from the estimates furnished by Stewart et al. (1978) – 520×10^6 tons C_{org} $year^{-1}$ and Schlesinger and Melack (1981) – $(370-410) \times 10^6$ tons C_{org} $year^{-1}$. It is inferior to the value given by Richey et al. (1980) – 1×10^9 tons C_{org} $year^{-1}$, but is 2 – 3 times as high as the estimates arrived at by Uspensky (1956), Skopintsev (1961), Kemp (1979) – about 200×10^6 tons C_{org} $year^{-1}$.

Subterranean Discharge of Organic Matter. The data on the subterranean discharge into the seas and oceans are very scarce and conflicting even if furnished by one and the same author: 1600 and 230 km^3 $year^{-1}$ (Nace 1970). Zektser and Kalinin (in *The World Water Balance and the Earth's Water Resources,* 1974, p. 112) employed the hydrogeological method and determined in an independent manner (rather than from the difference in the water balance) the underground runoff from the continents into the world ocean as being roughly equal to 2200 km^3 $year^{-1}$. This value gives an idea of the underground runoff from the zone of the upper 200 m. Djamalov et al. (1977) made use of the hydrologic-hydrogeologic method to calculate the underground runoff from the continents and large islands as being equal to 2460 km^3 $year^{-1}$, including that into the Pacific Ocean – 1340 km^3 $year^{-1}$, the Atlantic – 850 km^3 $year^{-1}$, the Indian Ocean – 220 km^3 $year^{-1}$, the Arctic Ocean – 50 km^3 $year^{-1}$ (only from the territory of Europe and partly from Asia). According to Shvets' data (1973), the average C_{org} content in ground waters or, more accurately, in the upper hydrodynamic zone (the volatile component being taken into account) is equal to 27 mg l^{-1}.

As much as 66.4 mln tons C_{org} $year^{-1}$ arrives in the world ocean with underground runoff, bypassing rivers; this includes 36.2 into the Pacific Ocean, 6.0 into the Indian Ocean, 23.0 into the Atlantic, 1.3 into the Arctic Ocean. These values give an idea of the quantity of C_{org} being supplied to the oceans from the upper hydrodynamic zone.

The total of C_{org} in the Earth's subterranean waters is put at 2.5×10^{12} tons C_{org} (Shvets 1973), that is, about three orders higher than the former figure of 10^9 tons (Skopintsev 1950). The much higher estimate of the quantity of OM present in subterranean waters is mainly due to the fact that these calculations were made for a much greater thickness of deposits (5 km) and larger volume of waters (about $5-6 \times 10^7$ km^3). Furthermore, account was taken of the C_{org} contained in volatile compounds which formerly had been disregarded.

The supply to the ocean of strongly-mineralized waters from deep-lying layers is obviously small, but on account of their high C_{org} content, on averagely about 50 mg C l^{-1} (Shvets 1973), they may be additional sources of allochthonous OM. It is yet impossible to assess the contribution of this OM in the total for the ocean.

1.2.2 Eolian Transfer

It has been figuratively observed that the ocean is a trap for atmospheric dust. After the works of Udden (1894), Thoulet (1908), Free (1917) (cited by Bruyevich and Gudkov 1954) it was thought for a long time that the quantity of eolian material carried by air currents exceeded the input of particulate matter with river runoff. Although with respect to organic matter this was not later confirmed (Romankevich 1977), the concept of a great importance of atmospheric dust in sedimentation, particularly in pelagic oceanic areas, has found support in the extensive material accumulated by now (e.g., Lisitzin 1974).

Data on the content of OM in the eolian suspended material are few today. Ketseridis et al. (1976), Simoneit (1977) discovered that atmospheric suspended material in European and Atlantic air masses contained fatty acids, phenols, amines, aliphatic and aromatic hydrocarbons. It was also observed that certain groups of OM in atmospheric suspended material maintain rather constant relationships. This could be due to a common origin and/or to a similar transformation of these constituents in the eolian material of the explored areas.

In the moderate and tropical regions of the Pacific Ocean, north and south of the equator, we used glass fibre filter sheets and Petryanov filters to collect airborn suspended material, and determined the content of OM and certain organic constituents in it.

As revealed by the analyses, the C_{org} content in eolian suspended material varies from 8.7 to 53.0%, i.e., some individual samples within the accuracy of

Table 10. The content and composition of organic matter in eolian suspended material over the Pacific Ocean area[a]

Organic matter constituents	Number of samples	Limit values	Median	Average
C_{org}, % dry mass	22	8.68 – 53.03	31.71	33.32
N_{org}, % dry mass	22	1.34 – 10.28	6.05	5.91
N/C, %	22	11.2 – 21.7	17.4	17.7
Carbohydrates, % dry mass	10	1.92 – 8.18	4.03	3.99
Carbohydrates/C_{org}	10	4.22 – 8.85	6.16	6.02

[a] Samples taken in the eastern half of the Pacific Ocean between 34°N – 26°S and 179°W – 77°W

C_{org} determination were found to consist entirely of organic matter. The OM content in the eolian material off the sea coasts in the arid zone was found to be lower than the average value (about 33%). A very high content of OM in some individual samples of eolian suspended material is, apparently, due to the fact that OM had partly formed from seawater by bubbling. Such suspended material contains a greater quantity of OM than could be expected on the basis of the OM content of the initial seawater.

The annual input of eolian material to the world ocean is put at 1600×10^6 tons of dry mass (Lisitzin 1974). If, despite the processes of chemical transformation, fractionation and physico-chemical recycling of OM which, obviously, occur on a large scale at the ocean-atmosphere interface, we assume that the average C_{org} content in the eolian transports from dry land is 20%, this will give the value of the annual C_{org} supply to the ocean through the atmosphere (disregarding the volatiles and the majority of anthropogenic substances) of 320×10^6 tons C_{org} year^{-1}. According to the latest data, the flow of petroleum hydrocarbons from the atmosphere onto the world ocean's surface is put at about 3.3×10^6 tons, which is five times as high as the calculated values disregarding the formation of aerosol hydrocarbons from the vaporous phase in the oceanic atmosphere.

The material of the eolian genesis is distributed quite differently from dissolved and particulate OM of the river runoff. Its greatest part is deposited with atmospheric precipitation on the periphery of the arid zones and in the adjoining parts of the moderate humid zones and the equatorial zone. About 10% of eolian OM is removed from the atmosphere with dry matter during its gravitational precipitation and collision with the ocean surface. Tentative estimates of the quantity of C_{org} of the eolian genesis falling out in the Pacific (160×10^6 tons), Atlantic (93×10^6 tons) and Indian Oceans (67×10^6 tons) were obtained on the assumption that the input of this material is proportional to the area of each of the oceans.

Tropospheric transfer of eolian suspended material can be either local or distant. In the case of local transfer, the eolian material falls out close to its source, that is, some tens or hundreds of kilometers away from it. In the case of distant tropospheric transfer the distance of propagation may exceed 6000 km. Eolian material plays a significant role in supplying OM onto the sea bed on the periphery of humid zones and in the arid regions where the material of eolian origin makes up from 25 to 75% of the content of pelitic sediments.

Even a rough estimate of the quantity of OM falling out with rains onto the ocean surface can be of much interest. Such information is indispensable in studying the migration of elements in the atmosphere and hydrosphere, the biological productivity, the viscosity of water, the potential and capillary properties of the water – air interface. As regards the total balance of C_{org} in the ocean, we need not take into account the larger part of this OM since its source is, in the first place, the ocean itself, as well as the eolian suspended material (which has already been included in the balance calculations) and volatile organic compounds of a different genesis.

According to our measurements made in the tropical part of the Pacific (VII – X, 1972), the samples of non-filtered rain water contained $1.3-3.2\ \text{mg}\ C_{org}\ l^{-1}$. Rather similar values, $2.5-3.5\ \text{mg}\ C\ l^{-1}$ were obtained by Menzel (1974). For the total amount of atmospheric precipitation falling out on the world ocean of $458{,}000\ \text{km}^3$, (*The World Water Balance and the Earth's Water Resources,* 1974, p. 569) and its average C_{org} content of $2-3\ \text{mg}\ l^{-1}$, there arrives annually on the ocean surface as much as about $900-1400\times10^6$ tons C_{org}. A most tentative estimate gives the amount of cyclic organic carbon in the system ocean-atmosphere-ocean as being of the order of $500-1000\times10^6$ tons C_{org}.

1.2.3 The Material of Glacial and Wave Abrasion

Ice-floes and waves carry out into the ocean the products of decomposition of sedimentary and igneous rocks and, along with them, a certain amount of OM. In some oceanic regions, for example in the Ross Sea, the OM of such genesis can make up as much as 90% of all organic compounds in the sediments (Sackett et al. 1974).

The annual solid runoff with glaciers into the world ocean is put at 1500×10^6 tons (dry mass), while the quantity of wave abrasion material is estimated to be 500×10^6 tons (Lisitzin 1974). The average C_{org} content of the material being carried out from the Precambrian Antarctic platform and the north-eastern part of the Canadian shield may be assumed to equal 0.10%. This value differs little from the C_{org} content of the sedimentary rocks of the later Proterozoic of the Russian platform (0.18%, Ronov and Migdisov 1970) and from the average C_{org} content of the igneous rocks of the Mid-Indian ridge which was 0.09% according to our determinations in 22 samples.

The average C_{org} content of sedimentary material of wave abrasion is, apparently, similar to that of sedimentary rocks on the continents (including effusives) – 0.4% (Vassoyevich et al. 1973, citing Ronov and Yaroshevsky's data 1967). These assumptions enable us to estimate the annual supply of C_{org} with solid glacial runoff material as being equal to 1.5×10^6 tons and in the case of coastal abrasion – 2.0×10^6 tons.

Thus the supply of OM to the ocean due to the destruction of old deposits (as part of the sources considered) is at present not large, though by no means equal to zero. We shall make no great error if we assume that the supply of OM to each of the oceans as part of the material of wave abrasion as well as of the Antarctic glacial abrasion is equal in quantity. Furthermore, the Atlantic Ocean receives an additional supply of about 0.45×10^6 tons C_{org} as part of the glacial abrasion material from Greenland.

In the period of Quaternary glaciation the supply of OM due to glacial abrasion was approximately three times as high. Moreover, the bulk of the material was supplied not to the southern hemisphere as it is today, but to the northern hemisphere where 60% of all glaciers were concentrated at the time.

1.2.4 Other Sources

Submarine Volcanism and Hydrothermal Activity. During volcanic eruptions, solid, liquid and gaseous products, among them organic compounds, find their way into the ocean. The supply of the products of land volcanism to the ocean was included as a constituent part in eolian transfer, river and underground runoff, glacial and wave abrasion and, consequently, has been taken into account. The quantity of organic matter entering the ocean due to the eruptions of submarine volcanoes cannot be easily estimated at present. From the available data on land volcanism, we can make certain deductions.

Active volcanoes on dry land annually supply to the ocean as much as $1-1.5$ km^3 of solid and liquid (lava) products, i.e., about $2-3 \times 10^9$ tons of material. No quantitative estimates of gaseous eruptions from the ocean floor are available, but according to Lisitzin (1974), they produce far less than solid eruptions. The average C_{org} content in this material being $0.2-0.3\%$, the volcanoes supply to the Earth's surface as part of the solid phase as much as $4-9 \times 10^6$ tons C_{org} year^{-1}. Taking account of the distribution of volcanoes and of the relationship between the number of active land (436–483) and submarine (61–69) volcanoes it would seem unlikely that the input of DOC + POC due to subaquatic eruptions exceeds 10×10^6 tons.

Oil Outflows Into the Sea. There exist in the world ocean more than 250 points where natural oil outflows have been recorded. Most of them are located in near-continental areas with very thick bottom sediments and high tectonic activity. The total amount of oil which finds its way into the ocean due to seeping from the sediments is put at about 0.6×10^6 tons year^{-1} (Wilson et al. 1974).

Organic Pollution. In the past decade it became particularly manifest that the biosphere has much suffered from industrial and other wastes with which a great many organic substances are finding their way into the ocean (Vinogradov 1973, Nesterova and Simonov 1979, Monin and Nelepo 1982). They comprise about 6 million tons of oil (according to different estimates, from 2 to 10 million tons) and about 6 million tons of "non-natural" organic compounds which likewise derive their origin from oil, gas and other combustible minerals, as well as from the phytomass. It is predicted that in the future this situation will continue and the world ocean will annually receive as much as 4–6 million tons of oil.

The means by which anthropogenic organic matter gets into the ocean are different: runoff with fluvial waters, discharge of industrial wastes, sewage and ballast waters from ships, oil leakage during off-shore mining and transportation. Some of these compounds have been taken into account in estimating the river runoff and eolian transfer, but their larger part (some 5 million tons) apparently reaches the seas and oceans from unrecorded sources.

The annual supply to the world ocean of organic pollutants amounts to n 0.01% from the quantity of photosynthesis production. Some of them are not easily susceptible to biochemical oxidation (alkylbenzene sulphonates, aromatic hydrocarbons, amines, lignin, humic acids, etc.) and may selectively persist in the water and in the bottom sediments.

Input of Organic Matter from Outer Space. In the total balance of OM in the ocean this material occupies last place. According to different authors, $4\times10^3-10^7$ tons of cosmogenic material arrives annually from outer space. In the most widespread stone meteorites the C_{org} content is about 0.04% (Vinogradov 1962). If we proceed from these values, the maximum input of outer space carbon to the ocean is 4000 tons. It is quite possible that this organic matter attains the seabed in about the same quantity. In this case, the outer space carbon in the eupelagic red clays accumulating at the rate of 1 mm or less in a 1000 years will amount to no less than 1% of the total C_{org}.

1.2.5 Relationship of the Various Sources and Forms of Supply

The above materials permit us to evaluate the annual photosynthesis production in the world ocean as being equal to 25.6×10^9 tons C_{org} and the input of C_{org} from dry land – $1.0\cdot10^9$ tons. This makes up a total of 26.6×10^9 tons C_{org} (about 50×10^9 tons of OM) or about 2.6×10^{17} Cal.

The allochthonous component of OM constitutes about 3.8% of the annual total OM being formed and supplied to the ocean. It is some 4–5 times as high as was formerly assumed: about 1% of phytoplankton production (Skopintsev 1950, Bogdanov et al. 1971). A significant increase in the proportion of allochthonous OM in the total OM reproduction and input into the ocean is mainly due to the fact that the various sources and routes of organic matter supply have been more adequately taken into account.

The comparison of literary data on the supply of allochthonous (fluvial) OM to the ocean [Vernadsky (1934) – about 100×10^6 tons C_{org}; Uspensky (1956) – of the order of 180×10^6 tons; Skopintsev (1961) – 360×10^6 tons; Menzel (1974) – 260×10^6 tons] with those cited above indicates that we have until now underestimated the effect of dry land and its role in the supply of OM to the world ocean.

The overwhelming part of OM is annually supplied or formed in the ocean in the particulate form. The stable form of organic compounds in the ocean is the dissolved state; the factors governing its distribution are discussed in the next chapter.

2. Carbon of Dissolved Organic Matter in the Ocean

Organic matter is present in the world ocean in diverse dispersive and phase states. Because of methodological limitations of isolation and quantitative analysis, OM is classified at present in the majority of works only as dissolved and particulate (suspended). Definition of the term "dissolved organic matter" (DOM) means distinguishing between these two arbitrarily identified forms of organic compounds in natural waters. No uniform opinion prevails on this point. However, the majority of research workers in the field of geochemistry and ecology share the opinion that DOM includes the sum of organic compounds passing through filters with $0.45-1$ μm pore size. Owing to colmatage, i.e., washing of finely-dispersed particles into filter pores, all of them retain particles of a lesser diameter. This form of OM includes the truly dissolved (particle size below 0.001 μm) and colloidal ($0.001-0.1$ μm) OM, as well as a small proportion of larger particles.

In natural waters DOM occupies, in the biogeochemical and trophodynamic aspects, an intermediate position between living organisms and biogenic inorganic salts, being the product of incomplete decay of the organisms and, in certain cases, a direct predecessor of living organic matter. The quantity of dissolved carbon in the ocean exceeds that of particulate carbon and is the prevailing form of C_{org}.

DOM exerts a significant influence on the chemistry of waters. Participating in the system of decomposition principally as a donor of electrons, OM brings down Eh and pH, contributes to the dissolution of $CaCO_3$ and to raising the alkalinity of water. The effect of OM on the carbonaceous system is particularly strong below the photosynthetic zone. The compounds liberated in OM decomposition become part of the reserve of biogenic substances and participate in the formation of the small- and large-scale irregularities of the chemical energy field in the ocean. The capacity of numerous microelements to form complex, chelate and clathrate compounds with the products of incomplete OM decomposition changes their solubility, geochemical behaviour and utilization by the organisms. All these processes derive their origin from life and in their turn, through O_2, Ca^{2+}, HCO_3^- and microelements, influence the vital activity of marine organisms by either facilitating or suppressing the respiratory and nutritive systems.

The different OM forms and microorganisms in the ocean are integrated into a single metabolic and biogeochemical system which combines some conjugate physico-chemical and biochemical processes, for instance OM sorption $\rightleftarrows$ desorption, extracellular hydrolysis of organic macromolecules, resynthesis of organic and mineral-organic compounds. Therefore the processes of OM synthesis-destruction associated with the transformations of mineral constituents have in-

Table 11. Analytical methods for determining the DOC content of sea water

Sample pretreatment	Sample (ml)	Method of OM oxidation	Method of carbon dioxide determination	Accuracy, according to author's data	Source
Adding phosphoric acid and thallium sulphate; removing the precipitate by filtration; evaporating the solution until dry	50	Adding sulphuric acid, potassium bichromate, silver bichromate, cerium sulphate to the dry residue; oxidation by heating to 120 °C	Titration of barium hydroxide solution in Roeberg vessels	0.1 mg C l^{-1}	Krogh and Keys (1934)
Acidifying with a mineral acid and evaporating until dry	–	Adding potassium saltpetre to the dry residue; oxidation by heating to 420 °C	Same	7%	Datsko and Datsko (1950)
Acidifying with sulphuric acid and drying in vacuo	5 – 10	Adding sulphuric acid, potassium and silver bichromates in equal volumes to the sample; oxidation by heating to 130 °C	Titration of barium hydroxide solution	6% for 3 mg C l^{-1}	Kay (1954)
Acidifying with muriatic acid and evaporating until dry	30	Pyrolysis and oxidation in air flow at 900 °C	Titration of barium hydroxide solution in Roeberg vessels	–	Skopintsev and Timofeyeva (1961)
Acidifying with a mineral acid and air blowing	50	Adding sulphuric acid, potassium and silver bichromates in equal volumes to the sample and heating to 130 °C	Coulometric titration with barium hydroxide	3% for 1 mg C l^{-1}	Duursma (1961)
Acidifying with phosphoric acid; adding potassium persulfate; nitrogen blowing; sealing in an ampule	5	Autoclaving (130 °C, 30 min)	From absorption in the infrared region	10% for 1 mg C l^{-1}	Menzel and Vaccaro (1964)

Acidifying with a mineral acid and air blowing	15 – 50	Oxidation with a mixture of sulphuric acid, potassium and silver bichromates while heating to 130 °C	Automatic coulometric titration with barium hydroxide	5% for 2 mg C l^{-1}	Liutsarev and Smetankin (1980)
Acidifying with phosphoric acid and nitrogen blowing	10	Irradiation with short-wave light in the presence of potassium persulfate	Automatic conductometric determination in sodium hydroxide solution	0.06 mg C l^{-1} for 1 – 10 mg C l^{-1}	Ehrhardt (1969)
		Dry burning in oxygen flow at 900 °C	From absorption in the IR region	5%	Sharp (1973)
Acidifying of the sample with 85% phosphoric acid (5 drops)	25	Dry burning at 730 °C in oxygen flow	From a change in thermal conductivity of gases by using a catharometer	8.5%	Gordon and Sutcliffe (1973)
Adding 0.25 ml 0.1 N hydrochloric acid	10	Ultraviolet irradiation in potassium persulfate solution	From absorption in the IR region	2.5%	Collins and Williams (1977)
Acidifying with phosphoric acid until pH 3; freezing; evaporating in vacuo with purified air blowing	15	Dry burning at 850 – 900 °C in a quartz boat	Same		MacKinnon (1978)

disputably determined in the past and determine at present the chemical image of the ocean which has changed but little in the past 500 million years.

The principal trends in the study of dissolved OM in the seas and oceans are at present connected with the problems of the trophodynamics of the ecosystems, the geochemistry of the elements and the pollution of natural waters with anthropogenic wastes.

The sources of DOM in the seas and oceans are the products of bacterial decomposition and autolysis of plants and animals, their metabolic excretions, as well as OM supplied from the land.

The first studies of DOM took place eighty years ago in the Mediterranean. Netterer found that concentrations of C_{org} off the Greek coast and in the open Mediterranean were correspondingly 5 – 10 and about 1 $mg\,l^{-1}$ (Netterer 1892, cit. Duursma 1965).

The high content of chlorides in the ocean, the dissolved carbon content being ten thousand times lower, hindered the application of the classical methods of elementary analysis and was largely responsible for a long period of absence of any more or less reliable methods for the determination of dissolved carbon. Until the mid-sixties, therefore, determinations of dissolved carbon were done by using the methods of oxidizability in neutral, acidic and alkaline media. At present these methods are considered to be less reliable than the direct (after $CO_{2\ org}$) determinations of dissolved carbon. The results obtained with their aid indicate a varying resistance of OM to oxidation rather than its total content.

A great many direct determinations of dissolved carbon in the seas and in different oceans areas have now been made. They are discussed in some monographs and review works, (for example: Skopintsev 1950, 1971, Datsko 1959, Duursma 1961, Koyama 1962, Menzel and Ryther 1970, Starikova 1970, Khailov 1971, Liutsarev 1973, Williams 1975, Romankevich 1977, Wangersky 1978, Skopintsev et al. 1979, MacKinnon 1981). In these works the data on dissolved carbon were obtained by different methods (Table 11). As revealed by special methodological studies, the results of DOC determinations by the persulphate oxidation method (Menzel and Vaccaro 1964) were underestimations by 10 – 50% compared to the dry oxidation (MacKinnon 1978) and photo-oxidation methods (Collins and Williams 1977) which yield similar values. This in many (though not in all) cases may be the reason why Soviet data on the DOC content (e.g., Skopintsev et al. 1966, 1979, Starikova 1967, 1970, Liutsarev et al. 1972) are higher than those obtained by foreign researchers, who in the 1960's and 1970's mostly employed the Menzel-Vaccaro method.

At present research workers investigating DOM in the ocean are unanimous in only one thing, that the DOC content is decreasing with depth, which occurs principally in the upper 200 – 400 m. On other issues, opposing and irreconcilable viewpoints are not infrequent. Contradictory data have been obtained on DOC variations below 300 m; it is not quite clear whether there are regional differences in the DOC content, or whether seasonal variations of DOC and its relationship with biological productivity can be clearly identified.

With respect to DOC distribution in deep waters, two different views have been expressed. One of them presumes that DOC is transferred by currents and gradually disintegrates as it is being carried away from the supply source. The

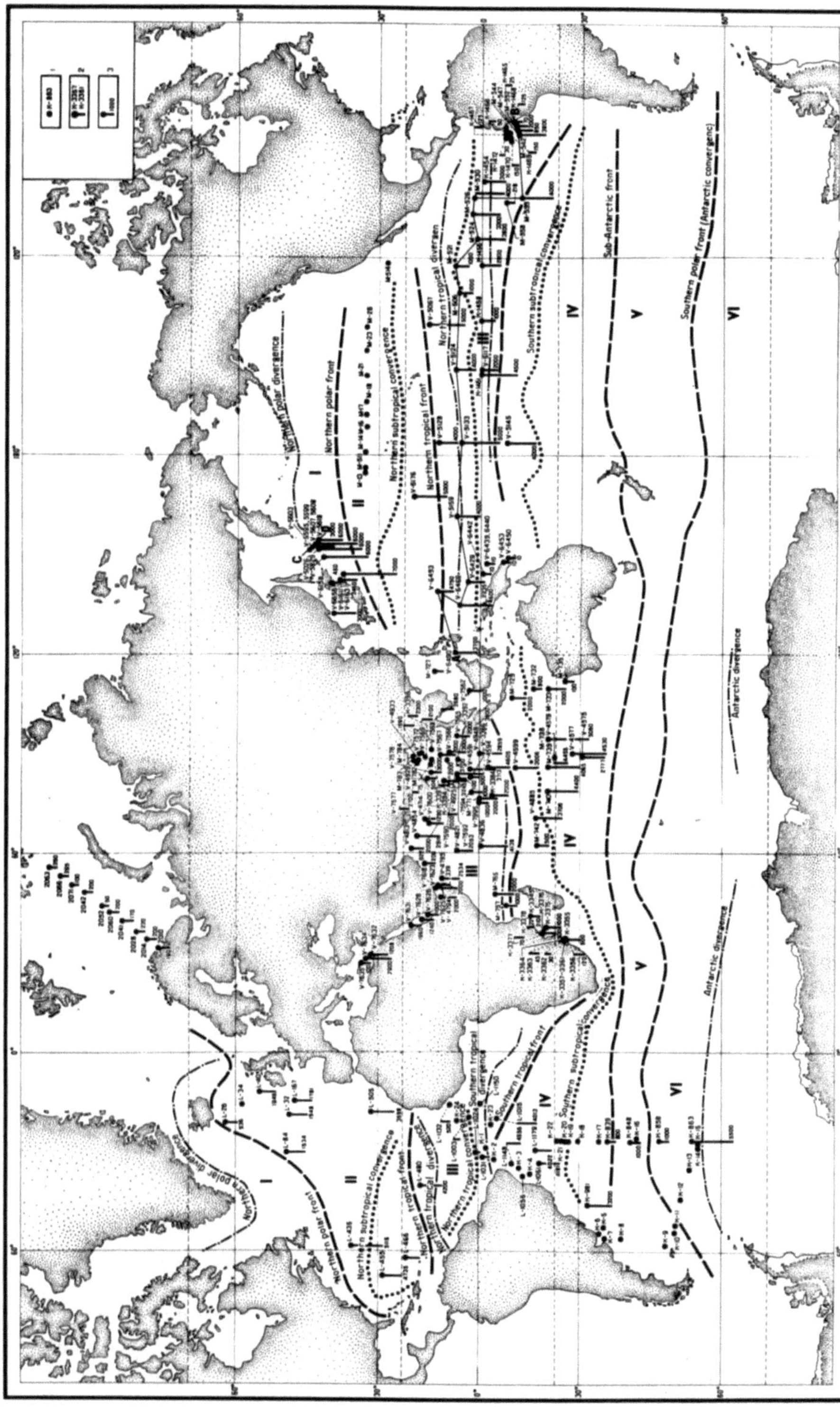

Fig. 2. The sites of samples in which the C_{org} content was determined: Station Nos: *V* Vityaz; *K* Akademik Kurchatov; *L* Mikhail Lomonosov; *M* Dmitry Mendeleev. Thickness of the explored mass, *m*. Macrocirculation systems: *I* sub arctic; *II* northern subtropical; *III* equatorial; *IV* southern subtropical; *V* subantarctic; *VI* antarctic. *Dots:* sampling sites in Soviet cruises alone. Sampling sites in inland seas not shown. *AB* section line

DOC distribution in ocean depths may thus manifest regional differences (Sverdrup et al. 1942, Redfield et al. 1963, Pytokowicz 1968). The other view presupposes that the supply of DOC, as well as of POC, to deep waters is negligible and their distribution is homogeneous (or nearly homogeneous) both in space and time (Menzel and Ryther 1970, Menzel 1974, MacKinnon 1981).

The present chapter contains material on the determination of the C_{org} content in approximately 1500 samples of sea water, performed by the methods of dry and/or wet burning developed in the Soviet Union (Skopintsev and Timofeyeva 1961, Liutsarev and Smetankin 1980). Most of the analyses for the C_{org} content involved 2–3 parallel determinations. The investigations covered different areas in the Pacific, Atlantic and Indian Oceans (Fig. 2).

A comparison of the results obtained by the two methods has indicated that they yield similar absolute values of the C_{org} content in sea water down to about 1000 m depth. At greater depths the method of wet burning (Liutsarev and Smetankin 1980a) yields lower values (about 0.7–1.2, average about 1.1 mg $C\,l^{-1}$). The values which are nearest to these were obtained by some western workers by the method of dry burning (Sharp 1973, Gordon and Sutcliffe 1973, MacKinnon 1978). However, it should be noted that the results of DOC determinations in the same area by one and the same researcher using the wet oxidation method (e.g., Liutsarev and Smetankin 1980a) may differ by 20–30%, occasionally even more. The reasion for this has not yet been clarified. Seemingly, it stems from the chemico-analytical difficulties involved in the quantitative microanalysis of DOC in a system as intricate as oceanic waters.

2.1 Total Quantity

Table 12 summarizes data on the C_{org} content in the water mass in the open part of the Pacific, Atlantic and Indian Oceans, which were obtained beyond the limits of the coastal shallow-water area, i.e., out of direct influence of fluvial waters, by the two aforementioned methods. It was compiled on the basis of the results from published works (Skopintsev and Timofeyeva 1962, Skopintesev et al. 1966, 1979, Starikova 1966, 1967, 1970, Liutsarev et al. 1972, Starikova and Yablokova 1973, Artemiev and Melnikov 1974, Artemiev and Scherbinin 1975, Liutsarev and Romankevich 1975, Romankevich 1977), as well as from data recently obtained at the P.P. Shirshov Institute of Oceanology, USSR Academy of Sciences. Only some of the water samples had been filtered prior to C_{org} determination to remove the bulk of suspended material. However, in the case of oceanic waters the POC content below the photic zone is so small as practically to have no effect on the accuracy of DOC determinations. Thus, the data in Table 12 are helpful in forming an idea of the overall content of C_{org} in oceanic waters.

The weighted average contents of DOC in the waters of the Pacific, Indian, and Atlantic Oceans are rather similar: 1.26, 1.28, 1.51 mg DOC $\cdot\,l^{-1}$, respectively. The water volumes of the various oceans being taken into account (*The World Water Balance and the Earth's Water Resources,* 1974), the average C_{org} content in the world ocean is 1.33 ± 0.20 mg DOC l^{-1} and its total quantity is

Table 12. The DOC content in the Pacific, Indian, and Atlantic Oceans (mg l^{-1})[a]

Depth, m	Number of samples	Extreme values	Average from all samples	Median	Max/min ratio		Occurrence frequency of concentrations, %					Number of stations	Weighted average content	
					cf.[b]	cf.[c]	<0.5	0.51–1.00	1.01–1.50	1.51–2.00	>2.00		extreme values	average
Pacific Ocean (707.1 × 10^6 km^3)														
0–100	196	0.54–2.72	1.40	1.35	5.04	1.59	0	23	30	30	17	35	0.80–2.62	1.39
101–200	47	0.49–2.52	1.21	1.07	5.14	1.42	0	40	34	17	9	30	0.69–2.45	1.22
201–1000	76	0.32–2.18	1.11	0.95	6.81	1.46	3	42	39	13	3	27	0.60–1.75	1.15
>1000	58	0.57–2.63	1.12	1.07	4.61	1.36	0	36	57	5	2	23	0.52–2.01	1.08
Total for the explored mass[d]	377	0.32–2.72	1.28	1.22	8.50	2.00	1	28	43	21	7	41 (23)	0.59–1.88 (0.72–1.88)	1.22 (1.26)
Indian Ocean (282.6 × 10^6 km^3)														
0–100	224	0.34–3.20	1.54	1.47	9.40	1.61	1	14	36	33	16	57	0.69–2.94	1.58
101–200	38	0.70–2.50	1.32	1.27	3.57	–	0	26	45	24	5	52	0.80–2.50	1.47
201–1000	88	0.69–2.70	1.38	1.30	3.91	1.21	0	24	31	41	4	45	0.80–2.50	1.34
>1000	67	0.46–1.80	1.26	1.26	3.91	1.26	2	19	55	24	0	38	0.58–1.90	1.25
Total for the explored mass[d]	423	0.34–3.20	1.42	1.40	9.40	1.88	1	18	38	33	10	65 (36)	0.63–2.12 (0.63–2.08)	1.32 (1.28)
Atlantic Ocean (329.7 × 10^6 km^3)														
0–100	59	0.66–3.88	1.79	1.76	5.88	1.30	0	7	27	37	29	21	1.09–2.56	1.83
101–200	23	0.69–2.76	1.60	1.65	4.00	1.61	0	9	26	56	9	21	0.77–2.08	1.63
201–1000	51	0.36–2.63	1.55	1.60	7.30	1.32	4	4	25	63	4	21	0.50–1.86	1.56
>1000	65	1.01–2.40	1.43	1.46	2.37	1.22	0	0	68	29	3	18	1.06–1.64	1.40
Total for the explored mass[d]	198	0.36–3.88	1.59	1.55	10.78	1.72	1	4	40	43	12	21 (9)	0.59–1.88 (0.50–2.56)	1.50 (1.51)

[a] Table was compiled from the results of DOC determinations made at the stations shown in Fig. 2
[b] The ratio of max and min values
[c] The average value from the ratio of max and min values, as established at each station
[d] Bracketed figures refer to the whole surface – bottom interval

Table 13. Total quantity of organic matter carbon in certain natural formations

Object of study	C_{org}, 10^{12} tons	Source
World ocean waters	1.8 – 2.2	Skopintsev (1965, 1971), Romankevich (1977), Skopintsev et al. (1979)
Subterranean waters	2.5	Shvets (1973)
Soils	1 – 3	Uspensky (1970), Vassoyevich et al. (1976), Bolin et al. (1977)
Peat	0.3	Vassoyevich et al. (1976)
Coals	30	Vassoyevich et al. (1976)
Combustible and coal shales	10	Vassoyevich et al. (1976)
Oil in deposits	2.5	Vassoyevich et al. (1976)
Hydrocarbon gases	2	Vassoyevich et al. (1976)
All rocks in the stratisphere (sidementites)	15,000	Vassoyevich et al. (1976)
Including those of the oceanic sector	5,000	Vassoyevich et al. (1976)

$(1.5-2.0)\times10^{12}$ tons, or, on average, 1.8×10^{12} tons DOC. This corresponds to about 5 kg C_{org} over each square metre of the ocean floor.

This value is inferior to those arrived at by Vernadsky (1934) – 5.0×10^{12} tons OM, or by Skopintsev et al. (1979) – 2.19×10^{12} tons C_{org}, but is 2 – 3 times the estimates given by some foreign authors – $(0.66-0.82)\times10^{12}$ tons C_{org} (Williams et al. 1969, Menzel 1974). It should be noted that in view of such significant discrepancies in the estimates of the total quantity of DOC in the world ocean (a three-fold difference between the extreme values) there is a need for undertaking an international intercalibration of sampling techniques and sample analysis methods for DOC content.

The total of 1.8×10^{12} tons C_{org} is of the same order as the quantity of C_{org} contained in subterranean waters, soils, peat (Table 13). So we may repeat: "that the impression of the negligibility of scattering disappears in reality and is replaced by a reverse one: the impression of immensity" (Vassoyevich 1973, p. 45).

2.2 Time of Residence

The residence time of organic matter carbon in the ocean is a fundamental biogeochemical characteristic of the system of OM decay, which was called by Vernadsky (1965) the second geochemical function of OM. Residence time indicates the length of the period of renovation of DOC in the ocean, and this gives an idea of its turnover rates and its conservatism.

In the ocean the bulk of DOC is contained in the deep and near-bottom waters which account, for instance, in the Pacific Ocean for 73% of the total water volume. The highly active surface layer (0 – 200 m) makes up about 5% of the water volume in the ocean and contains about 6% of total DOC. The cycle of OM in this layer proceeds at a fairly high rate. If we, like Vernadsky (1960, p. 75), consider the organisms as "semiliquid, occasionally liquid aqueous colloidal systems", then the residence time in the photic layer of this living "dissolved" OM (biomass) will be equal approximately to a single year (Khailov et al. 1973). The residence time of dissolved OM in the active oceanic layer, where its mass is only partly composed of the trophically valuable OM components, is considerably longer.

In water taken from the depths of 1880 m and 1920 m west of California, by means of a direct method using $^{14}C_{org}$, the residence time of DOC was established to be 3470 ± 330 and 3350 ± 300 years respectively (Williams et al. 1969). In the Black Sea the residence time of water determined after $^{14}C_{carl}$ at 300 – 700 m depths was found to be about 900 years, and at the 2000 m depth, 2000 ± 200 years (Östlund 1974).

Experimental investigations have also furnished indirect evidence of the great age of DOC in ocean depths of 500 – 2900 years (Postma 1968, Skopintsev 1971, Khailov et al. 1973).

Proceeding from gross phytoplankton production (which he assumed to be equal to 38.4×10^9 tons $C_{org}\,yr^{-1}$), the annual addition to the DOC reserve of the organic matter of plankton (3% of the primary production of 1.15×10^9 tons DOC yr^{-1}), and allochthonous production (5 mg l^{-1}; 180×10^6 tons DOC), as well as from the total OM quantity ($1.3 - 2 \times 10^{12}$ tons C_{org}), Skopintsev (1971) has calculated that the residence time of DOC in the ocean is 1000 – 1500 years.

The age of stable DOC calculated by a different, "trophic" method was found to be equal to about 2900 years (Khailov et al. 1973). The authors made use of the OM, with labelled carbon, from macrophyte thallomes which has been disintegrating in the sea water for about a year. Given additional assumptions concerning the use of particulate carbon by bacteria for plastic exchange, we may use the data cited in the above work to calculate that the age of stable DOC is about 2000 years.

Residence time of the bulk of DOC in the ocean can be estimated if the annual in situ consumption of oxygen for OM oxidation is known. Unfortunately, such data are contradictory (cf., for instance, Skopintsev 1981). The methods of determining biochemical oxygen demand result in overestimation of oxygen consumption compared to indirect estimates from hydrophysical parameters. According to Craig's calculations, the oxygen consumption rate in the Pacific deep waters is estimated at 0.004 ml $l^{-1}\,yr^{-1}$ (Craig 1971). In the northern part of the Atlantic the oxygen demand in deep waters is 0.002 ml $l^{-1}\,yr^{-1}$. The average biochemical oxygen demand in deep oceanic waters may apparently amount to 0.002 – 0.003 ml $l^{-1}\,yr^{-1}$ (Packard et al. 1971, Skopintsev 1981).

For the oxygen equivalent of 2.10 ml O per 1 mg C_{org} this corresponds approximately to 0.001 mg $C_{org}\,l^{-1}\,yr^{-1}$. Oxygen consumption for POC oxidation in deep zones is small. Only in zones of high biological productivity and transport to the seabed of a significant quantity of large organic particles can it be significant. Therefore, if we attribute the oxygen demand to the oxidation of DOM alone, the content of which in deep waters averages 1.1 mg $C\,l^{-1}$, the residence time of this OM will amount to 1100 years.

In some works the OM age in ocean depths is estimated (or can be calculated on the basis of the data supplied by the authors) at significantly lower values, namely, 20 – 140 years (e.g., Sorokin 1971a).

Given below are the calculations made by the author on the quantity of annual supply of carbon of the relatively stable DOM into the ocean and its residence time. The calculations included materials on the supply to the ocean of both allochthonous and autochthonous OM.

3% of gross OM production by phytoplankton	900×10^6 t	1109×10^6
6% of OM production by phytobenthos	36×10^6 t	
50% of DOM input from land	137×10^6 t	
5% of POM input from land	36×10^6 t	
Total quantity of DOC below 200 – 300 m	$(1.4 - 1.9) \times 10^{12}$ t	
Residence time of the bulk of DOC in the ocean	1300 – 1700 yr	
Contribution of allochthonous DOC to the formation of total DOM in the ocean	about 16%	
The quantity of DOC annually removed from the cycle	1109×10^6 t	

An important and very difficult problem is the one related to the estimation of the quantity of OM which fails to disintegrate in the active upper oceanic layer and goes over from the chains of primary producers, consumers and reducers into the reserve of deep-water DOC. For phytoplankton and phytobenthos we may adopt the estimates made by Skopintsev (1950, 1981) – 3% (for plankton) and Khailov (1971) – 6% (for plankton and phytobenthos). Menzel (1974) believes that the annual supply into the DOC reserve from the chains of primary producers (phytoplankton), consumers and bacteria is 4% of gross OM production by phytoplankton.

In river waters about 30% of OM (11 – 50%) is highly unstable to the effect of decay and is oxidized to CO_2 within 40 – 70 days. The rest of OM (70%) disintegrates within a year at 15 – 20 °C by about 50%. Under lower temperatures (8 °C and 0 °C) about 23% and 10% of this relatively stable OM will disintegrate in a year (Skopintsev 1950). An experiment with the decay of "humic water" (flowing out of a swamp) has demonstrated that at 20 °C as much as 30 – 40% of this stable OM is subject of decay over a year (Romanenko 1972). The bulk of fluvial OM arrives in the tropical zone of the ocean where decay processes proceed at a rapid rate. OM here partly coagulates and goes over into the particulate material which, in turn, partly dissolves. It may be assumed that about 50% of DOM from river discharge replenishes the reserve of stable OM in the ocean. A similar value has been arbitrarily adopted for the OM of the subterranean inflow. No data are available on this score. However, the principal processes that compel us to introduce the factor of 0.5 are those of coagulation and sorption.

At present, we may only tentatively estimate the quantity of OM forming in the course of dissolution and incomplete decay of allochthonous particulate OM coming into the ocean from land and going over into the reserve of stable OM. The adopted value for the OM of the solid fluvial and glacial runoffs, eolian transfers and wave abrasion (5% of the total) was obtained by extrapolating the results of three-year-long observations of decomposition of POM in a number of rivers of the Arctic Ocean basin (Romankevich 1977).

In deep oceanic waters about 16% of DOC is of the allochthonous origin. This value was obtained as a result of balance calculations. In the future it should be verified by means of direct determinations of lignine decomposition products in different types of waters. However, the determination of aromatic aldehydes alone in sea water will not be able to supply an unambiguous answer to this question. There is a need for a complex utilization of numerous tracers of allochthonous OM and a comparative study of its polymer composition and of dissolved OM in sea water.

In the surface waters 45 – 70% of C_{org} synthesized by planktonic algae and macrophytes changes over during their lifetime and postmortem into DOC (Khailov 1971).

In the photic layer the allochthonous DOC accounts for not more than 5% of total DOC. This is true for the photic layer of the ocean as a whole. Near river mouths this relationship may be different. A different ratio between the allochthonous and autochthonous DOC in the photic layer and in deep waters is principally the result of the rapid utilization of the trophically valuable metabolites and decay products of marine organisms.

The presence in the ocean of a large quantity of DOC slowly withdrawn from the cycle indicates that the bulk of OM contained below the photic layer is resistant to enzymatic and chemical oxidation. Direct observations of the variation of its content in ocean depths corroborate these views (cf. Sect. 2.3). Microbiologic, chemical and isotopic studies likewise give evidence of the conservatism of the bulk of DOC in oceanic depths.

Below 500 – 800 m the content of saprophytic bacteria growing on different (mainly proteinic) substrates is extremely small, and the total biomass of the bacteria, as well as their production and heterotrophic activity are low (Sorokin 1973, 1977). The isotopic ratio $DO^{13}C/DO^{12}C$ in the mass of oceanic waters varies but little (Williams 1968, Williams and Gordon 1970). The weak decreasing trend of $DO^{13}C$ with depth points, firstly, to OM decay that is taking place, and, secondly, to the preferable mineralization of less conservative OM components that are heavier with respect to $DO^{13}C$ (Galimov 1968, Parker and Calder 1970).

A relatively high resistance of the bulk of DOC to oxidation may be caused by a number of factors. They can be grouped under two categories based on different opinions.

One view is that the main factor determining the long-term residence and the large quantity of DOC in the ocean is the nature of the OM itself. Following the decay in the upper ocean layers of the bulk of primary production, at depths below 200 – 500 m there are preserved stable, biologically inert organic compounds very slowly undergoing transformation (e.g, Skopintsev 1971, Menzel 1974). Low DOC concentrations, the presence of antibiotics in sea water, low temperatures and high hydrostatic pressure are not chief inhibitors of the OM decay processes.

The other view insists that environmental factors, such as temperature, pressure, the presence of interfaces, as well as OM concentration and the formation of high-molecular organometallic complexes have a crucial effect on microbiological activity and OM preservation in ocean depths (e.g., Duursma 1961, Jannasch et al. 1970, 1971, Jannasch and Wirsen 1973, Sorokin 1973, 1978, Dawson and Duursma 1981). In some cases the substrate specificity of the bacteria proves very important (Jannasch et al. 1974).

To be able to understand the cycle of the bulk of OM in the ocean neither the geochemists nor the ecologists can be indifferent to whether the deep-water OM is stable only in situ (low temperature, high pressure, a lower number of interfaces) or whether it is due to the nature of OM (chemical composition, structure) which fails to be utilized by bios even where deep waters upwell to the photic layer. In the latter case the probability of OM being removed from the cycle due to slow coagulation and transition into suspended material, sorption and settling onto the bottom becomes greater.

Deep DOM resistance to destruction and its low trophic value due to its chemical composition and structure have found proof in the researches on the destruction of DOM by the microflora of natural waters. In some experiments it was observed that a natural population of microorganisms oxidizes up to 50% of OM in the surface water layer and does not utilize at all (or utilizes very slowly) the OM from deep waters even at 20 °C and normal pressure (e.g., Barber 1968). In contrast to this, the method of apparent oxygen demand[2] (e.g., Ogura 1970) points to a great effect of environmental factors, particularly, temperature and pressure, on decomposition of deep-water DOM. A comparison between oxygen demand and DOC content has shown that in surface waters about 50% of OM is capable of mineralizing to CO_2, whereas in deep waters (below 500 m) no decomposition of OM has been recorded. A temperature decrease alone is hardly capable of increasing 50 – 100 times the presence of DOC in deep waters, as compared to particulate carbon, (materials on particulate OM are given in Chap. 3), or to prolong 20 – 30 times in an experiment the presence of resistant DOM as compared to the trophically valuable DOC components the rate of bacterial consumption of which is 4 – 20 times as high for the same temperature (Khailov et al. 1973). A high hydrostatic pressure will in many cases lower the activity of enzymatic systems and suppress the growth of heterotrophic microorganisms (Kriss 1973, 1976). Long-term experiments at the depth of 1830 m have demonstrated that the rates of OM decay in situ are 1 – 3 orders below those observed under the laboratory conditions (Jannasch and Wirsen 1973). At the same time, the

2 The apparent O_2 demand is a difference between the O_2 content in the conditions of an equilibrium with the atmosphere under in situ temperature and salinity, and the measured O_2 concentration. This method, notwithstanding certain assumptions, gives an idea of the in situ oxidation principally of DOC

dehydrogenase activity of bar-tolerant bacteria in deep-water layers may in certain cases be high and comparable with bacterial activity in surface water layers (Stupakova 1975).

An analysis of the data on OM decay has led the author to the conclusion that the conservative nature of DOM is an important, though not the chief factor, and only when combined with low temperature and high pressure in deep waters will it determine the great stability and age of OM. Proceeding from the total amount of DOC in the ocean and its residence time, we may calculate that about 1×10^9 tons DOC is being withdrawn annually from the ocean. The fate of deep-water DOC is different: the greater part of it disintegrates down to the end products, but some portion is adsorbed on suspended particles and goes over into the fossil state. This is evidenced by a lower degree of the co-precipitation of deep-water OM with hydroxides under experimental conditions and by specific OM composition in red clays, pointing to the fact that this OM was, apparently, largely adsorbed from sea water. A quantitative assessment of these quantities is a task of future research.

2.3 Distribution Patterns

Figure 3 shows a general pattern of distribution of DOM carbon in the water mass of the Pacific, Indian and Atlantic Oceans, and Figs. 4–9 gives DOC distribution at the specific stations.

The DOC content at the majority of stations is observed to decrease with depth. In the Pacific the weighted average DOC content in the layers 0–100, 100–200, 200–1000 and below 1000 m is 1.39, 1.22, 1.15, 1.08 mg DOC l^{-1}, in the Indian Ocean – 1.58, 1.47, 1.34, 1.25 mg DOC l^{-1}, in the Atlantic – 1.83, 1.63, 1.56, 1.40 mg DOC $\cdot$ l^{-1}, respectively (see Table 12). In the Pacific the most frequent concentrations in the 0–100 m layer are 1.01–2.00 mg DOC l^{-1} (60%) and deeper – 0.51–1.50 mg DOC l^{-1} (74–93%). In the Indian and Atlantic Oceans the rate of occurrence of high DOC concentrations, as well as average weights for all of the above depth ranges, are higher than those in the Pacific.

The variability limits of extreme values of DOC content are great: in the Pacific 0.32–2.72 mg l^{-1} (0–7000 m layer), in the Indian Ocean 0.34–3.20 mg l^{-1} (0–5060 m layer), in the Atlantic 0.35–3.88 mg l^{-1} (0–5565 m layer). The values for the areas with a great influence of river runoff have been omitted. However, the deviation of these values from the average in 70–90% of cases does not surpass ± 0.3 mg DOC l^{-1}. The deviation from the weighted average concentrations for the 0–200 m layer amounts to ± 0.3, and at greater depths, to ± 0.2 mg DOC l^{-1}.

In each of the three oceans the highest DOC values are characteristic of the uppermost layers, while low values occur at all depths. A steep slope of the curves limiting the region of maximum concentrations points to a more abrupt decrease of DOC in the Pacific and Indian Oceans in the 0–400 m layer (see Fig. 3). In the Atlantic this is less distinctly manifested.

In the vertical direction a decrease is observed, though not a smooth one, in DOC content, in the occurrence rate of high concentrations and of the ratio of

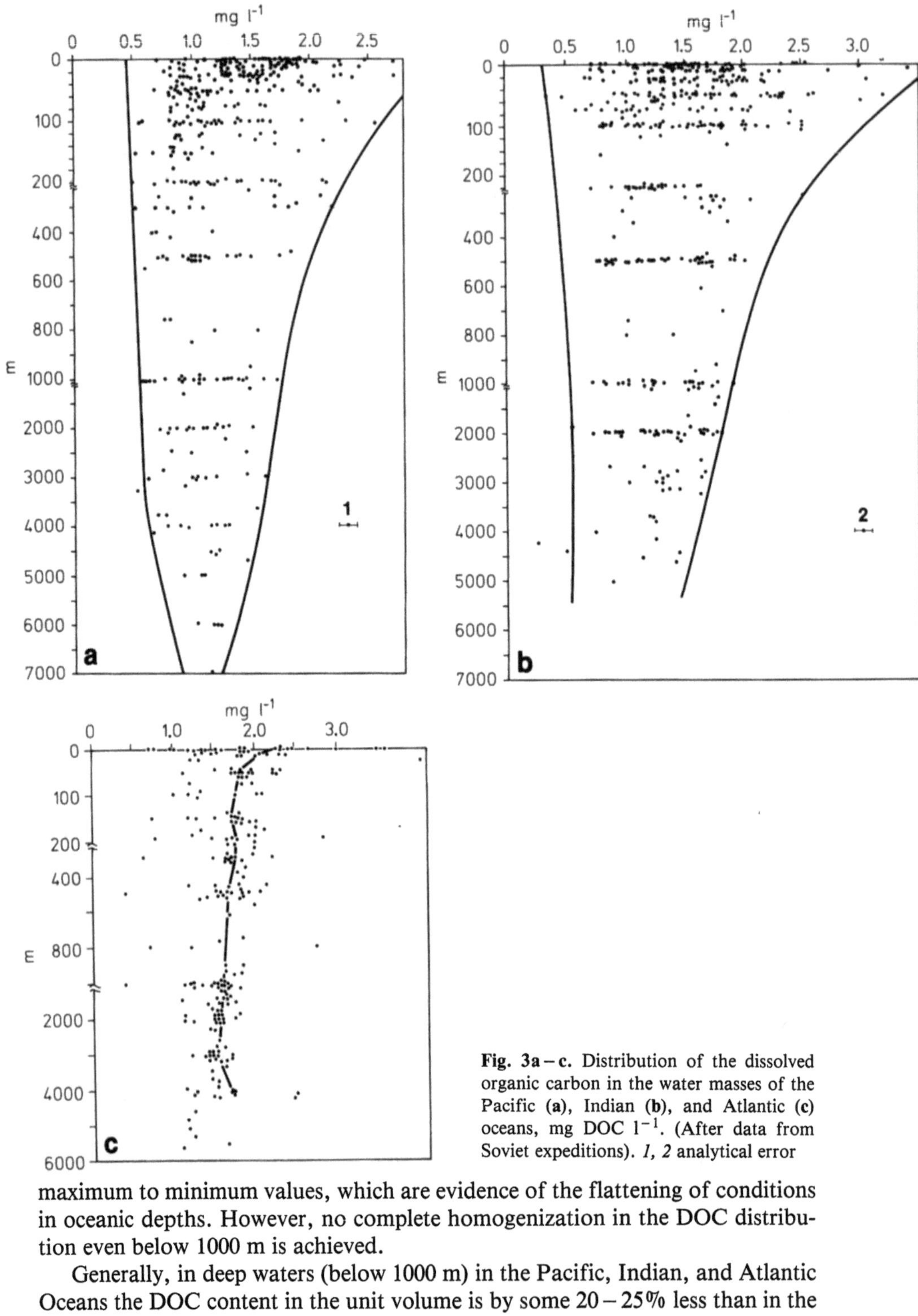

Fig. 3a–c. Distribution of the dissolved organic carbon in the water masses of the Pacific (**a**), Indian (**b**), and Atlantic (**c**) oceans, mg DOC l^{-1}. (After data from Soviet expeditions). *1, 2* analytical error

maximum to minimum values, which are evidence of the flattening of conditions in oceanic depths. However, no complete homogenization in the DOC distribution even below 1000 m is achieved.

Generally, in deep waters (below 1000 m) in the Pacific, Indian, and Atlantic Oceans the DOC content in the unit volume is by some 20–25% less than in the

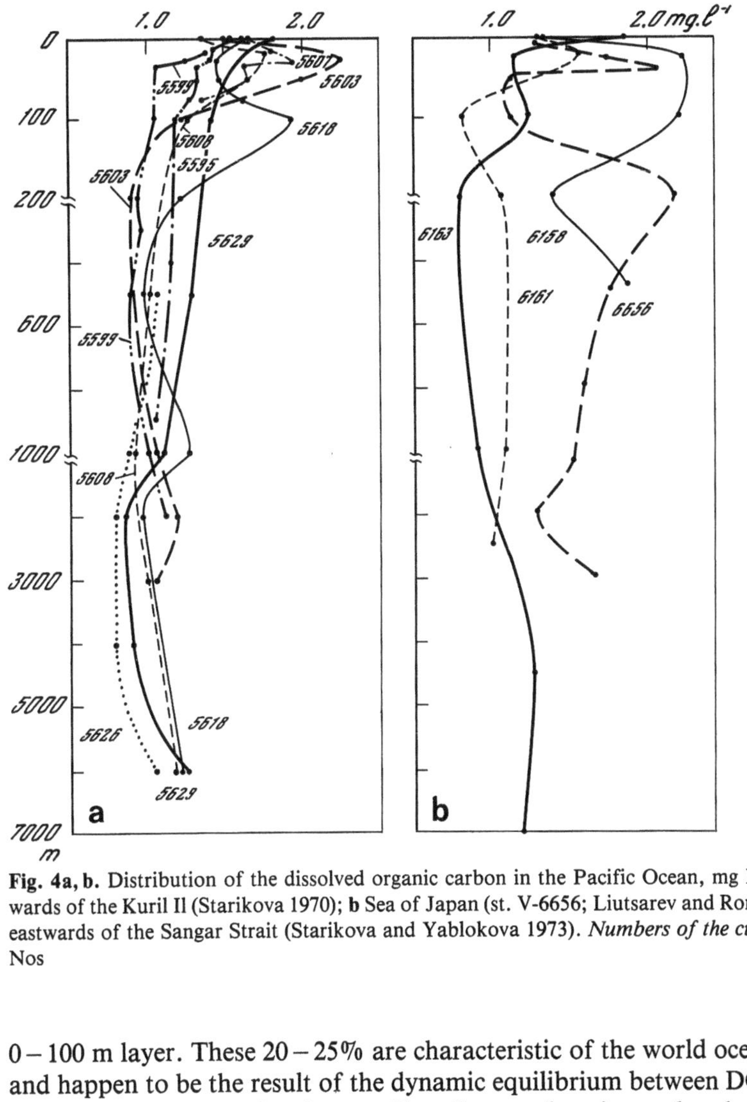

Fig. 4a, b. Distribution of the dissolved organic carbon in the Pacific Ocean, mg DOC l^{-1}. **a** eastwards of the Kuril Il (Starikova 1970); **b** Sea of Japan (st. V-6656; Liutsarev and Romankevich 1975), eastwards of the Sangar Strait (Starikova and Yablokova 1973). *Numbers of the curves* are Station Nos

0 – 100 m layer. These 20 – 25% are characteristic of the world ocean as a whole and happen to be the result of the dynamic equilibrium between DOC input into ocean depths (convective downwelling flows, advective and turbulent transfer, particulate carbon changing into dissolved carbon) and its loss (mineralization due to bios, chemical decomposition to CO_2, aggregation, flocculation, utilization and precipitation of aggregates). In contrast to any surface region, hydrodynamics affects this quantity only indirectly, determining the residence time of OM, the degree of aggregation at interfaces, the DOC input to the photosynthesis zone where the probability of DOC utilization becomes greater.

An essential factor is the presence in the oceans of the areas with generally increased DOC content. In the upper layer (0 – 200 m) where occurs the reproduction and decay of the bulk of plankton the regional differences in the DOC content do not evoke much surprise. However, the fact of their existence in deeper layers requires additional confirmation.

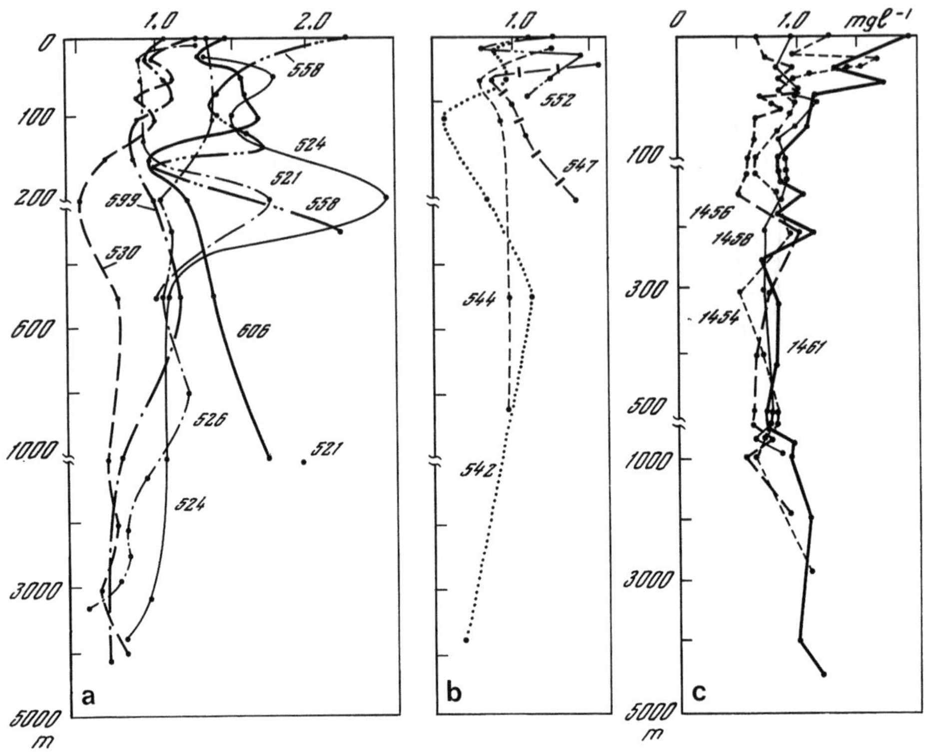

Fig. 5a – c. Distribution of the dissolved organic carbon in the Pacific Ocean, mg l^{-1}. **a** eastern equatorial-tropical area, data of 1972 (Romankevich 1977); **b** area off Peru, data of 1972 (Romankevich 1977); **c** equatorial region and eastern part of the ocean. (Liutsarev et al. 1975). *Numbers on the curves* are Station Nos

The Pacific. In the explored part of the Pacific such areas are the western equatorial-tropical region and the Sea of Japan (Liutsarev and Romankevich 1975, Romankevich 1977). With an overall vertical decrease of DOC, higher DOC values are preserved here down to the maximum investigated depths (2200 – 4700 m) compared to other areas of the Pacific. Particularly high values were recorded in the low-production area of water convergence (the western part of the anticyclonic subtropical structure), as well as in the Sulu Sea.

Water samples taken from the surface layer off New Guinea (Langemak Bay), near atolls in the western part of the ocean (Admiralty, Ninigo Is.) did not show an increased DOC content (1.36 – 1.75, average 1.64 mg l^{-1}).

In the Kuril-Kamchatka area, east of the Tsungaru Strait and in the Sea of Japan, as well as in the Peruvian area and in the eastern equatorial-tropical region the DOC content at all the horizons was found to be on the average lower than in the western and central equatorial-tropical area of the Pacific.

Differences in DOC distribution are distinctly manifested on the latitudinal profile from the Sulu Sea to Peru, running in the open part of the Pacific nearly entirely within the confines of the equatorial water structure (Fig. 10). In the surface layers a change-over of DOC content is observed here in the west-east

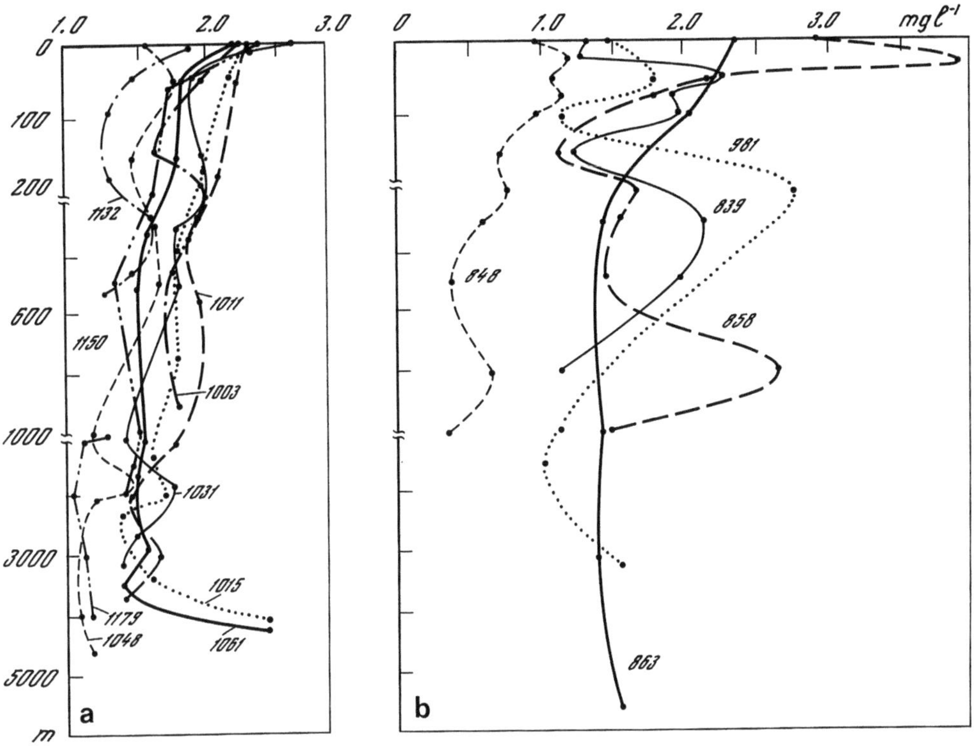

Fig. 6a, b. Distribution of the dissolved organic carbon in the Atlantic Ocean, mg l^{-1}. **a** tropical area South of equator (Skopintsev et al. 1966); **b** southern part of the ocean (Artemiev and Melnikov 1974). *Numbers on the curves* are Station Nos

direction from over 2.00 to $1.00-0.76$ mg l^{-1}. In the intermediate layers the pattern is more complicated. With the overall decrease of DOC in the eastern direction there are observed some spots of increased DOC concentrations at the depths of about 200, 500, 1000 m. An increase of DOC in the intermediate waters is particularly characteristic of the eastern equatorial region and is, possibly, associated with DOC supply from the west by intertrade and equatorial deep countercurrents, as well as by OM penetration from the high-production photic zone. Such a distribution of DOC agrees with the presence here of an intricate system of zonal opposite-directed currents, as well as of local upwelling and downwelling zones.

In the waters below 1000 m the DOC content is likewise decreasing in the eastern direction from $1.50-1.01$ to $0.75-0.51$ mg l^{-1}. Consequently, a decrease of DOC content from west to east takes place in the surface, intermediate and deep waters.

During a later cruise of the research vessel "Akademik Kurchatov" (1974) it was once again demonstrated that the eastern high-productivity equatorial zone and the area off Peru are not distinguished by a high DOC content (Fig. 5c). However, other workers cite high values of DOC content in these areas (up to 4 mg l^{-1} and more, average >2.5 mg l^{-1}, for the $0-200$ m layer) and assume

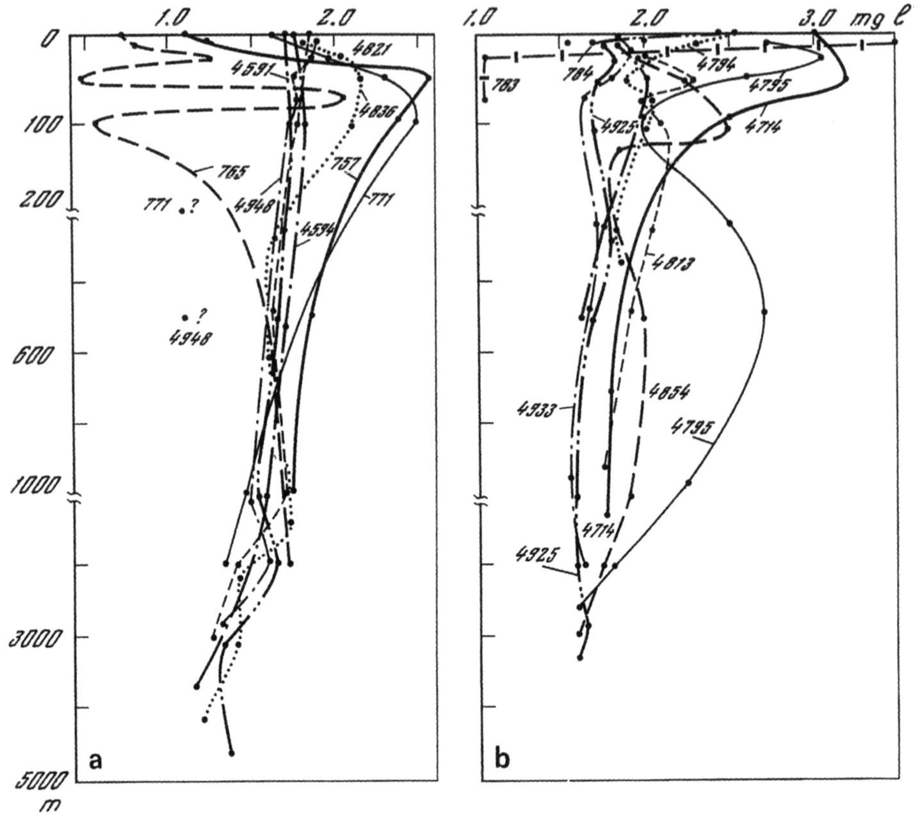

Fig. 7a, b. Distribution of the dissolved organic carbon in the Indian Ocean, mg DOC l^{-1} (Starikova 1967, Artemiev and Shcherbinin 1975, with additional data). **a** equatorial waters; **b** Bay of Benghal, Arabian Sea. *Numbers on the curves* are Station Nos

that there exists a direct correlation between primary phytoplankton production and DOC content both in the surface and in deep waters (Skopintsev et al. 1979). Our data do not substantiate this thesis.

The materials used to arrive at these conclusions were obtained directly in sea cruises on the basis of a uniform methodology (Liutsarev and Smetankin 1980). Every possible precaution was taken to avoid the contamination of the samples in the process of sampling with water bottles (to this end, the 200-litre water bottles made of stainless steel were specially cleansed), during the storage in the coolers (not more than 1 – 5 days when necessary) and while determining the blank correction. The identified differences in DOC content in deep waters are a few times as great as individual errors of determination and seem quite trustworthy.

In the Pacific equatorial-tropical zone a decrease of DOC content from west to east is accompanied by an increase of phytoplankton production (see Fig. 1), zooplankton biomass and POC content. Heterotrophic microorganisms growing on nutritive substrates exhibit a different distribution. Their content in the Pacific western tropical humid area, oligotrophic with respect to phytoplankton,

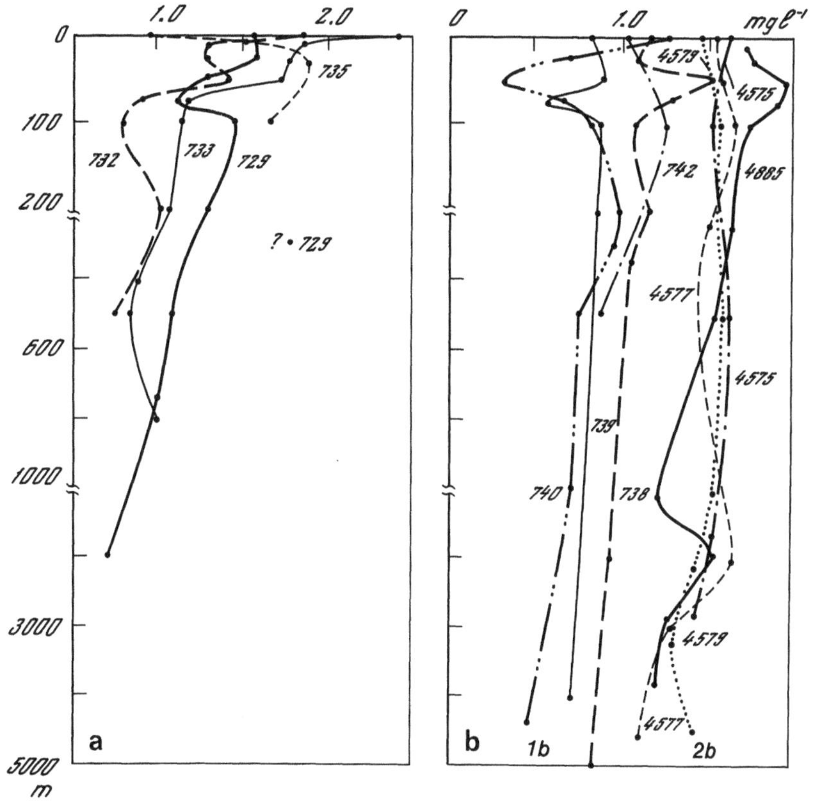

Fig. 8a, b. Distribution of the dissolved organic carbon in the Indian Ocean, mg DOC l^{-1}. **a** Java-Australian region; **b** subtropical waters; *1b* after data by Artemiev and Shcherbinin (1975); *2b* after data of 1959 – 1960 by Starikova (1967). *Numbers on the curves* are Station Nos

is observed to be higher than in the high-production south-eastern area, off Peru and Chile (Kriss et al. 1972, Kriss 1976), this being in agreement with the DOC distribution pattern.

It is worthwhile to note an asymmetry in OM production in the ocean and on land. The bulk of OM on land (biomass about 520×10^9 tons C_{org} and production about 54×10^9 tons C_{org} per year) is produced in tropical and subtropical humid zones (80%) where phytoplankton production is low and DOC content is high. Part of the stable terrigenous OM inavoidably finds its way into the ocean, the input in the equatorial-tropical humid oceanic areas being larger than in the arid ones.

Atlantic Ocean. Most of the data on DOC in the Atlantic have been obtained between 62°N to 16°S by the method of dry burning. Skopintsev with co-authors (Skopintsev et al. 1966, 1979) after examining these materials noted that the DOC content at the same depths in different ocean areas was rather similar. DOC content, as a rule, declines with depth. Increased DOC concentrations were

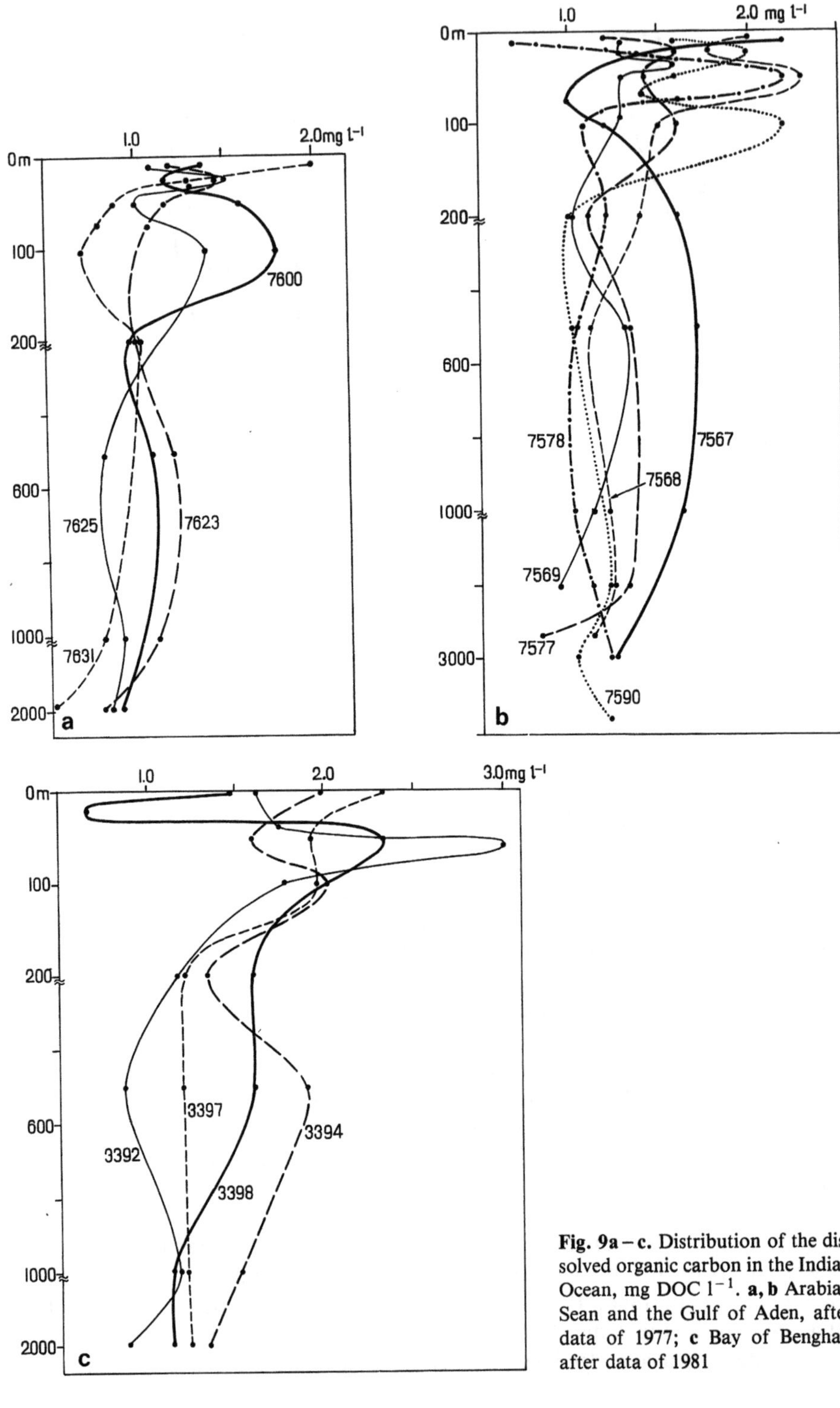

Fig. 9a – c. Distribution of the dissolved organic carbon in the Indian Ocean, mg DOC l^{-1}. **a, b** Arabian Sean and the Gulf of Aden, after data of 1977; **c** Bay of Benghal, after data of 1981

recorded in the upper layer at the equator, as well as at the lowest horizons at some of the stations (L-1015, 1062; cf. Fig. 6a). The weighted average DOC content in the water mass from 0 to 4000 m was, according to these data, 1.54 mg l^{-1}. From the surface down to the depth of 3000 m the average DOC concentration decreased by a factor of 1.5. The greatest variations (1.3 fold) took place in the upper 150-metre layer.

In November – December 1971, mainly on a meridional profile along 26°W between 6°N and 56°S, there were taken water samples in which DOC content was determined after the Liutsarev and Smetankin (1980) method (cf. Fig. 6b).

A comparison of all these materials indicates that the weighted average concentrations of C_{org}, as obtained by both methods in different parts of the Atlantic Ocean, are quite similar. For the 0 – 100 m layer they were found to be 1.83 mg l^{-1}. For the 100 – 200 m, 200 – 1000 m and 1000 – 4000 m layers the weighted average content of C_{org} obtained by the methods of dry and wet burning were 1.65 and 1.49 mg l^{-1}, 1.60 and 1.38 mg l^{-1}, 1.48 and 1.31 mg l^{-1}, respectively.

Arctic Ocean. Data on DOC distribution in this area are very scarce (Kinney et al. 1971, Melnikov and Pavlov 1978, Melnikov 1980, Beliayeva and Romankevich 1983). Direct measurements of DOC made in May – October 1975 and in April – May 1976 at the drifting station "Severnyi Polius (North Pole) – 22" by the method of wet burning (Liutsarev and Smetankin 1980) yielded relatively high values of 2 – 3 mg DOC l^{-1} in the surface and subsurface waters and 0.5 – 1.0 mg DOC l^{-1} at the depth of 500 m. Consequently, despite a short vegetation period and low primary plankton production and biomass values, DOC concentrations obtained in the Arctic Ocean's pelagial are comparable to those in more productive areas of the Antarctic.

DOC concentrations in the ice, measured during the drift of the station, showed inhomogeneity of DOC distribution in the ice mass and of DOC content in the autumn type of ice – 1.1 – 2.1 mg DOC l^{-1}, in the winter type – 1.8 – 3.1 md DOC l^{-1}, and in the perennial type – 1.9 – 2.4 mg DOC l^{-1}. Variations of the DOC content in the ice mass of different types are associated with the peculiarities of the ice-formation process, with sharp changes of the ambient temperature and ice growth rate, with the penetration of dissolved and particulate substances into intercrystalline space, with their squeezing out into the water and the redistribution of brines (Melnikov 1980).

Determination of DOM in the Barents Sea at the Murmansk-Franz Josef Land profile (non-filtered water samples were analyzed) performed by the same method in August – September 1976 yielded for the 0 – 270 m layer – 1.0 – 4.8, average 2.2 mg C_{org} l^{-1} (Beliayeva and Romankevich 1983). Maximum concentrations of C_{org} in the sea water were recorded in the upper water layers in the southern part of the profile. The content of C_{org} was observed to decrease with depth and in the direction of Franz Josef Land.

The waters of the North-Cape current are characterized by high C_{org} concentrations. The waters of the cold Polar current penetrating into the Barents Sea from north-east can be distinctly traced in the form of a tongue with low DOM

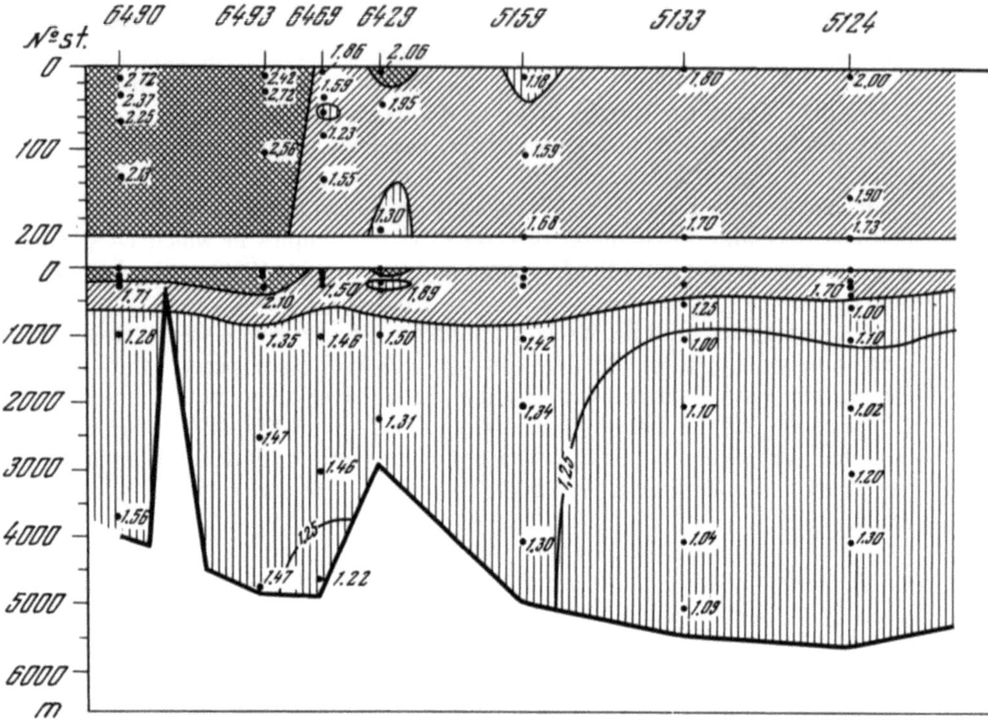

Fig. 10. Distribution of the dissolved organic carbon on a latitudinal profile in the equatorial-tropical area of the Pacific, mg DOC l^{-1}. *1* = <0.50; *2* = 0.51 – 0.75; *3* = 0.76 – 1.00; *4* = 1.01 – 1.50; *5* = 1.51 – 2.00; *6* = >2.00. The location of the profile is shown in Fig. 2 (line *AB*)

concentrations. In different parts of the profile the average DOM concentration is 1.5 – 2 times that in the 0 – 200 m layer of the world ocean. This is due to high production indices of phytoplankton and phytobenthos in the Barents Sea, reduced decay rates of OM and the resultant of the balance: input – decay – outflow into adjacent marine areas.

Indian Ocean. First data on DOC in the Indian Ocean were obtained in 1959 – 1961 (Starikova 1967). The samples taken during this cruise were filtered through a glass filter covered with a layer of barium sulphate and preserved by adding muriatic acid. 1 – 3 months later the water was evaporated and the residue was subjected to thermal decomposition after the Skopintsev method (Skopintsev and Timofeyeva 1961).

In 1973 these data were supplemented with DOC analyses (14 stations) performed directly on board (Artemiev and Scherbinin 1975) after a special procedure (Liutsarev and Smetankin 1980).

In the deep waters of the equatorial structure, similar values were obtained in both cases (cf. Fig. 7a). The weighted average concentration below the depth of 1000 m in the first case was 1.43 – 1.50, average 1.45 mg l^{-1}, and in the second case – 1.40 – 1.69, average 1.54 mg l^{-1}. At the same time, for the halistatic re-

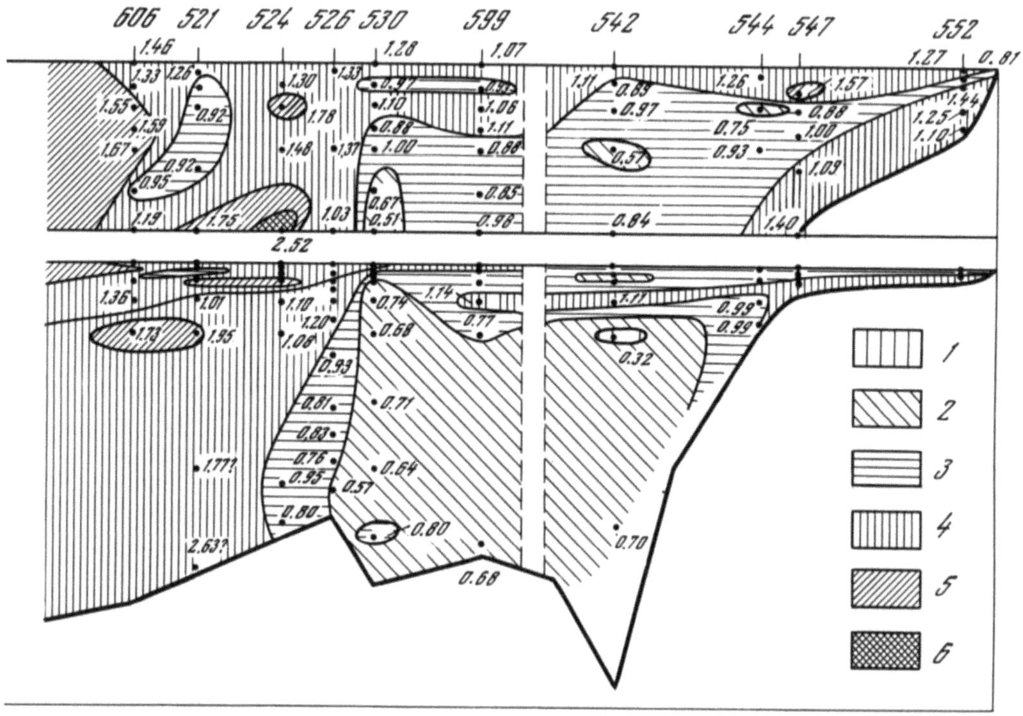

Fig. 10 (continued)

gion south of the southern tropical front (cf. Fig. 8b) the DOC values at all depths, obtained in July 1973, proved to be 1.5 – 2 times as low as the ones cited earlier.

According to the 1973 data the waters of the subtropical structure were found to be the poorest in DOC content in the Indian Ocean (for the 1000 – 5000 m layer the weighted average value was 0.70 mg DOC l^{-1}). Here there is a clear relationship between salinity and the DOC content. These waters were also among the most transparent in the world ocean.

In the Javan-Australian region DOC content in the upper layers was higher than in the area (the 1973 data are compared in Fig. 8a). Below 500 – 600 m the differences disappeared and the values approached those encountered at the same depths in the halistatic area with central-Indian water (0.71 – 1.09 mg DOC l^{-1}).

In 1977 and 1981 the DOC content was determined in the mass of waters in the northern part of the Indian Ocean (Fig. 9) at 20 and 18 stations, respectively, by using a uniform methodology of sampling and analysing by the method of wet oxidation (Liutsarev and Smetankin 1980a). The weighted average concentrations for all the stations in the Bay of Bengal, the Gulf of Aden and the Arabian Sea was found to be lower than those obtained earlier in the same areas: about

1.8 mg DOC l^{-1} (0 – 100 m) and about 1.1 mg DOC l^{-1} (below 1000 m). These values are closer to the ones cited by MacKinnon (1981) for the Atlantic and Pacific.

Thus, the new data do not corroborate the earlier recorded decline of DOC content in transition from the highly productive waters of the Arabian Sea and the Gulf of Aden to the waters of equatorial and subtropical structures. At present it is yet too early to speak of regional differences in DOC content in deep waters or of the existence of a direct correlation between phytoplankton production and DOC concentration at all depths.

The problem of regional differences in DOC content in the mass of oceanic waters is, in effect, part of a more comprehensive and complex issue concerned with the character of oceanic water circulation, which is as yet far from being solved. In accordance with the existing concepts, the intermediate waters (200 – 2000 m) are characterized by a system of equatorial counter-currents, by the spreading of subtropical intermediate waters towards high and low latitudes, and by the intensification of meridional circulation. Deep and near-bottom waters are being formed in the North Atlantic and in the Weddell Sea where the conditions facilitate the downwelling of the coldest waters, richest in oxygen. The transformed deep waters of Atlantic origin, by-passing Africa, become replenished with deep Antarctic waters and penetrate far to the north into the Indian and Pacific Oceans.

The transfer of deep and near-bottom waters has not been adequately investigated, but generally it appears that for the surface waters of greatest importance is the latitudinal component of transfer, whereas in the case of deep and near-bottom waters – the meridional component. However, the average long-period transfer in the entire mass of the Pacific waters appears to retain a latitudinal direction (Stepanov et al. 1977, Burkov 1980).

At present we do not possess enough data obtained by reliable methods yielding comparable results to be able to give an unambiguous answer to the question whether there exist regional differences in DOC content in the waters below 500 – 1000 m.

It can be presumed that its important source in these waters are not the stable transformation products of planktonic OM alone, but also stable allochthonous OM supplied from the land. The latter, apparently, is principally responsible for the differences in DOC concentrations in the waters of the explored areas of the western and eastern parts of the Pacific equatorial-tropical region.

A decrease of DOC in the waters of the Pacific northern equatorial-tropical zone in the west-east direction may be due to the slow ageing and mineralization of autochthonous und allochthonous OM transferred in this direction in accordance with the existing concepts of the average long-period transfer of waters. However, the decay of this OM with a great proportion of conservative allochthonous components is, apparently, unable, at intermediate depths, to effect the reduction of oxygen, the content of which in the Pacific western equatorial-tropical zone is higher at corresponding depths than in the eastern zone. This is in agreement with the lower activity of enzymic systems in tropical waters and with the turnover rate of DOM which is several times as low in the Pacific western tropical waters compared to subtropical and Kuroshio waters (e.g., Sieburth 1971, Seki et al. 1972). This is a prerogative of a more labile OM the oxidation of which in the high-production upwelling zones contributes to the formation here at intermediate depths of a layer of oxygen minimum.

One of the ways to establish a relationship between DOC and phytoplankton production involves an analysis of correlations between the average values of these parameters for large ocean regions. Regional DOC distributions indicate that the relationship between plankton, the principal OM source in the ocean, and DOC may take various forms. The type of relationship in a specific place and at a specific time depends on the sign of the resultant component in the supply-loss system. This is determined by the syngenetic production and the input of allochthonous DOC, the mineralization of OM and its outflow with the currents. Owing to the conservative nature of the bulk of DOC and its long residence time, the effect of currents on DOC distribution in the ocean is extremely strong, so that it determines the type of relationship with biological productivity. The absence of significant correlation between these indices appears to be quite common.

This was noted in 1972 while carrying out biogeochemical research in the Peruvian upwelling zone. The abundant development of phytoplankton in the area did not lead to any perceptible increase of DOC content either in the deep or surface waters. The causes behind this are, apparently, the rapid mineralization of labile OM, the transition into the particulate form through aggregation, sorption and bioutilization, the outflow of decay products of DOM with the Humboldt current waters and the reduction DOC concentration due to the inflow of deep Antarctic waters which are poor in DOC. At the same time, some researches working in the area of the Peruvian upwelling observed in other years increased DOC concentrations which were found to be in direct correlation with phytoplankton production (Skopintsev et al. 1979).

Winding up the discussion of problems involved in the DOC distribution in the ocean, we would like to call attention to the possible role of plankton pellets and other large detrital particles, skeletal remains of foraminifera and, to a lesser extent, coral detritus and siliceous remains of plankton as additional sources of deep-water DOC. Below the critical depth, principally on the seabed surface, carbonaceous exoskeletal formations are practically fully dissolved. The OM contained therein, which was practically inaccessible to the microflora, is relased and, on account of the low activity of deep-water bacterio-plankton, fails to be utilized and goes over into the truly-dissolved or colloidal state. In fact, at some of the stations in the Pacific basins increased DOC concentrations were recorded in deep water-layers. The content of C_{org} of deep-water pelagic carbonaceous sediments amounts to $0.3-0.5\%$. In the future by consulting the difference between the annual production of planktonic foraminifera and the absolute mass of deposited $CaCO_3$ it may well be possible to estimate the quantity of OM going over from this source into the near-bottom waters.

All that has been said above about DOM distribution refers, as implied in the definition of DOM, to the sum of truly-dissolved, colloidal OM and part of larger particles. The methods of their separation are based on the use of molecular membranes, coagulation and sorption on hydroxides, carbonates, coal.

The two most commonly used methods of DOM separation are gel filtration and ultrafiltration. The latter presents certain advantages in studying diluted OM solutions, such as oceanic, marine and fluvial waters. Its application involves little transformation of the concentration and ionic composition of inorganic dissolved substances, no significant change of the physico-chemical conditions of existence of colloidal OM, while enhancing the latter's concentration. Therefore, most of the data obtained in the past decade (Tables 14 and 15) have been obtained by the method of ultrafiltration (Gjiessing 1970, Sharp 1973, Ogura 1974, 1977, Baturina et al. 1975, 1979, Maurer 1976, Liutsarev et al. 1984).

The DOM proper is usually assumed to comprise that part of DOM which passes through membrane ultrafilters with the nominal retainable molecular mass in excess of 500. The apparent pore diameter in such filters, according to the method of determining the pore size, lies within the range of 10 – 50 Å. As far as is known at present, the colloidal part of DOM can be subdivided into coarse, medium and fine colloids. The molecular masses of these colloidal fractions and the apparent diameter of ultrafilter pores retaining them lie within the following ranges: for coarse colloids – above 100 thousand, >500 Å; for medium colloids – 100 thousand – 20 (10) thousand, 500 – 200 (150) Å; for fine colloids – 20 (10) thousand – 1.0 (5.0) thousand, 200 (150) – 100 (50) Å. Table 16 presents average data on the relative content of different DOM fractions.

An analysis of available data has demonstrated that in the open ocean areas DOM is principally represented by the truly-dissolved (Table 16). It was observed that DOM in bays, seas, high-production oceanic areas contains a perceptibly higher proportion of colloidal OM. Presumably, it is the variation of this quasi-conservative component of DOM that causes the regional differences in the content of DOC. A still higher relative content of colloidal OM occurs in the zones of mixing of fluvial and sea waters, as well as in fluvial waters themselves. In the series: rivers – mixing zones – sea, the relative proportion of truly-dissolved OM increases and that of colloidal OM decreases. According to the data available at present, in the mixing of fluvial and sea waters it is predominantly the medium colloids that are subject to coagulation.

The DOC content in interstitial water is about one order higher than in oceanic water; in four investigated samples it amounted to 12 – 20 mg C l^{-1} (Table 16). The proportion of colloidal OM in terrigenous calcerous pelitic sediments proved to be higher than in terrigenous near-continental muds of the fans of the Limpopo and the Ganges. In two samples of interstitial water from red deep-water clays the dominant OM mass was represented by the truly dissolved fraction.

Table 14. The distribution of the dissolved organic carbon by sea water fractions in the Pacific Ocean

No. of station	Coordinates	Depth (m)	Samples and filter[a]	Molecular weight of matter retained	C_{org} in each sample, $mg\,C\,l^{-1}$	Concentration factor		Relative C_{org} content (% of the total)
						calculated (after volume)	actual (after C_{org})	
1454	01°04′ S 97°00′ W	30	Initial sea water	6	1.08	–	–	–
			XM-300	300,000	2.03	11.08	1.98	8.70
			PM-10	10,000	1.54	10.90	1.67	5.80
			UM-2	1,000	3.13	7.20	3.44	19.30
			Filtrate	<1,000	1.27	–	–	65.80
		3000	Initial sea water	–	1.03	–	–	–
			XM-300	300,000	1.33	28.00	1.29	1.10
			PM-10	10,000	1.77	20.20	1.74	3.80
			UM-2	1,000	9.00	19.20	9.18	42.70
			Filtrate	<1,000	0.43 – 0.66	–	–	52.40
1456	00°03′ S 122°05′ W	45	Initial sea water	–	0.93 – 1.13	–	–	–
			XM-300	300,000	2.00	28.46	1.94	3.41
			PM-10	10,000	1.40	20.60	1.44	8.75
			UM-2	1,000	5.90	12.87	6.05	21.80
			Filtrate	<1,000	1.00	–	–	66.00
		1000	Initial sea water	–	1.00	–	–	–
			XM-300	300,000	1.10	14.55	1.10	0.75
			PM-10	10,000	1.00	13.55	1.01	0.06
			UM-2	i,000	8.86	10.60	8.90	48.10
			Filtrate	<1,000	0.93	–	–	51.06
1458	00°04′ S 139°02′ W	23	Initial sea water	–	0.82 – 1.20	–	–	–
			XM-300	300,000	2.20	24.80	2.20	5.03
			PM-10	10,000	1.40	24.70	1.47	1.21
			UM-2	1,000	5.70	13.20	5.53	25.35
			Filtrate	<1,000	1.17 – 1.20	–	–	70.43
1465	07°28′ S 79°37′ W	9	Initial sea water	–	2.20	–	–	–
			XM-300	300,000	3.65	19.10	1.66	3.63
			PM-10	10,000	3.82	15.40	1.80	5.36
			UM-2	1,000	23.00	24.00	11.50	37.37
			UM-0.5	500	6.00	17.60	3.33	6.10
			Filtrate	<500	1.13	–	–	48.00

[a] Apparent filter pore sizes: XM-300 – ca. 100 Å; PM-10 – ca. 16 Å; UM-2 – ca. 13 Å; UM-0.5 – ca. 10 Å

Table 15. The distribution of C-DOM by seawater fractions

No. of station site	Depth (m)	Sample and filter[a]	Molecular weight of substance retained	C_{org} (mg C l^{-1})	Concentration factor		C_{org} of colloidal OM (mg C l^{-1})	C_{org} of colloidal OM (% of DOM)
					calculated in terms of volume	in terms of C_{org}		
3397, Bay of	0	Initial sea water	–	3.67	–	–	–	–
Bengal	(thin	UAM-500	⩾100,000	4.33	35.5	1.99	0.06	1.6
15°21′8 N	layer)	UAM-300	⩾75,000	4.50	23.2	1.67	0.08	2.1
87°53′0 E;		UAM-200	⩾30,000	3.50	16.2	1.52	0.07	2.0
Bottom depth		UAM-100	⩾1,000	4.86	14.2	2.31	0.19	5.3
2800 m		UAM-50	⩾500	3.00	3.0	1.69	0.41	11.2
		Filtrate	<500	–	–	–	–	77.8
	24	Initial sea water	–	1.73	–	–	–	–
		UAM-500	⩾100,000	2.40	27.6	1.66	0.04	5.5
		UAM-300	⩾75,000	4.40	35.5	2.24	0.07	3.8
		UAM-200	⩾30,000	4.26	16.7	1.97	0.13	7.3
		UAM-100	⩾1,000	6.27	16.5	2.24	0.21	12.1
		UAM-50	⩾500	3.30	6.0	2.33	0.31	18.0
		Filtrate	<500	–	–	–	–	53.3
	2000	Initial sea water	–	1.57	–	–	–	–
		UAM-500	⩾100,000	2.56	45.6	1.72	0.02	1.5
		UAM-300	⩾75,000	2.46	24.1	1.58	0.04	2.4
		UAM-200	⩾30,000	2.94	18.1	1.57	0.06	3.8
		UAM-100	⩾1,000	6.10	15.8	3.45	0.27	17.5
		UAM-50	⩾500	2.86	3.6	3.62	0.55	35.2
		Filtrate	<500	–	–	–	–	39.6
Seychelles,	–	Initial sea water	–	1.67	–	–	–	–
Curieuse I.,		UAM-500	⩾100,000	1.87	45.0	1.27	0.01	0.5
coral lagoon,		UAM-300	⩾75,000	5.80	60.2	4.15	0.07	4.4
40 m from the		UAM-200	⩾30,000	4.07	37.3	2.66	0.07	4.1
coastline,		UAM-100	⩾1,000	4.60	23.6	3.46	0.14	8.3
hor. 0		UAM-50	⩾500	7.30	7.5	1.35	0.25	15.2
		Filtrate	<500	–	–	–	–	67.5

[a] Apparent diameters of filter pores: UAM-50 – 44 Å, UAM-100 – 125 Å, UAM-200 – 220 Å, UAM-300 – 280 Å, UAM-500 – 490 Å. Pore diameters determined after the flow rate of distilled water according to Poiseuille's equation (p. 1.5 atm, to 25 °C)

Table 16. The distribution of C-DOM by fractions in marine and interstitial water, % of total

Area, type of sediments		Colloidal OM				Proper-dissolved OM	Source
		Coarse colloids	Medium colloids	Fine colloids	Total		
Oceanic, marine and estuarine waters							
Open ocean		0	2	6	8	92	this volume, Gorshkova (1983), Ogura (1977), Sharp (1973), Bordovsky et al. (1981)
Seas, gulfs, high-productive ocean areas		10	10	30	50	50	Baturina et al. (1975, 1979), this volume, Gorshkova (1983), Starikova et al. (1976), Khailov (1968, 1971), Maurer (1976), Ogura (1974, 1977), Gagosian and Stürmer (1977), Brown (1975)
Zones of intermixing of river and seawater	rivers	15	40	25	80	20	this volume, Liutsarev and Pashkova (1983), Moore et al. (1979)
	mixing zones	6	54	17	77	23	this volume
	seas	10	20	30	60	40	this volume, Liutsarev and Pashkova (1983)
Interstitial waters, Indian Ocean							
Terrigenous pelite muds of the Limpopo and Ganges fans (St. 3357, 3397)		3	7	35	45	55	this volume, Gorshkova (1983)
Terrigenous calcareous (30–50% $CaCO_3$) coccolith-foraminiferal ooze of the continental rise (St. 3368, 3386)		3	16	54	73	27	this volume, Gorshkova (1983)

3. Carbon of Particulate Organic Matter

At present the particulate material, one of the components of which in natural waters is OM, is assumed to comprise that part of OM which is retained in the filters with 0.45 – 1 µm pore size or in separators. Large OM particles in natural waters are present in an extremely scattered state and are usually not to be found in filtration particulate suspensates. Their quantity in the ocean, compared to the bulk of OM is insignificant, but they play an important role in the supply of labile OM to the sediments and in nutrition of the benthos.

The OM contained in particulate material consits of (a) living planktonic algae and other living green cells, microzooplankton, including eggs and larvae and the greater part of aggregated bacterioplankton, (b) the remnants of various organisms and of their fecal pellets, (c) OM in skeletal structures and in terrigenous and eolian components, (d) OM which has been coprecipitated, sorbed from the solution or aggregated.

Thus particulate material contains both living and non-living OM. The two forms may be present in different proportions and this, as can be expected, affects the composition and properties of particulate material.

Owing to its generally high activity, particulate OM plays an important role in geochemical processes at the sedimentogenesis stage. It is, in effect, an indicator of a basin's productivity, its nutrient supply and of the intensity of biochemical processes.

Vernadsky (1921, 1934, 1965) was one of the first to have studied particulate organic matter (POM) in the biogeochemical aspect and to have called attention to the importance of estimating the total quantity of particulate carbon in the ocean as one of the "planetary constants". Moreover, scientists have long been aware of the significant trophic function of POM (Mobius 1880, Lohmann 1908). An intensive study of POM (not living plankton alone, but living and non-living OM combined) did not, however, begin until the middle to late 1950s. Lisitzin (1955, 1974) was one of the first to have carried out comprehensive investigations of particulate material and he rightly noted that its study enables us to follow the progress of formation of the chemical composition of bottom sediments. This fully applies to OM as a constituent of particulate material.

Although much progress has been made in studying the composition and distribution of POM, there still exist great difficulties in its quantitative estimation in the seas and oceans. These difficulties are due mainly to a very wide dimensional spectrum of suspended particles, to seasonal and daily variations of the concentrations, the relationship between vertical and horizontal transfer, resuspension of the bottom sediments and their change-over into a particulate state and the scattered and occasionally mosaic distribution of particulate material. All this necessitates, depending on the object of the investigation, the use of different water volumes (from 0.5 l to dozens and hundreds of tons), of different sampling methods (filtration, separation, sediment traps) and calls for further improvement and standardization of sampling techniques. Thus methodological difficulties are at present associated, not with the analytical procedure of POC determination (its accuracy being usually ±5–10% of the quantity determined) but rather with the differences in sampling methods involving the use of water bottles of different designs, immersion pumps, autofiltration in situ and sediment traps. Very important are filtration and separation procedures, the magnitude of the applied pressure or suction, the character of the filtering material, the time elapsing between collection of the sample and filtration, the shape and size of sediment traps, and so on.

The development of inorganic filters (glass fiber and silver) capable of withstanding calcination and complete removal of OM as well as improvement of filtration and sampling techniques, reducing the effect of the surface film in which OM is concentrated, have both contributed to obtaining reliable data on the distribution of POC and, in most cases, on its composition. It is to be noted that because of sorption onto filters of various DOM components, for instance fatty acids (Saliot 1975, cit. by Cauwet 1981), and because of the dependence of the quantity and composition of POC on the type and porosity of filters, only results obtained with use of identical filters and similar techniques for isolation of particulate material are truly comparable. Even in this case, however, slight differences in the methods may be responsible for large differences in quantitative evaluations of POC (Wangersky 1976).

Recent progress in sampling techniques with the use of sediment traps has introduced practically no changes in estimates of total POM in the ocean, although it has furnished a great deal of new information on the distribution and composition of large particles (mostly fecal material), their role in sediment accumulation and in the activity of benthonic heterotrophs (e.g., Knauer and Martin 1980, Suess 1980, Tanoue and Handa 1980, Gagosian and Lee 1981, Iseki 1981, Tseitlin 1981, Degens and Ittekkot 1983).

A summary of data on particulate material in marine and oceanic waters is contained in several works, including (Riley 1970, Bogdanov et al. 1971, Nakajima 1973, Parsons 1975, Romankevich 1977, Cauwet 1981).

3.1 POC Distribution from Separation Data

There is no unanimous opinion today as to whether separation or filtration yield a better recovery of particulate material. This mostly relates to different designs and performance characteristics of the separators and supercentrifuges in use, as well as to the operational procedures adopted and the methods employed for collecting the separated material. No standardization of the available methods exists. Our data obtained at the Institute of Oceanology, the USSR Academy of Sciences, by means of "De Laval" separators (types VIB-1929 and LAB 201-21A) and Soviet-built types HCM-2, HCM-3121 (Zhivago 1980), did not substantiate the opinion that separation may produce a 30% better recovery than filtration

Table 17. The content of POC in the surface layer (about 0–7 m) of the ocean according to separation data

Ocean	Number of samples	Extreme values	Mean from sample number	Occurrence rate of concentrations (%)				
				<5	5–10	10–50	50–100	>100
POC, $\mu g \cdot l^{-1}$								
Pacific	76	1.4–140.8	36.7	11	9	54	21	5
Atlantic	54	6.9–620	61.2	0	13	52	22	13
Indian	81	3–389	63.1	1	0	60	19	20
World ocean	211	1.4–620	53.1	4	7	56	20	13

Ocean	Number of samples	Extreme values	Mean from sample number	Occurrence rate of concentrations (%)			
				<5	5–10	10–20	>20
POC, % of dry mass							
Pacific	80	0.5–21.5	7.2 ± 5.3	51	20	28	1
Atlantic	55	3.2–21.5	11.1 ± 5.1	5	50	38	7
Indian	91	1.4–22.0	10.3 ± 4.7	11	41	46	2
World ocean	226	0.5–22.0	9.4 ± 2.9	24	35	38	3

(Breck 1978). Our average values for the content of particulate material ($\mu g\ l^{-1}$) turned out to be 30% below that for POC isolated by filtration through a GF/C filter at the same pelagic areas in the tropical regions of the Pacific and Indian Oceans.

In the surface water layer (0–5 m) of the ocean, on the evidence of 211 samples taken with separators on a number of cruises made by the "Ob", "Vityaz", "Akademik Kurchatov", research vessels of the Institute of Oceanology, the USSR Academy of Sciences, the POC content varied between 1.4 and 620 $\mu g\ l^{-1}$, averaging $53.1 \pm 26.4\ \mu g\ l^{-1}$ (Table 17, Fig. 11). The quantity of organic carbon in all these samples was determined by the classical Knopp method involving separate gravimetric measurement of C_{CaCO_3} and C_{org}. The variability of POC content in the Pacific, Atlantic and Indian Oceans is great, but the most frequent concentrations in all the three oceans are 10–50 $\mu g\ l^{-1}$ (52–60% of all cases). As regards average values, the POC content the Pacific is less than that of the Atlantic and Indian Oceans on account of the vast areas of oligotrophic water within it: in the Pacific – 55%, in the Atlantic – 31%, in the Indian Ocean – 27%. Differences in POC content between the Indian and Atlantic Oceans in surface waters were not clearly significant.

Chester and Stoner (1974) analysed 66 samples of surface particulate material (0–5 m layer), using with a supercentrifuge. These samples came mostly from the marginal parts of the Pacific, Atlantic and Indian Oceans. They found that POC content varied between 6.6 and 211 $\mu g\ l^{-1}$, with the average value being

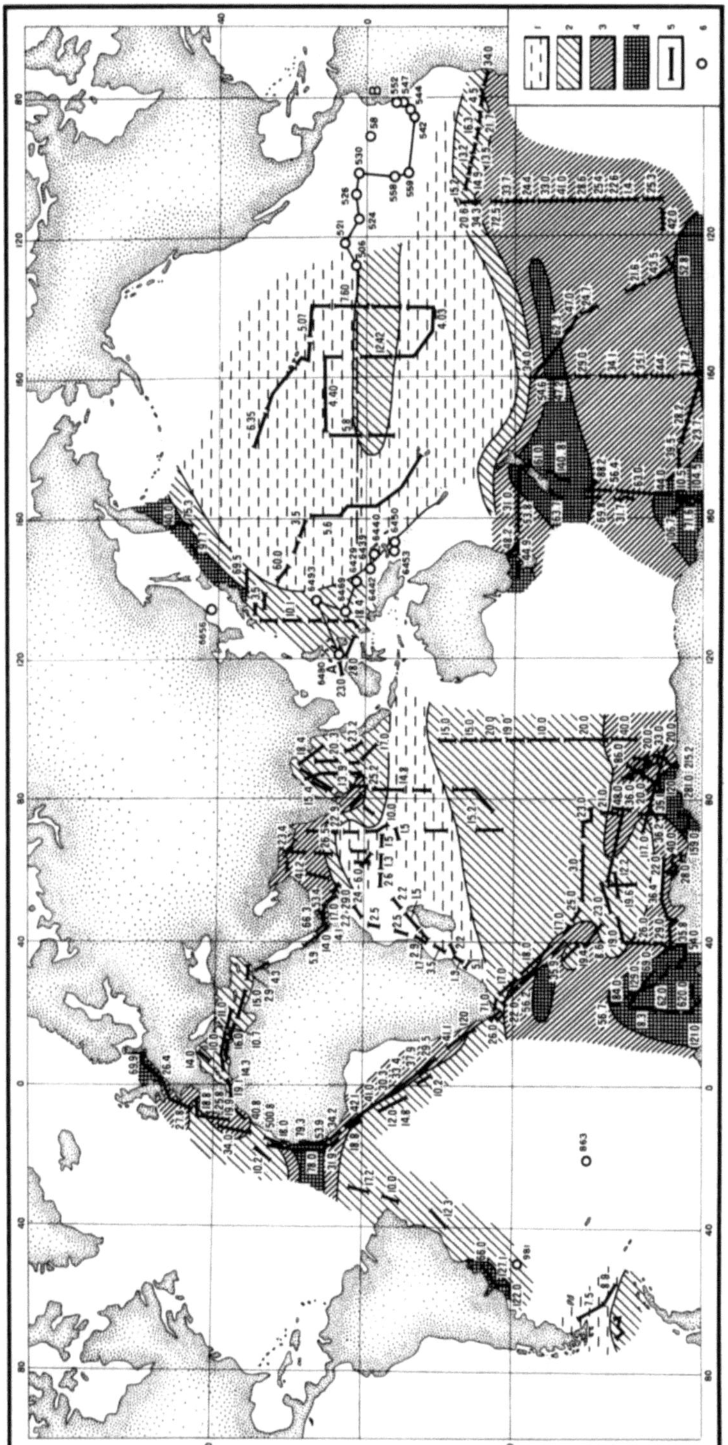

Fig. 11. Distribution of POC in the surface water layer of the World Ocean according to separation samples analysis data, μg POC l^{-1}: *1* = <10; *2* = 10 – 25; *3* = 25 – 50; *4* = >50; *5* = samples collected by separators; *6* = location of suspended material samples collected by filtering methods (filter Whatman GF/C) on the Vitiaz and Akademik Kurchatov cruises. *AB* is the pro file line

about 52 $\mu g\,l^{-1}$. Comparison of separation and filtration samples demonstrated that in separation as much as 99% of particulate material was recovered by being retained during water filtration in membrane filters of 0.45 μm pore size. A decreasing series of POC concentration established by the authors ran north-east Atlantic > China Seas > South Atlantic > Indian Ocean: this is clearly valid only for those areas where the samples had been taken.

The results obtained by now show that the highest POC concentrations are typical of neritic ocean areas (shelves, continental slopes and adjoining parts of deep basins), whereas the lowest POC concentrations are a characteristic of pelagic areas of the ocean floor in tropical and subtropical latitudes. In marginal and particularly inland seas POC content according to separation and also filtration data is generally higher than in the oceans outside their coastal areas, though the variability limits of concentrations there are equally significant (Table 18). A high POC concentration is characteristic of the Pacific equatorial zone. A spot of high POC concentrations in the Pacific Ocean north of 50°S has found no explanation as yet. A high POC content near the Antarctic is partly associated with the fact that the samples were mostly taken here in the spring – summer season, when a brief, though abundant development of phytoplankton was replaced with a bloom of zooplankton and other OM consumers. Given enough seasonal observations, POC content here might be lower.

The average POC values obtained in the Pacific and Indian Oceans (36.7 and 61.2 $\mu g\,l^{-1}$), which have been more uniformly studied compared to the Atlantic Ocean, are in agreement with the average values of total phytoplankton production namely 46 and 81 $gC\,m^{-2}\,yr^{-1}$, respectively. Comparison of POC and primary production distribution maps (cf. Figs. 11 and 12) shows that while there are significant differences between the two, such regularities as circumcontinental and latitudinal zonation are common to both.

This inference can be illustrated by the distribution pattern of particulate and precipitated OM at the profile from the Aleutian Islands to the Antarctic (from 54°22′N to 66°29′S, Fig. 12). As shown in Fig. 12, the highest concentrations of particulate OM are observed in the 0 – 100 m layer, declining progressively in the deeper layers. In the moderate latitudes of the northern hemisphere the content of POM at practically all depths is higher than in the Antarctic region of the Pacific. The sharper vertical gradient of POM decrease in the Antarctic waters is associated with the peculiarities of their biological structure, as well as with their generally relatively modest biological productivity (Koblentz-Mishke 1977). The abundant blossoming of phytoplankton near the Antarctic is of brief duration and does not lead to the creation of large annual masses of POC. Zooplankton utilizes a significant proportion of this OM and contributes to the development of even sharper vertical gradients of its distribution. This situation recurring year after year becomes reflected in the sediments in the form of low C_{org} content. In the sediments this is further intensified by the diluting effect of the terrigenous material of glacial abrasion, which is poor in C_{org}. Thus, at the already mentioned profile of the Aleutian Islands – Antarctica, increased concentrations of particulate and precipitated OM were observed in the boreal, near-Antarctic and equatorial zones, as well as in the region of the northern tropical divergence of waters, i.e., in the areas of high biological productivity.

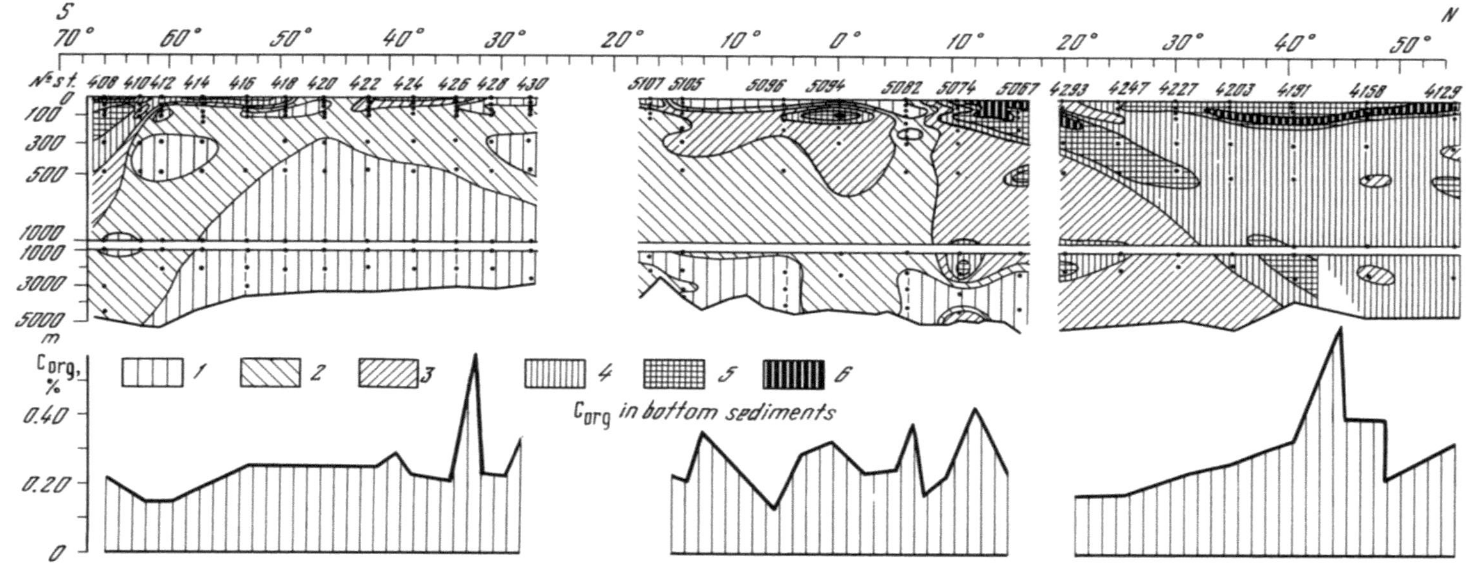

Fig. 12. Distribution of POM, on the profile from the Aleutians to the Antarctic (from microscopic investigation data), µg OM l^{-1} (Bogdanov et al. 1971, with supplements): *1* = <10; *2* = 10 – 25; *3* = 25 – 50; *4* = 50 – 100; *5* = 100 – 250; *6* = >250; *below:* C_{org} distribution in bottom sediments, % of dry mass

Table 18. Content of particulate organic carbon in seas and oceans (in $\mu g\ l^{-1}$)

Sampling area	Layer (m)	Extreme values	Source
Pacific Ocean			
Coastal waters off San-Diego	6	550 – 3250 [a]	Fox et al. (1953)
Coastal waters off Vancouver Island	0 – 21	50 – 531 [b]	Parsons (1963)
Northeastern Pacific	10 – 3000	36 – 196 (50) [b]	Parsons and Strickland (1962a)
	15 – 3000	32 – 254 [b]	Parsons and Strickland (1962a)
	250 – 400	30 – 110	Hobson (1967)
Northeastern Pacific	0 – 1000	12 – 67 [c]	Handa (1970)
	0 – 50	15 – 450 [c]	Nakajima (1973)
	50 – 1500	0 – 155 [c]	Nakajima (1973)
Western Tropical Pacific	0 – 3000	23.6 – 65 [b]	Finenko and Ostapenia (1971)
	0 – 4750	5.5 – 86.4 (31.4) [d]	Liutsarev et al. (1972)
Kuril-Kamchatka Basin	0 – 6000	54 – 468	Bogdanov and Shaposhnikova (1970)
Gulf of California	Euphotic layer	40 – 397 (192) [c]	Zeitschel (1970)
North of the Hawaiian Islands	0 – 4000	(0.8 – 25.7) [c]	Gordon (1971)
Bering Sea	0 – 50	60 – 80 [c]	Handa and Yanagi (1969, quoted by Parsons 1975)
	0 – 300	~25 – 200 [c]	Nishizawa and Nakajima (1971)
Area of the Aleutian Islands	0 – 300	25 – 125 [c]	Nishizawa and Nakajima (1971)
East of Hokkaido	0 – 300	30 – 293 [c]	Nishizawa and Nakajima (1971)
Phillipine and Carolina Basins	0 – 300	< 15 – 165 [c]	Nishizawa and Nakajima (1971)
Oyashio Current	0 – 50	60 – 160 [c]	Handa and Yanagi (1969), Parsons (1975)
Kuroshio Current	0 – 50	60 – 80 [c]	Handa and Yanagi (1969), Parsons (1975)
Japan Sea	0 – 2000	14 – 56 (29.8) [d]	Liutsarev and Romankevich (1975)
Tasman Sea	23 – 4210	36 – 207 (100) [c]	Dal Pont and Newell (1963)
Santa Catalina Basin	0 – 1300	17 – 130 [c]	Holm-Hansen et al. (1966)
Near California	10 – 1025	20 – 120 (52)	Holm-Hansen and Booth (1966)
Area of Peruvian Current	0 – 20	250 – 1250	Menzel (1967, quoted by Parsons 1975)
	20 – 100	40 – 500	Menzel (1967, quoted by Parsons 1975)

Table 18 (continued)

Sampling area	Layer (m)	Extreme values	Source
	200–3000	<15	Menzel (1967, quoted by Parsons 1975)
Atlantic Ocean			
Atlantic Ocean	0–5	7.5–129[e]	Lisitzin et al. (1975)
	0–5	3.2–21.5 (12.1)[f]	Lisitzin et al. (1975)
North Atlantic	0	87[c]	Gordon (1970), Riley (1970)
	100–5300	(19.7)[c]	Gordon (1970), Riley (1970)
	10–3000	60–130[c]	Szekielda (1968a)
	0	28–88 (51)[b]	Ostapenia and Kovalevskaya (1965)
Equatorial Atlantic	0–5285	40–320[c]	Skopintsev et al. (1966)
South Atlantic	0–5500	7.2–60.5 (21.9)[d]	Artemiev and Melnikov (1974)
Nova Scotia and Newfoundland shelf	0	(41–176)[c]	Gordon (1970a), Riley (1970)
Irminger Sea	0	(44)[c]	Gordon (1970a), Riley (1970)
	100–3300	(12.7)[c]	Gordon (1970a), Riley (1970)
Chesapeake Bay	–	1000–3000[c]	Biggs and Flemer (1972, quoted by Parsons 1975)
North Atlantic coastal waters	–	570–1700[c]	Manheim et al. (1970, quoted by Parsons 1975)
Between Recife and Buenos Aires	0–5200	<2–12[c]	Hobson and Menzel (1969)
Sargasso Sea	0–50	20–60[c]	Menzel (1967), Parsons (1975)
	100–4000	<10[c]	Menzel (1967), Parsons (1975)
	0	(29–142)[c]	Gordon (1970), Riley (1970)
	100–4000	(8.8–47.3)[c]	Gordon (1970), Riley (1970)
	0	14–677[c]	Riley et al. (1965)
	6–700	4–30[c]	Hobbie et al. (1972)
	0–4720	0–850[c]	Wangersky and Gordon (1965)
Bermuda Islands – Liberia near Western Africa	0	300–800[c]	Wangersky and Gordon (1965)
5–20°N, 50–60°W	1–1000	19–86[c]	Menzel and Goering (1966)
Gulf of Guinea and Romanche Basin	0–6500	2–100[c]	Copin-Montegut (1973)
Near Southwestern Africa	0–200	6–970[c]	Hobson (1971)

Table 18 (continued)

Sampling area	Layer (m)	Extreme values	Source
Near Hatteras Inlet	4–70	10–152[c]	Hobbie and Holm-Hansen (1972)
Azov Sea	0–11	120–2650[g]	Datsko (1959)
	0–11	740–3760[h]	Datsko (1959)
Black Sea	0–2000	50–910[g]	Datsko (1959)
	0–2000	210–1030[h]	Datsko (1959)
	10–1500	20–240[c]	Skopintsev et al. (1971)
Bosporus area	0–10	74–185 (121)[b]	Suschenia (1961)
	0–75	190–1090[b]	Finenko (1965)
Western halistatic region	0–75	220–610 (262)[b]	Finenko (1965)
Sevastopol Bay	0–14	264–1022 (750)[b]	Finenko (1965)
Sea of Marmara	0	70–81 (75)[b]	Suschenia (1961)
Aegian Sea	0	54–99 (67)[b]	Suschenia (1961)
Ionian Sea	0	44–107 (86)[b]	Suschenia (1961)
Adriatic Sea	0	50–171 (97)[b]	Suschenia (1961)
Mediterranean Sea	0	61–72 (67)[b]	Suschenia (1961)
	0–5	1–15 (13)[e]	Lisitzin (1964)
Coastal waters	0	50–157 (81)[b]	Ostapenia and Kovalevskaya (1965)
Bay of Biscay	0	(87)[b]	Ostapenia and Kovalevskaya (1965)
English Channel	0	(104)[b]	Ostapenia and Kovalevskaya (1965)
North Sea	0	91–132 (113)[b]	Ostapenia and Kovalevskaya (1965)
Baltic Sea	1–230	(23.3–50.9)[c]	Erhardt (1969)
	0–1	30–2070 (540)[e]	Emelianov and Romankevich (1977)
Arctic Ocean	0–20	7–13	Kinney et al. (1971)
Indian Ocean			
Western part of Ocean	0	57–157	Menzel (1964)
Western and central parts of Ocean	15	46.5–138 (78.5)[b]	Mullin (1965)
Java-Freemantle profile	0–5000	(7.0–16.1)[c]	Newell and Kerr (1968)
Gulf of Aden	0–100	80–180[c]	Szekielda (1968b), Parsons (1975)
Java-Australia profile	0	17–32[c]	Newell (1969)
Red Sea	0–5	14–66.6 (40.3)[e]	Lisitzin (1961, 1964), Gordeyev (1968)
North of 50°S	0–5	~(16–30)[e]	Lisitzin (1961, 1964), Gordeyev (1968)
South off the Polar Front	0–5	3–389[e]	Lisitzin (1961, 1964), Gordeyev (1968)

Table 18 (continued)

Sampling area	Layer (m)	Extreme values	Source
Discrete areas of the world ocean	0–5	4–389	Bogdanov et al. (1971)
	0–5	0.5–21.48[f]	Bogdanov et al. (1971)
	0–5	6.6–211	Chester and Stoner (1974)
	0–5	4.6–29.9[f]	Chester and Stoner (1974)

Figures in parenthesis show mean values
[a] Estimated from the difference between the dry weight and ash-residue
[b] Bichromate oxidation. In most cases the C_{org} content was calculated from the ratio: $C_{org} \simeq 0.67\ O_2/2$
[c] Various dry combustion methods, with recording of released CO_2
[d] Wet combustion, with culonometrical recording of released $CO_{2(org)}$ (Liutsarev 1973)
[e] Wet combustion, with gravimetrical or titrimetrical measurement of $CO_{2(org)}$ (Zalmanzon 1957 and Sokolova 1975)
[f] Percent of dry particulate matter
[g] Water settling of month duration. Combustion in fusion cake of potassium saltpeter. Measurements expressed in $CO_{2(org)}$
[h] Total particulate and colloidal-dissolved C_{org}, precipitated with aluminium hydroxide

Circumcontinental and latitudinal zonation are characteristic of numerous phenomena in the ocean (Bezrukow 1964a). This owes its origin to an irregular distribution of heat on Earth (circumcontinental and latitudinal) and to the location of the continents (circumcontinental), which determine the pattern of water circulation and biogeochemical processes associated with it.

3.2 POC Distribution from Filtration and Sediment Traps Data

At the Institute of Oceanology of the USSR Academy of Sciences, a considerable amount of information has been amassed on the distribution of POM in different areas of the world ocean, the water samples being mostly collected with the aid of 200–500 l stainless-steel water bottles (Bogdanov and Shaposhnikova 1970, Bogdanov et al. 1971, Romankevich 1971, 1979, Pustelnikov 1973, 1975, Lisitzin 1974, 1978, Lisitzin et al. 1975, Emelianov and Romankevich 1979, Artemiev et al. 1981, Romankevich et al. 1982, Artemiev 1983, Beliayeva and Romankevich 1983).

By analysing these data and by comparing them with the C_{org} distribution in bottom sediments it became possible in the late 1960's to furnish evidence in support of the view that the distribution of POC in oceanic depths and that of OM in the surface sedimentary layer are determined by the distribution of POM and primary production in the phytosynthetic layer (Bogdanov et al. 1971).

In his comprehensive work, Riley (1970), after summarizing and critically analyzing the data obtained mostly by foreign research workers, discussed the biological aspects of the problem of POM in the ocean (sources, total quantity, regional differences in the distribution, the morphological, elementary and biogeochemical composition, seasonal cycles, etc.). It was demonstrated that, in the surface waters, the seasonal and regional differences in POC content were principally associated with the development cycles of the plants. At various interfaces, DOM may change into the particulate form and this complicates POC distribution in the mass of waters and its relationship with the biological productivity. It was noted that POC content in deep waters in an ocean area and phytoplankton production are generally directly correlated. This confirmed the opinion that vertical transport of POM is more rapid than horizontal transfer by currents of diffusion. It was noted that to achieve an unambiguous interpretation of POC distribution in ocean depths and of its role in biological processes, there is need for many more observations of seasonal POC variations, evaluations of the relative importance of sedimentation of planktonic OM, advective turbulent transfer, resuspension of the material, as well as of the conversion of the dissolved into particulate OM.

Extensive data on the distribution of particulate carbon, as well as of particulate nitrogen in the water mass of the Pacific Ocean were obtained by Japanese researchers. Along the meridional profile 155°W from 50°N to 15°S six types of waters were identified differing in their average content of particulate carbon and particulate nitrogen. In each type of water a close correlation was observed between POM content in the 0 – 150 m layer and in the 150 – 4000 m layer (Nakajima 1973). In the nortwestern part of the Pacific several hundred determinations were made of the total quantity of particulate material, and of the content of particulate carbon and nitrogen, and it was shown that there existed regional differences in their concentration both in the surface and in the deeper-lying waters. The highest average concentrations for all those areas were found to be typical of the surface layer (0 – 50 m) in which the content of particulate material varied within the range 0.10. – 3.95 mg l^{-1}, particulate carbon 15 – 450 µg l^{-1}, particulate nitrogen 0.8 – 65.0 µg l^{-1}. High POC contents were observed in the upper several dozen metres.

The vertical decrease of POC (outside the oligotrophic areas) has a stepped character. From the upper layer (300 m), within each of the steps it was almost exponential and can be described by the equation

$$c_i = c_{i_0} e^{-k_i(z_i - z_{i_0})},$$

where c_{i_0} is the maximum concentration at the depth z_{i_0} in the step i; c_i is the concentration at the depth z_i.

The concentration reduction rate (k_i) varied between 0.002 and 0.04, higher values being observed in the surface layer. Below the photic layer two POC and N-POM maxima were found. One of these coincided with the main halocline and the other with intermediate subarctic waters. The average POC and N-POM content in the surface layer (0 – 50 m) correlated linearly with the average POC content in the lower-lying mass where it was 0 – 155 µg l^{-1} (Nakajima 1973).

Menzel and Ryther examined POC distribution in deep ocean waters and arrived at conclusions which differ radically from those described above (Menzel and Ryther 1979, Menzel 1974). Menzel considers that particulate material in deep ocean waters has neutral buoyancy and does not settle, that POM is resistant to decomposition and is not utilized in situ by organisms. POC fluctuations at great depths are small (4 – 5 μg l^{-1}) and its distribution is uniform and homogeneous. The POC content below 200 – 400 m is not influenced by the magnitude of primary production.

These conclusions have not been confirmed by abundant new data obtained in subsequent years on primary production, O_2 consumption in the water mass, the distribution and flow of POM, including result derived from sediment trap data. By summarizing the the available data (Wangersky 1976, Romankevich 1977) it became possible to plot the nomogram of the absolute C_{org} masses reaching different depths in hemipelagic and pelagic ocean areas in the primary phytoplankton production – depth coordinates (Fig. 13; Bogdanov et al. 1979). The nomogram an estimate to be made of what percentage of POM reaches any given depth (cf. Fig. 13. inset), and makes possible the determination of the absolute C_{org} masses reaching deep water layers and the ocean floor.

On the basis of sediment trap data an empirical relationship between annual mean net primary production rate, depth (for depths below 50 m) and the POM flow was deduced (Fig. 14; Suess 1980).

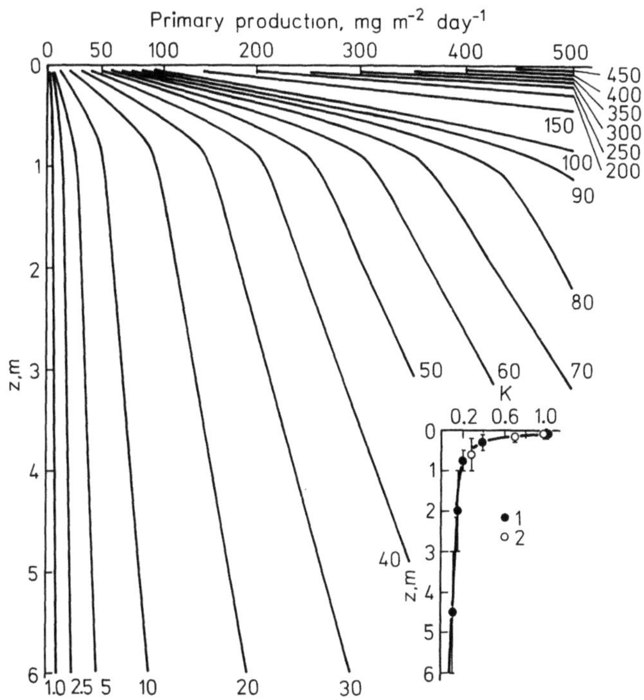

Fig. 13. A nomogram of absolute POC masses (g cm^{-2} 1000 yr^{-1}) reaching deep water layers and the ocean floor against the magnitude of primary production and the depth of the ocean (Bogdanov et al. 1979). *K* the ratio of POC at depth to POC at surface; *1* after data by Wangersky (1976); *2* after data by Romankevich (1977)

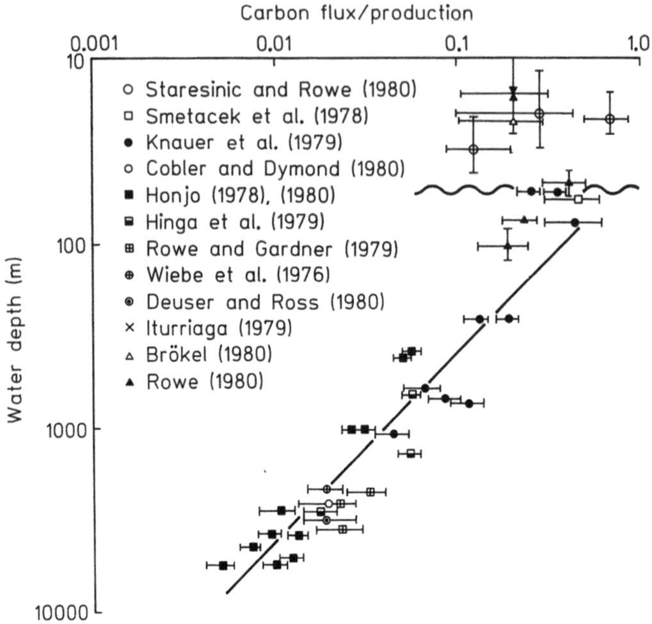

Fig. 14. Change with depth of POC flux (C_{flux}) normalized to average annual phytoplankton production ($C_{prod.}$) (Suess 1980). Analysis data of sediment trap samples. *Straight line* described by equation: $C_{flux(z)} = \dfrac{C_{prod.}}{0.0238z + 0.212}$; $n = 33, r^2 \simeq 0.79, z \geqslant 50$ m. *Undulating line* denotes the lower boundary of the euphotic zone. Actual values from data by various workers (quoted by Sness 1980) shown by different symbols

$$C_{flux(z)} = \frac{C_{prod}}{0.0238\,z + 0.212} \quad \text{for} \quad n = 33, r^2 = 79, z \geqslant 50,$$

where C_{flux} is the POC flow at a given depth (g m^{-2} yr^{-1}), C_{prod} the primary C_{org} production at the surface (g m^{-2} yr^{-1}), z is the depth at which POC is measured (below 50 m), n the number of measurements and r^2 the correlation coefficient.

The above formula (with certain assumptions, such as a similar pattern of variation of particle size in suspensions with depth during decomposition, a precipitation rate of large particles averaging some 50 – 100 m per day and a linear relationship between depth and half-life of POM) furnished the basis for calculating the decomposition rate constant for POM:

$$2.38\,t = \frac{1}{C_{flux}/C_{prod}} - 0.212,$$

where C_{prod} is production at t = 0 and k = 2.38 (i.e. z(0.0238)), having the dimensionality of day^{-1} and the character of a second-order rate constant (Suess 1980). Thus improved techniques for obtaining particulate material from different ocean depths and the latest information on POM distribution have confirmed the existence of a close direct correlation between primary phytoplankton production and POC content in the depths of the ocean. They make it possible to obtain a quantitative expression for this dependence for the world ocean hemipelagic and pelagic areas below the euphotic zone, as well as to estimate POM decomposition rate and the associated oxygen demand at different depths.

Between 1971 and 1981, on a number of cruises, water filtration through the glass fiber filters Whatman GF/C was used to obtain samples of suspended material from the surface down to the depth of 5500 m over the Peruvian shelf and continental slope, in the north-eastern, Peruvian, Carolina and West-Mariana basins, in the Sulu Sea, near atolls in western Pacific (Liutsarev et al. 1972, 1975), in the area of the Argentine basin and the South-Sandwich trench (Artemiev and Melnikov 1974), in the Sea of Japan (Liutsarev and Romankevich 1975), in the central part of the Arctic basin (Melnikov and Pavlov 1978), in the Barents Sea (Beliayeva and Romankevich 1983), in the Baltic (Artemiev et al. 1981), in the northern part of the Indian Ocean (Artemiev 1983). The POC content in these samples was determined by the same method and in the same device as the DOC content (Liutsarev and Smetankin 1980a, b). The analysis was performed by using part of a glassfiber filter (11 cm in diameter) cut into sectors (studies were usually made by using 10 – 20° filter sectors placed opposite each other). The analysis involved, as a rule, a parallel determination. The relative error of the chemicoanalytical procedure was below ± 10% for two parallel determinations (n = 150). According to these data, significant POC variations (13.7 – 402 $\mu g\,l^{-1}$) occur in the upper 100-metre water layer in the Pacific, the weighted average being 63.7 ± 32.2 $\mu g\,l^{-1}$ (Table 19, Fig. 15). In the 100 – 200 m layer both the range of POC variations and the weighted average content decrease, being 6.7 – 87.6 $\mu g\,l^{-1}$ and 41.8 ± 20.2 $\mu g\,l^{-1}$, respectively. Below 200 m

Table 19. Distribution of POC in the Pacific according to filtration data (filter GF/C), $\mu g\,l^{-1}$

Layer (m)	Number of samples	Extreme values	Mean from sample number	Median	Occurrence rate of concentrations (%)				
					<10	10 – 25	25 – 50	50 – 100	>100
0 – 100	75	13.7 – 402	69.4	47.0	0	11	41	30	18
100 – 200	22	6.7 – 87.6	31.1	29.2	14	27	45	14	0
200 – 1000	28	4.0 – 38.8	15.8	11.8	29	53	18	0	0
>1000	22	4.0 – 21.8	10.7	8.0	55	45	0	0	0

Layer (m)	Number of stations	Weighted average content			
		extreme values	average	max/min ratio	vertical gradient ($\mu g\,l^{-1}\,m^{-1}$)
0 – 100	17	21.8 – 150.9	63.7 ± 32.2	6.0	
					0.219
100 – 200	15	8.2 – 128.6	41.8 ± 20.2	15.7	
					0.052
200 – 1000	13	7.4 – 31.0	18.4 ± 7.4	4.2	
					0.004 [a]
>1000	11	5.5 – 20.1	11.2 ± 5.1	37	

[a] Calculations were made for the average depth of 3600 m

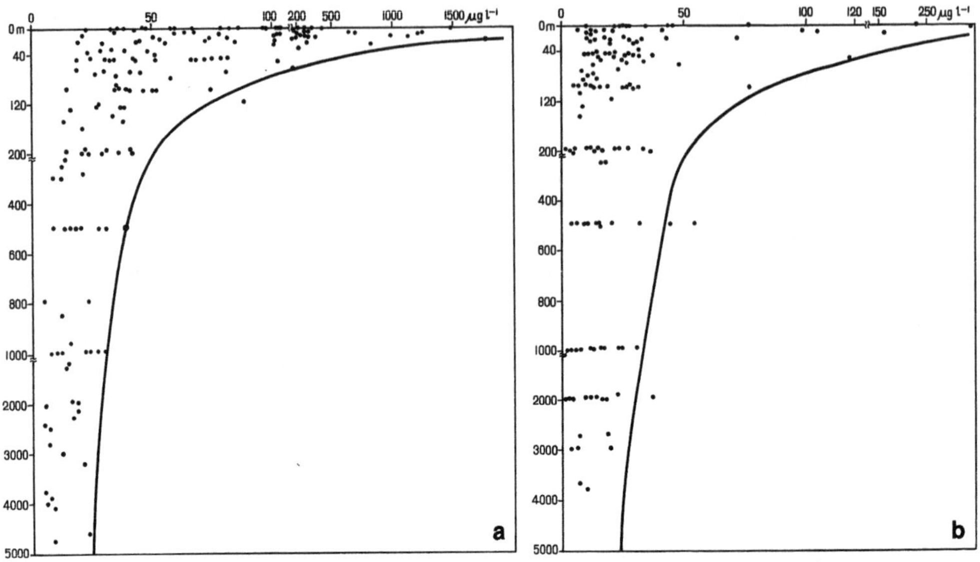

Fig. 15a, b. Distribution of POC in the Pacific (**a**) and Indian (**b**) water mass, µg POC l^{-1}. Analytical error of determination ±10%

depth the differences between the stations become even less significant, the weighted average amounting to 18.4 ± 7.4 µg l^{-1} (200 – 1000 m) and 11.2 ± 5.1 µg l^{-1} (>1000 m). The median values within the depth intervals discussed are less than the arithmetical means and weighted averages, pointing to a shift in all the series towards lower values. Correspondingly, the occurrence rates of different concentrations decrease in a vertical direction. Below 1000 m as much as 55% of all values relate to the POC content less than 10 µg l^{-1}. The vertical gradients calculated for the weighted average concentrations point to a high decomposition rate of OM in upper ocean layers (0 – 100 – 200 m) and exhibit only a tendency for decrease below 1000 m (gradient of 0.004 µg POC l^{-1} m^{-1}).

Within the depth range 200 – 1000 m there are quite distinct regional differences in POC content. High values were observed above the Peruvian slope, in the highly-productive eastern equatorial zone of the Pacific, whereas the lowest values were recorded in the low-productivity areas (West-Mariana and Peruvian basins, the Sulu Sea). Below 1000 m the differences in absolute values of POC content are not great (4.0 – 21.8 µg l^{-1}), but still they are a few times greater than the accuracy of the analyses.

The ranges of POC variability at depths below 1000 m can according to particular authors be very different (Table 20). They indicate the existence of regional differences in POC content at great ocean depths.

Analysis of our data obtained in the Pacific shows the following features in POC distribution (Fig. 16).

1. An eastward increase of POC content in transition from the stations in the western equatorial-tropical region to those in the eastern region and further to

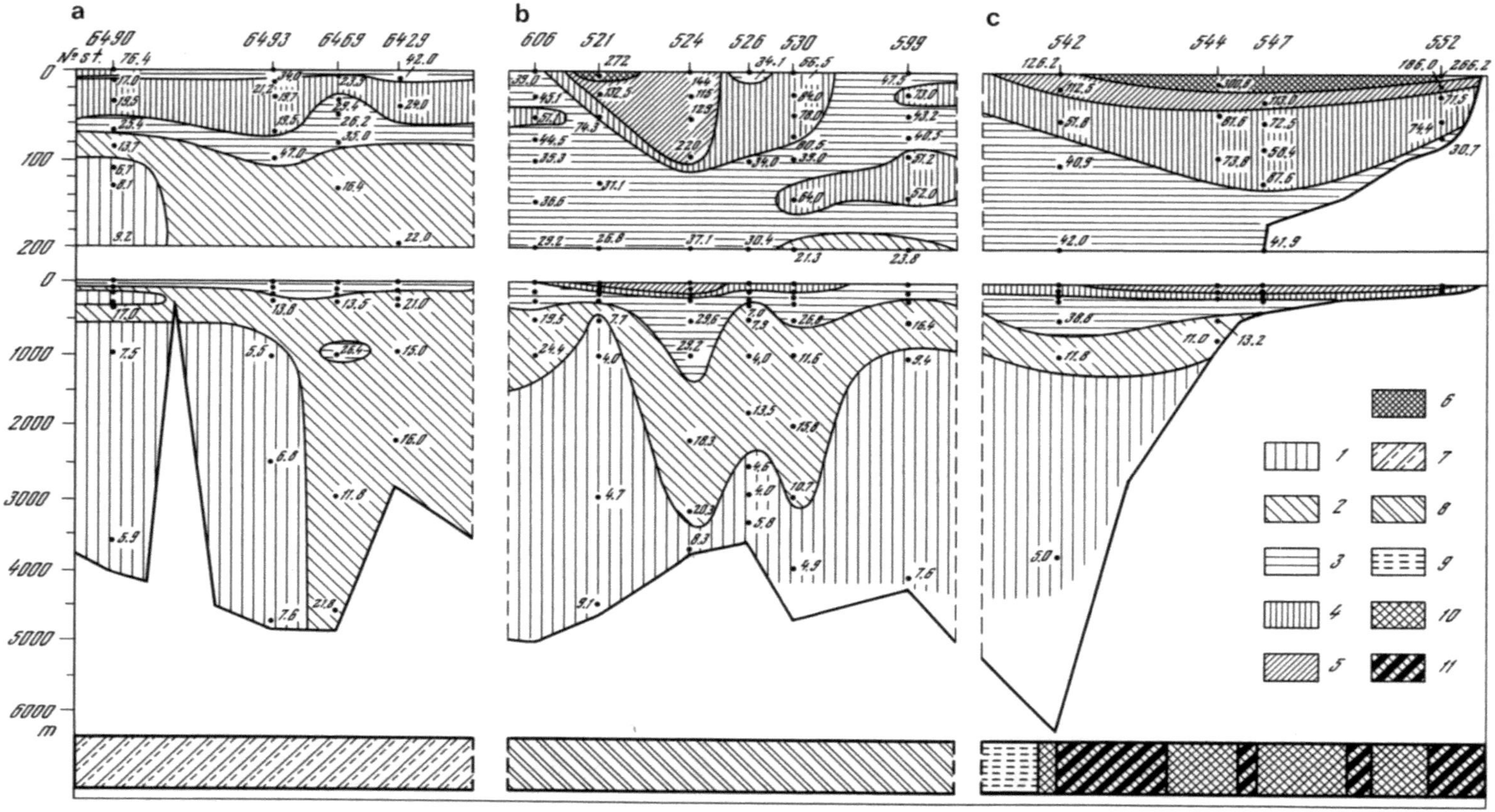

Fig. 16a – c. Distribution of POC in the Pacific water mass, μg POC l^{-1}: *1* = 4 – 10; *2* = 10 – 25; *3* = 25 – 50; *4* = 50 – 100; *5* = 100 – 200; *6* = >200. **a** western equatorial-tropical area; **b** eastern equatorial-tropical area; **c** area off Peru. *Below:* C_{org} distribution along the profile in the surface layer of the bottom sediments, % of dry mass; *7* = <0.3; *8* = 0.3 – 0.5; *9* = 0.5 – 1.0; *10* = 1 – 3; *11* = 3 – 5

Table 20. Regional differences in POC content below 1000 m, $\mu g\ l^{-1}$

Sampling site	Number of samples	Extreme values	Average	Source
The Pacific				
Sea of Japan	2	14.0 – 18.0	16.0	Liutsarev and Romankevich (1975)
West-Mariana and Caroline basins	5	6.8 – 21.8	12.8	Liutsarev et al. (1972)
Sulu Sea	1	5.9	–	Liutsarev et al. (1972)
North-Eastern basin	13	4.0 – 20.3	9.9	this volume
Peruvian continental slope	1	5.0	–	this volume
North-Eastern deep north of the Oahu i	90	0.8 – 4.0	2.7	Gordon (1971)
About 49 – 50°N, 132 – 145°W	4	32 – 68	46	Parsons and Strickland (1962)
Western equatorial part	6	23.6 – 40.3	30.6	Finenko and Ostapenia (1971)
Santa-Catalina trough	–	~17 – 29	–	Holm-Hansen et al. (1966)
Tasman Sea	8	36 – 207	95	Dal Pont and Newell (1963)
Ecuador area	–	5.18	–	Menzel (1967)
Sea of Japan	23	5.1 – 28.6	–	Nakajima (1973)
Bering Sea (near the Aleutian Is.)	3	36.0 – 78.5	62.0	Nakajima (1973)
Bering Sea (halistatic region)	4	12.0 – 20.5	16.0	Nakajima (1973)
Northern Pacific	2	4.7 – 30.0	11.9	Nakajima (1973)
Oyashio current area, east of Hokkaido	7	25.5 – 119.4	–	Nakajima (1973)
Kuroshio current area, south-east of Kyushu	4	4.0 – 8.7	–	Nakajima (1973)
East of Ryuku	4	6.0 – 12.7	–	Nakajima (1973)
Profile between New Guinea and Honshu i.	35	0.3 – 43.6	–	Nakajima (1973)
Profile along 150°W from 55°N to 63°S				
1000 – 3000 m	~62	0 – 21	4.9	Wangersky (1976)
below 3000 m	~84	0.2 – 2.5	3.6	Wangersky (1976)
Indian Ocean				
West-Australian basin	–	–	8.2	Newell and Kerr (1968)
The Atlantic				
Argentine basin	3	8.9 – 60.5	30.2	Artemiev and Melnikov (1974)
S. Sandwich trench's slope	3	8.1 – 8.7	8.4	Artemiev and Melnikov (1974)

Table 20 (continued)

Sampling site	Number of samples	Extreme values	Average	Source
Cape Verde basin	20	9–99	30	Wangersky and Gordon (1965)
Sargasso Sea	6	5–19	13	Wangersky and (Gordon 1965)
Atlantic Ocean (data by different authors)	–	13–74	–	Menzel (1974)
South Atlantic				
1000–3000 m	~125	0.70	3.0	Wangersky (1976)
below 3000 m	~108	0.50	3.0	Wangersky (1976)

the high-productivity waters of the Peruvian area; this is in agreement with the increase of primary and secondary bioproduction in this direction and is in sharp contrast with the DOC distribution described above.

2. The existence of an intricate pattern of POC distribution at some of stations in the eastern ocean zone, located near the equator. This complex POC distribution pattern is associated with the existence of a local system of zonal currents responsible for carrying POC down to great depths from the upper productive layers.

3. The presence at some stations, particularly in the oligotrophic western equatorial-tropical region, of a layer with increased POC content at the depth of 40–100 m. This layer adjoining the thermocline, is the zone where zooplankton and bacteria are concentrated and where a significant proportion of POM burns up in the metabolic furnace.

4. The presence of a vertical zonality in POC distribution, which is reflected in the bottom sediments as an increasing C_{org} content from the western to the eastern equatorial-tropical area and further to the region off Peru. Bottom sediments adequately reflect the existence of vertical zonality in POC distribution even when this cannot be recognized by collecting suspended material by any known method.

The decrease in POC content in the pelagic direction is illustrated in Table 21.

Wangersky (1976) investigated POC distribution at different depths in the Pacific and Atlantic Oceans and showed, using an uniform technique of sampling and analysis of suspended material (in this case the 5-litre Niskin sampling bottles and Selas Flotronic silver filter of 0.8 μm pore size), that values for both the oceans are similar for the layers 100–500 m (11 and 11.8 μg POC l^{-1}), 500–1000 m (4.6–6.3 μg POC l^{-1}), 1000–3000 m (3.0 and 4.9 μg POC l^{-1}), below 3000 m (3.0 and 3.6 μg POC l^{-1}, respectively).

In the upper layers of the Atlantic POC content was found to be higher than in the Pacific (0–100 m 54 and 30 μg POC l^{-1}, respectively); this agrees with the higher productivity of the Atlantic. POC content in both the oceans decreases

Table 21. POC content over Pacific shelf, slope, and ocean floor, $\mu g\ l^{-1}$ [a]

Layer (m)	Shelf	Slope	Ocean floor	Ocean floor basins
0 – 100	68	68	60	38
100 – 200	69	34	42	42
200 – 1000	–	21	17	13
>1000	–	10	12	8
For investigated water column	75(67)	21(42)	19(14)	10(10)
Average from samples	102	42	38	28

[a] The table was compiled from data obtained at the stations the location of which is shown in Fig. 2; weighted average values; bracketed figures refer to the surface-to-bottom interval

logarithmically with depth. The presence in the depths of the two oceans of spots of increased POC content permits us to speak of meso- and microscale variability of POC concentrations, which are associated with rapid sinking of part of the suspended material, as well as with its synthesis and utilization by migrating deep-water plankton and microflora.

In the central part of the Arctic basin, as shown by analysis of data from about 100 water samples tested for POM content between May 1975 and April 1976 during the drift of the research station "Severnyi polius (North Pole) 22" (83°55′N 166°59′W – 83°32′N 140°50′W), maximum POC concentrations amount to 25 – 45 μg POC l^{-1}. They were found in the surface Arctic waters during the ice thawing period (June – August). In other seasons, POC content in the mass of water down to the greatest depths explored (3000 – 3700 m) varied only a little and amounted to about 5 μg POC l^{-1}. The position of POC concentration maxima is unrelated to the vertical migration of zooplankton or water stratification and is, apparently, determined by a scattered distribution of total suspended material and to fluctuations of C_{org} concentration in it. In winter the characteristic features of POC distribution vary little and concentrations throughout the water column are low (about 5 – 10 μg POC l^{-1} in the surface and below 5 μg POC l^{-1} in the deep Atlantic waters) (Melnikov and Pavlov 1978).

Studies of POC distribution in the Barents Sea along the profile Murmansk to Franz Joseph Land (August – September 1976) showed POC concentrations decreasing in this direction, as well as from the surface towards the bottom (Fig. 17; Beliayeva and Romankevich 1983). This is also characteristic of the DOC distribution at this profile. The vertical POC concentration reduction rates (K_i) averaged 3.9×10^{-2} for the 0 – 50 m layer and 7.6×10^{-3} for the 0 – 200 m layer. As well as this decrease of K_i with depth, there was also observed a direct correlation between K_i and POC concentration in the surface water layer. This points to the presence there of a significantly larger proportion of labile, relatively quickly decaying OM. Anomalously low K_i values were recorded at the sites of inflow of allochthonous OM and of polluted waters. The distribution of K_{200} values along the profile points to lower POM decomposition rates in the more northern part

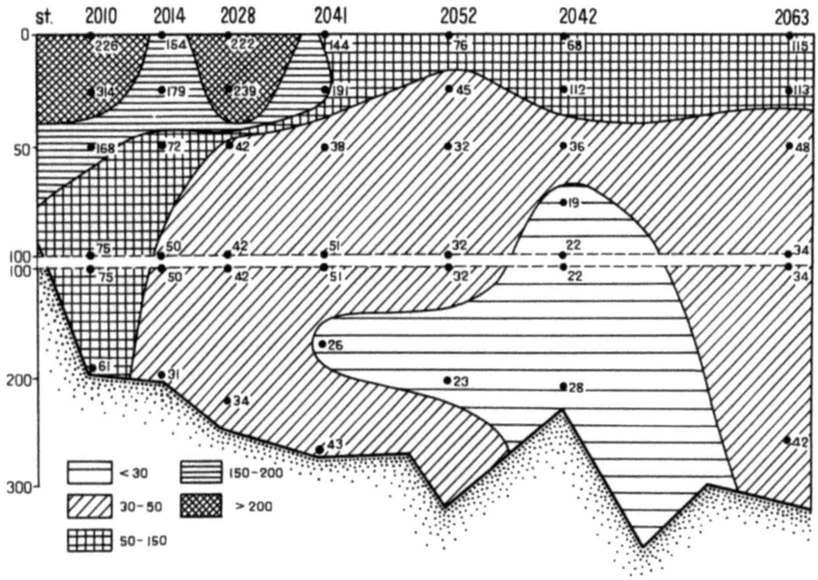

Fig. 17. Distribution of POC along the profile Rybachiy peninsula – Franz-Josef Land, μg POC l^{-1}

of the profile; this is associated with the effect of low temperature and the input of old OM into the water during ice thawing.

In most seas, POC content varies over a wide range and in many it is greater than that in the open ocean. Particularly high concentrations are observed in the intra-continental platform seas. For instance, in the Baltic Sea POC content is between 30 and 2070 μg l^{-1} (averaging, according to different data, 100 – 500 μg l^{-1}), in the Sea of Azov 120 – 2650 μg l^{-1} (average about 800 μg l^{-1}). In the intracontinental basin seas POC content is lower: in the Black Sea between 50 and 1090 (average 100 – 300 μg l^{-1}), in the Mediterranean 44 – 171 (average about 80 μg l^{-1}), in the Red Sea 14 – 67 (average 40 μl $^{-1}$). In basin-type marginal seas different POC concentrations have been recorded: in the Bering Sea 12 – 450, in the North Sea 90 – 130, in the Sea of Japan 5 – 56, and in the Sulu Sea 6 – 76 μg l^{-1}, although generally they appear to be lower than in the first two marine types.

POC distribution in the seas follows the general law typical for the oceans. Maximum content is observed near the coasts, in bays and gulfs (e.g. in the Sevastopol by 264 – 1022 μg l^{-1}, annual average – 750 μg l^{-1}; in Dalnezelenetskaya guba (inlet) (Barents Sea) – 250 μg l^{-1}). In more pelagic parts POC content usually declines. Most prominent in these seas is the vertical zonality of POM distribution, which finds expression in a decreasing POC content with depth (e.g., in the Baltic sea from 259.8 ± 38 at the 1 m depth to 91.3 ± 27.5 μg l^{-1} at the 200 m depth; Ehrhardt 1969). POC supply to the bottom is therefore greatly dependent on the particular depth. In the near-bottom layers POC concentration often increases due to rolling of settled particles, the local redistribution of which brings no changes into the general pattern of POC distribution in the water column.

The available data thus indicate that in the seas and oceans maximum POC content occurs in the upper 0 – 100 m. In the 100 – 500 m layer it tends to decrease and differences between the stations become less clear. Below 500 m a gradual decline of POC concentrations is observed, pointing to continuing decomposition and utilization of OM. Near the bottom (1 – 10 m from it) POC concentration at very many stations is higher because of resuspension of the sediments and their change into the particulate state. Regional differences in POC content distinctly shown in the upper 500 – 1000 m, are reflected in the bottom sediments even if they do not show in POC distribution below 1000 m.

A characteristic feature of POC distribution is the close relationship with biological productivity, the most important consequence of which is the decrease of POC content in the series: coastal upwelling zones > submarine continental margins > marginal parts of the ocean floor > central halistatic ocean regions. The reason for this global law is not only the decrease of primary production along this series but also a greater overall balancing between synthesis and decomposition of OM.

3.3 POC in Dry Matter of Suspended Material

Reliable data on POC content in particulate material dry matter were for a long time available only for surface samples obtained principally by means of separation. Recently data on POC content in particulate dry matter using the filtration method have become available. In samples of dry matter of suspended material, obtained with separators, POC content varies, according to our data, from 0.5 to 22%, averaging about 10% for the explored area of the world ocean (cf. Table 17, Fig. 18). The most frequent concentrations in the Pacific are below 5% and in the Atlantic and Indian Oceans from 5 – 20%. Owing to large variations in POC content in each of the oceans and the existence of blank spots on the map (cf. Fig. 18) differences in POC content in the Pacific, Atlantic, and Indian Oceans have not been adequately demonstrated.

According to other authors (Chester and Stoner 1974), POC content in the dry matter of separation samples of surface suspended material varies from 1.4 to 39.6%, averaging 16.6%.

From an analysis of 292 separation samples, an average POC content in the dry matter of suspended material was found to be about 11%, i.e., about 20% of suspended material in the ocean's surface layer is accounted for by OM.

The regional distribution of POC in the dry matter of suspended material in the surface layer is quite complex and is essentially different from the distribution of POC concentrations in a unit volume of water. This quantity is greatly affected by the degree of OM dilution with terrigenous material, as well as by the ash content of phyto- and zooplankton. The equatorial-tropical zone is distinguished by a markedly high OM content in suspended material. This may be due to a larger proportion in suspended material of the low-ash OM deriving from the consumers and to a larger proportion in the plankton of peridineans and in some places (for instance, off New Guinea and in the Red Sea) of blue-green

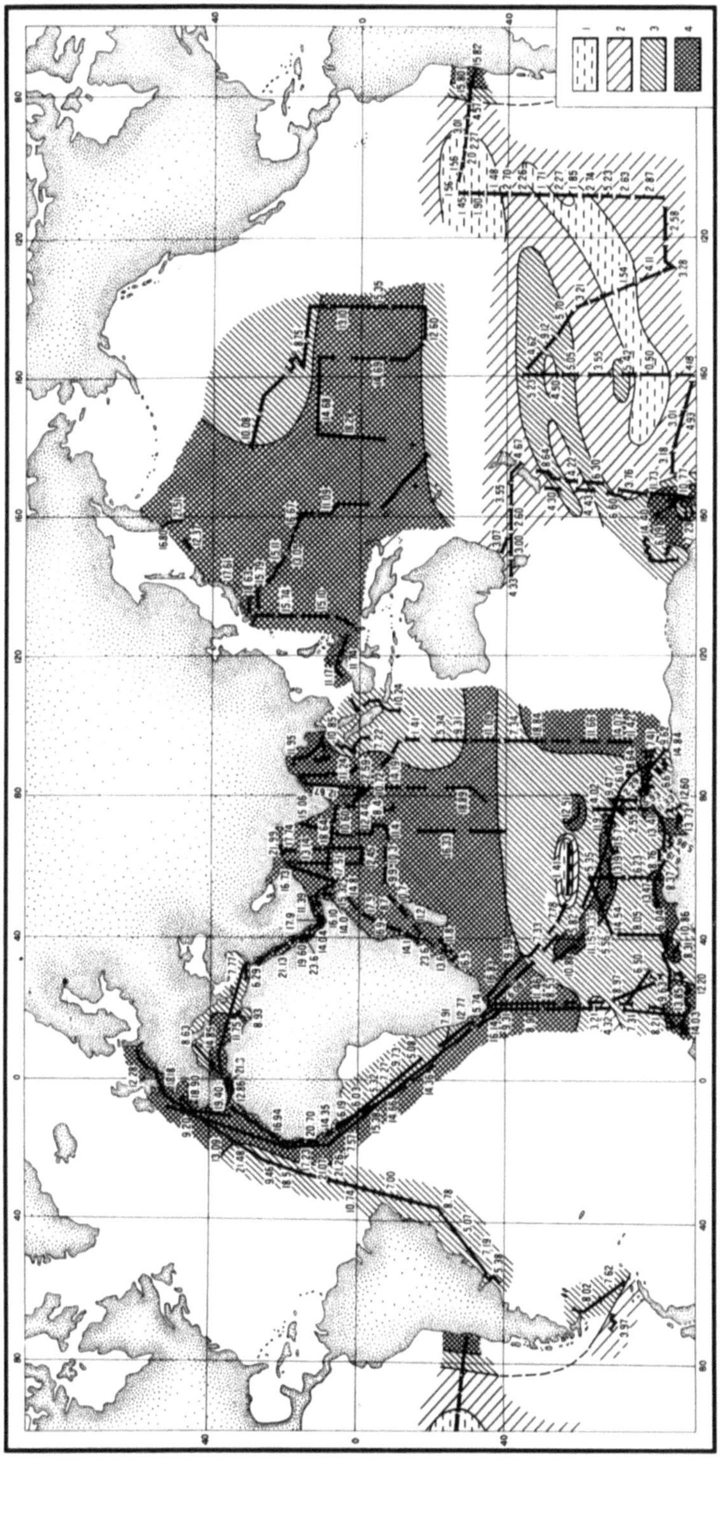

Fig. 18. The content of organic carbon in suspended material in the surface layer of ocean waters, % of dry mass: *1* = <2; *2* = 2 – 5; *3* = 5 – 10; *4* = 10 – 35

Table 22. Elementary composition of phytoplankton, phytobenthos and zooplankton

Object of study	C_{org}	Ash	N/C	P/C	H/C	Si/C	Source
	% of dry mass		wgt. %				
Algae							
diatom	11.7 – 49.5 (24.4)	7.6 – 76.5 (52.5)	12.2 – 22.6 (17.1)	1.9 – 5.4 (3.6)	13.3 – 15.4 (14.1)	0.7 – 333	Serenkov and Pakhomova (1959), Parsons et al. (1961), Vinogradova (1967), Strickland et al. (1969), this volume
peridineans	37 – 47.4 (42.6)	5.2 – 16.8 (11.1)	8.0 – 18.9 (12.8)	1.6 – 4.4 (2.5)	13.7 – 14.2 (14.0)	Traces – 0.4 (0.2)	Vinogradov (1953), Parsons et al. (1961), Vinogradova (1967), Strickland et al. (1969)
blue-green	44.9 – 47.6 (46.2)	4.0 – 10.7 (6.6)	10.7 – 12.9 (11.6)	2.8 – 3.1 (2.9)	14	2.9	Vinogradov (1953), Parsons et al. (1961), Vinogradova (1967)
Phytobenthos (zostera, Ruppia)	29.0 – 38.5 (33.2)	16.3 – 41 (33)	4.3 – 7.1 (5.3)	–	17.5	–	Vinogradov (1953), Weber et al. (1956), this volume
Crustaceous plankton	35.2 – 48.9 (43.0)	2.1 – 15.1 (10.0)	15 – 28 (23.1)	(3.0)	15.6 – 16.2 (15.9)	–	Vinogradov (1953), Vinogradova (1967), Beers (1966), this volume

Bracketed figures are average values

algae devoid of mineral skeleton. The effect of diluting terrigenous material, which usually contains little C_{org} is observed on the Antarctic coasts but is somewhat concealed off Kamchatka and the Kurils where samples had been taken in the period of plankton blossoming. The C_{org} content in suspended material there is the same as in diatom algae (12 – 20%).

A high POC content is typical for suspended material with predominant blue-green algae (14 – 21%, averagely 18.3%) and a low content for diatom suspensions (0.5 – 21.5%, averageing 7.5%). A similar variation of C_{org} content takes place in the series: blue-green algae > peridineans > diatoms (Table 22). Nearly in all cases the suspended material, as compared to phytoplankton, contains 2 – 3 times as little C_{org} in dry mass and 4 – 5 times less in comparison with zooplankton.

At the air-water interface and in the thin surface water layer there is an accumulation of living organisms and their decay products. Here the transition of some part of DOM into a particulate state also occurs. This necessarily leads to a change in the ratio between the mineral components and OM in suspended material in favour of the latter. It is shown not only in a sharp increase of POC (in $\mu g\ l^{-1}$) in the surface film but also in its greater content in the composition of dry matter. For example, data obtained in the Oshoro bay indicated that the POC content in the surface layer was 1.6 – 17.6 times greater than at the depth of 0.5 m. In suspended material dry matter, POC content generally made up 20 – 30% and was several times greater than in the subsurface layer (about 3 – 7%; Nakajima 1973).

On the evidence of data for filtration samples of suspended material, POC content in the surface layer (upper 50 m) varies from 2 to 45% of dry mass. The following values of POC content are given in literature: in the Gulf of California 34.3 ± 13.2% (n = 30; Zeitschel 1970); in the northern part of the Atlantic 14.4 – 21.1% (n = 7; Gordon 1970a), in the Gulf of Mexico 26.1 – 30.8% (Harris 1972), in the Sea of Japan 10 – 43% (n = 23), in the Bering Sea 6 – 37% (n = 30), in the northern part of the Pacific 12 – 45% (average 17%, n = 5), in the Oyashio current area east of Hokkaido 2 – 18% (average 6%, n = 6), south-east of Kyushu 21 – 27% (average 23%, n = 6), east of the Ryuku il. 7 – 33% (average 18%, n = 15), at the profile from New Guinea to the Honshu il. 5 – 37% (average 17%, n = 21) (according to Nakajima 1973), in the Mozambique area 5 – 32% (average 15%, Artemiev 1983), in the Gulf of Riga 2 – 42% (average 15%; Artemiev et al. 1981).

Since OM decays at a faster rate than the mineral portion, one could reasonably expect a decrease of POC in the dry mass of suspended material with depth. However, POC determinations in filtration samples of suspended materials obtained in the North Atlantic, in the northwestern part of the Pacific, in the Japanese and Bering Seas did not show a consistent decrease with depth of POC content in the dry mass of suspended material. At the same time, a comparison of the average values of POC content in water below 1000 m with those obtained for the 0 – 50 m layer generally show a lower POC content in the depths (usually 10 – 20%).

3.4 Relationship Between POC and DOC

The proportion of POC in the total sum of DOC + POC, as demonstrated by the data obtained by us in the Pacific, varies from 0.2 – 0.4% to 19.3%, i.e., by the factor 50 – 100 (Fig. 19). Maximum values were recorded in the high-production areas and the minimum ones, in low-production areas as well as in deep waters. In the Pacific upper layer (0 – 100 m) the weighted average of POC/DOC + POC was found to be 4.8 ± 2.9% and for the entire mass 0 – 3800 m – about 1.5%. In the explored Pacific areas in the layers 0 – 100 m, 100 – 200m, and below 1000 m the most frequent values of POC/(DOC + POC) are 3 – 10%, 1 – 3%, and less than 1%, respectively.

Table 23. POC content in total DOC + POC, %

Sampling site	Extreme values	Average	Source
Pacific (13°N – 12°S)	0.2 – 19.3	2.2	this volume
Sulu Sea (microlayer)	–	6.0	this volume
Off Peru – microlayer	~ 5 – 36	~21	Williams (1967)
– 24 – 74 m	~11 – 30	–	Williams (1967)
Equator, 155°W	–	~23	Nishizawa (1971)
Santa-Catalina trough	–	~10	Holm-Hansen et al. (1966)
Equatorial Atlantic			
0 – 164 m	6 – 18	9	Skopintsev et al. (1966)
215 – 5285 m	3 – 9	6	Skopintsev et al. (1966)
South Atlantic			
0 – 100 m	1.1 – 2.4	1.7	Artemiev and Melnikov (1974)
3000 – 5500 m	0.53 – 0.62	0.6	Artemiev and Melnikov (1974)
Black Sea			
0 – 100 m	1.7 – 20.9	9.0(22.6)[a]	Datsko (1959)
10 – 1500 m	2.0 – 4.9	~3.6	Skopintsev et al. (1971)
200 – 1000 m	2.8 – 22.9	10.8(19.7)[a]	Datsko (1959)
below 1000 m	5.6 – 13.2	8.3(9.4)[a]	Datsko (1959)
Sea of Azov, 0 – 10 m	2.3 – 32.2	13.0(34.3)[a]	Datsko (1959)
Baltic Sea, 1 – 230 m	2.0 – 6.5	~3.8	Ehrhardt (1969)
Sea of Japan			
0 – 100 m	1.2 – 3.1	2.4[b]	Liutsarev and Romankevich (1975)
100 – 200 m	0.7 – 1.2	1.0[b]	Liutsarev and Romankevich (1975)
200 – 1000 m	0.7 – 1.4	1.1[b]	Liutsarev and Romankevich (1975)
1000 – 2000 m	0.9 – 1.5	1.1[b]	Liutsarev and Romankevich (1975)

[a] Bracketed figures are the sum of particulate and colloidal dissolved OM

[b] Weighted average content

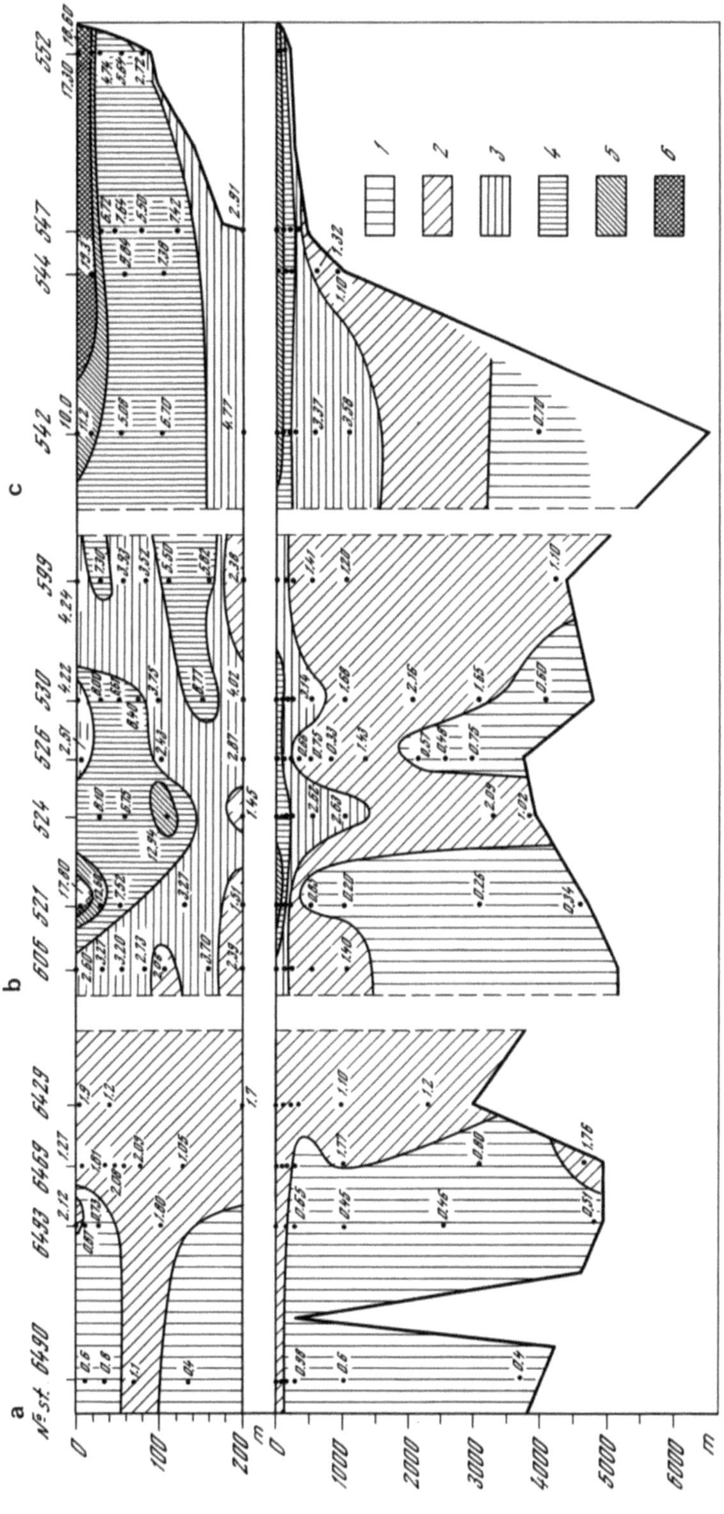

Fig. 19a – c. The proportion of POC in the Pacific waters in the total DOC + POC, %: *1* <1.0; *2* 1.0 – 2.5; *3* 2.5 – 5.0; *4* 5.0 – 10.0; *5* 10.0 – 15.0; *6* 15.0 – 20.0: **a** western equatorial-tropical area; **b** eastern equatorial-tropical area; **c** area off Peru

In the Pacific equatorial-tropical region the POC/(DOC + POC) ratio declines with depth and increases in an easterly direction. In the highly-productive Peruvian area the photic layer is observed to affect the magnitude of the POC/(DOC + POC) ratio in the shelf and upper continental slope areas. At some of stations in the eastern equatorial zone the elevated values of POC/(DOC + POC) occur at all horizons almost down to the bottom. In the areas of low biological productivity only the upper, as a rule, subsurface layer, where phytoplankton and its consumers accumulate, is distinguished by increased POC/(DOC + POC) values. The relative proportion of POM in total DOM + POM in the water mass changes but little here. Thus the distribution, in the part of the Pacific in question, of relative POC content with respect to the total C_{org} contained in sea water, correlates with biological productivity and likewise fails to render support to the concept of the ocean depths being allegedly isolated from the effect of the photic layer.

In the coastal and open high productively areas and in the seas the proportion of POC in DOC + POC is significantly higher than in pelagic ocean areas (Table 23). A particularly high POC/(DOC + POC) ratio, mainly due to elevated POC values is characteristic of the surface microlayer of water. In microlayer water from the Sulu Sea sampled by us in dead calm and with the slick visible to the naked eye, the POC/(DOC + POC) ratio was 6%, whereas at the depth of 0.2 m it was 2.7% and at the depth of 10 m – 0.6%.

According to Williams (1967), in microlayer water off Peru, POC content is 5 – 36% (average 21%) of the total DOC + POC, i.e., higher than in the deeper-lying waters. Differences in the values obtained by us and by Williams might be due to quite a number of factors, for example quite a different productivity of the areas studied: low in the first case and high in the second. Furthermore, the data given by Williams for the POC/(DOC + POC) ratio may be overrated, since DOC was determined by the Menzel and Vaccaro method. However, it is generally true that the concentration of POC at the water – atmosphere interface is a common phenomenon for zones of different productivity and the observed values are the result of complex biochemical and physico-chemical processes involving synthesis and destruction of different forms of OM at interfaces (Hunter and Liss 1981).

3.5 Living Organic Matter in Suspended Material

The most active and short-lived portion of suspended material is living OM. Drawing a line of demarcation between the living and non-living is a task that can be more easily coped with by geologists than by geochemists or oceanologists, while the latter can cope with it more easily than biochemists, since the criteria and levels at which the problem is considered are quite different (Rutten 1973). As regards organic geochemistry this problem is quite urgent, since it is the living matter that plays the most active part in the decomposition processes of OM and largely determines what part of it falls out from the cycle and accumulates in the sediments.

Efforts to determine the relationship between the living and non-living OM in suspended material were made a long time ago (Lohmann 1908). The methods being used at present for this purpose are based either on the direct microscopic calculation of phyto-, bacterio-, and microzooplankton cells or on the determination of some relatively unstable organic constituent of living OM which disintegrates rather quickly after the death of the organism, such as chlorophyll, protein. phosphorus, ATP (e.g., Nienhuis 1981). In the case of microscopic determination of living OM, account is taken of the green weight of the respective groups of organisms, or of their plasma weight, to be then recalculated in terms of living OM or "living" carbon, making use of some or other coefficients (e.g., Finenko 1965, Zeitschel 1970, Hobson 1971). In doing so, the phytoplankton biomass is often determined from the chlorophyll content, the coefficient taken for the recalculation into organic carbon being 35 – 70 and even 100.

The quantity of living OM in suspended material is estimated from the protein content which is determined after N_{org} either by the Lowry or biuret method or through a joint determination of chlorophyll and protein. However, chlorophyll and protein are contained not in the living matter alone. They are preserved in postmortem remains of the organisms and have been discovered by us in suspended material where the quantity of OM in terms of ATP was infinitesimal, as well as in the surface layer and in the mass of deep-water bottom sediments. Although the content of pheopigments and protein in suspended material sharply declines with depth, in larger pellets they are capable of reaching maximum ocean depths and cannot furnish unambiguous evidence about the proportion of living matter in particulate (as well as in precipitated) OM.

The results obtained yield a tentative idea of the relationship between living and non-living OM in suspended material. They indicate that non-living OM is, as a rule, sharply predominant in suspended material, accounting for 72 – 99.5% of total OM. Phyto-, zoo-, and bacterioplankton account respectively for 0.2 – 10%; 0.7 – 20% and 0.3 – 4% of OM in filtered suspensates (Finenko 1965, Mullin 1965, Krey 1967b, Odum and Cruz 1967, Zeitschel 1970, Melnikov 1975a).

In the microscopic study of suspended material from the Pacific and Indian Oceans it was demonstrated that in some individual cases OM of identified plankton remains may attain 40 – 70% and in the blossoming period more than 90%.

The content of living matter decreases with depth: at 200 m the cellular biomass accounts for less than 5 – 10% of OM contained in suspended material (Fig. 20; Hamilton et al. 1968, Fournier 1971, Melnikov 1975).

The method of estimating the quantity of living OM after the ATP content in suspended material (e.g., Holm-Hansen and Booth 1966) is based on the fact that the organisms with an active exchange make use of ATP in the energy processes taking place in the living cell. In the plasma and in organelles ATP serves as an accumulator and carrier of chemical energy which is spent on the processes of photo-, chemosynthesis, breathing, etc. The macroerg bonds between ATP phosphates are stable only under certain conditions. Outside the cell they are rapidly hydrolyzed. It has been found that in microorganism cultures ATP accounts for about 0.4% of C_{org}, and in the copepod plankton for about 0.9% (Holm-Hansen

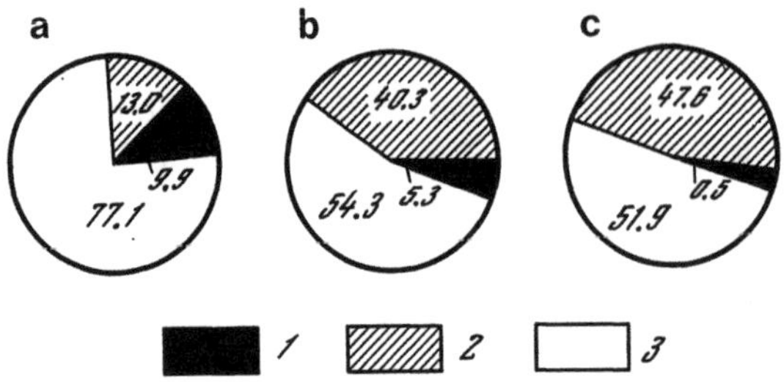

Fig. 20a–c. Relationship between OM microplankton and organic detritus in suspended material of the south-eastern Pacific, %: **a** eutrophic waters; **b** mesotrophic waters; **c** deep waters; *1* microplankton (less than 200 μm); *2* coarsely dispersed (aggregated) fraction of non-living OM (from 200 to about 1 μm); *3* finely dispersed fraction of non-living OM (<1 μm) (Melnikov 1975a)

1970, Eppley et al. 1971, Balch 1972). By assuming that the ATP and C_{org} relationship in the organisms is roughly constant we can use certain coefficients (usually $C_{org} = 250$ ATP) to calculate the carbon which is a constituent of living matter.

Determinations of the ATP content were performed by us for the eastern equatorial-tropical area and in a region off Peru (Fig. 21). The ATP content was determined by the chemoluminescent method (Strickland and Parsons 1968) in newly-obtained seawater samples (0.3 – 2.0 l). Filtering was carried out through a membrane filter with pore size of about 0.4 μm (Sinpor, Czechoslovakia). The extract luminescence was measured at a device securing displacement stability of luciferin-luciferase and the extract studied. Reagent supplied by "Du Pont" (for deep-water samples) and "Sigma" companies were used by way of enzyme – substrate complexes. The relative error of determination in the case of two parallel experiments with ATP content of 0.050 $\mu g\,l^{-1}$ was ±5%. The carbon of living OM was calculated on the base of ratio $C_{living} = 250$ ATP. The comparison of ATP and POC content determined in the same samples has shown the two to be generally in close direct correlation. Living OM was found at all depths, but it was present in significant amounts only in the upper layers. In the area of the Peruvian shelf the carbon content in living OM was found to be 100 – 38% of the total carbon in POM, the weighted average for the 0 – 70 m water column being 66%. Over the continental slope and the Peruvian-Chilean deep-water trench the weighted average content for 0 – 200 m at different stations was 16 – 41%, and in the eastern equatorial zone, 15%.

Below 200 m the carbon content of living matter was generally 1 – 15% of total POC, the median value being 7%. In ocean depths an increased content of living OM was recorded in the areas of high biological productivity, over the Peruvian continental slope and in the equatorial zone (7 – 42%), whereas a low content was observed mainly in the low-productive areas (in meso- and oligotrophic ocean-floor areas, 0.2 – 3.6%).

In the oceanic water masses layers (clouds) of condensed living matter occur. They may play an important role in the formation of the biological structure of the waters and may influence the biogeochemistry of the elements (Vinogradov 1968, Vinogradov and Lisitsin 1981). According to tentative estimates for the world ocean, the carbon of living matter accounts in the 0 – 300 m layer for 5% of POC and at greater depths (300 – 3800 m) for less than 3% of POC (Williams

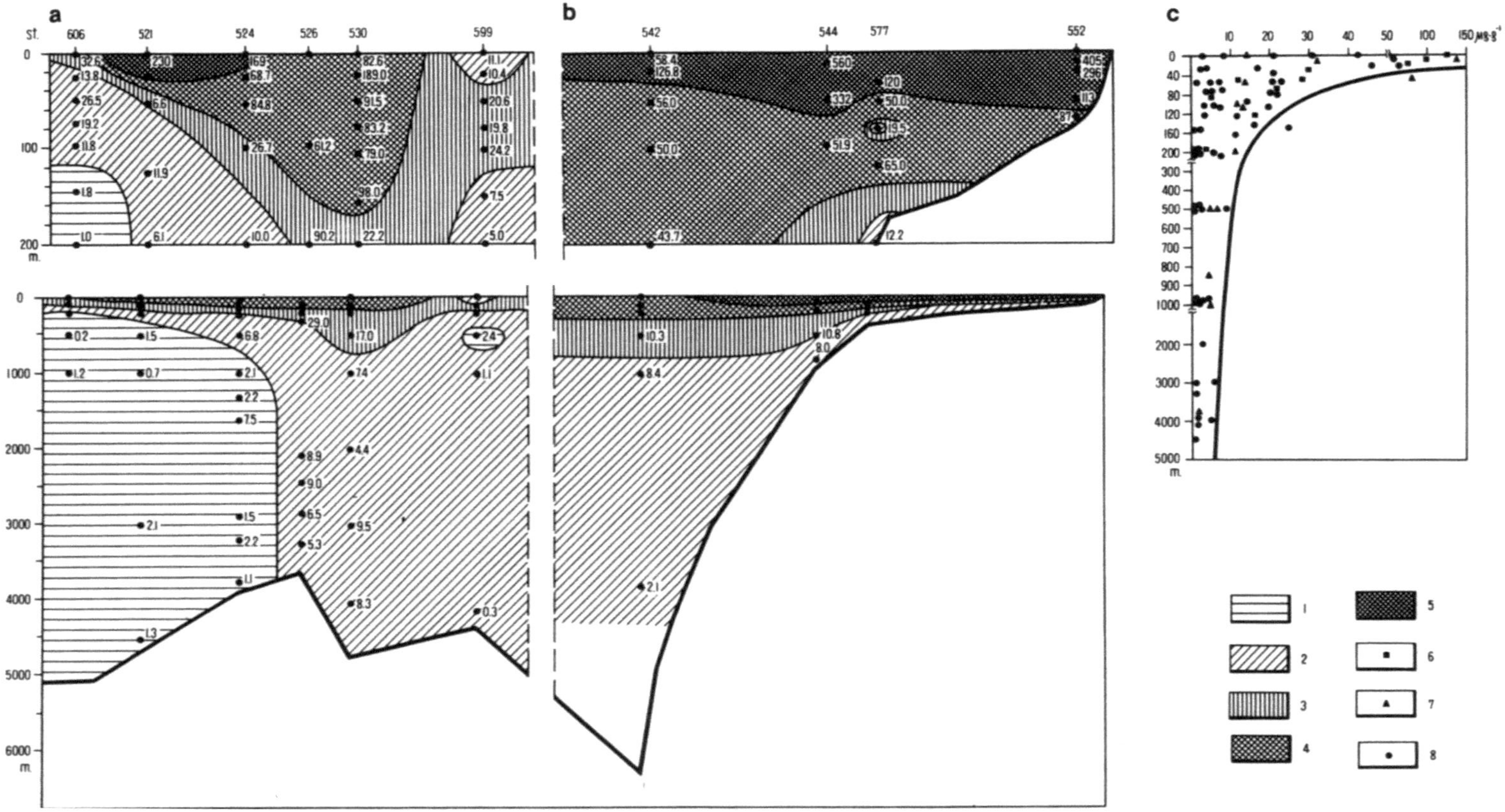

Fig. 21a – c. Distribution of adenosinetriphosphate (**a, b**) and living matter (**c**) in the Pacific water mass: **a** eastern equatorial-tropical area; **b** area off Peru; **c** distribution of living organic matter carbon in the water mass, μg C_l l^{-1}; *1* = <4 10^3 μg ATP l^{-1} (<1 μg C_L l^{-1}); *2* = (4 – 20) 10^{-3} μg ATP l^{-1} (about 1 – 5 μg C_L l^{-1}); *3* = (20 – 40) 10^{-3} μg ATP l^{-1} (about 5 – 10 μg C_L l^{-1}); *4* = (40 – 100) 10^{-3} μg ATP l^{-1} (about 10 – 25 μg C_L l^{-1}; *5* = 100 10^{-3} μg ATP l^{-1} (25 μg C_L l^{-1}); *6* = shelf region; *7* = continental slope area; *8* = ocean-floor area

1975, Parsons 1975, Whittle 1977). These appear to be underestimates since they disregard the proportion of living matter in the composition of fecal pellets and other large particles of suspended matter.

It would seem that in future one can hardly expect an alteration of the following established trends:

a) the high content of living OM in suspended material in the shelf and slope areas, particularly in upwelling zones;
b) a declining proportion of living OM in POM in transition from submarine continental margins towards the pelagial (except possibly for the areas with local upwellings and downwellings of large water masses, for instance at the equator);
c) the general character of vertical and seasonal distribution of ATP and living OM. However, the absolute values of these parameters may change according to the coefficients employed to calculate living OM in terms of ATP.

3.6 Total Quantity and Residence Time of Particulate Organic Matter in the Ocean

In 1971 the total quantity of particulate OM in the world ocean was estimated by us to be $(30-140)\times 10^9$ tons or about $(15-70)\times 10^9$ tons C_{org} (Bogdanov et al. 1971).

At present, relying on the new materials on POC distribution in different layers we can more accurately calculate the quantity of POM contained in the world ocean. The expression "more accurately" implies only that the calculations are based on more data on POC than in the late 1960's and that they were obtained on the basis of a more correct quantitative analysis of samples taken with special safeguards against contamination. However, regional and seasonal variations occurring in the ocean have, as formerly, been little accounted for. There are not sufficient data to permit a confident evaluation of the total number of large suspended particles not retained by the filtering method.

We have found the weighted average POC content for the 0 – 4000 m layer in the Pacific and Indian Oceans to be about 15 μg POC l^{-1}. Extrapolating this value for the water volume in the ocean (1299×10^6 km^3 without intracontinental and some highly-productive marginal seas of the world ocean, Stepanov 1982) yieds about 19.5×10^9 tons of POC. A low POC content in the Arctic Ocean, averaging 6 μg POC l^{-1} (without seas) because of the small volume of its waters, viz. 11.2×10^6 km^3 (without seas, Stepanov 1982), does not significantly change the above value. Taking account of the water volume of highly productive seas (37.2×10^6 km^3) and of the average POC content in them of 200 μg POC l^{-1} (7.4×10^9 tons POC), the total POC content in the world ocean amounts to about 27×10^9 tons. This estimate does not include a part of POM with particles exceeding the size of 1 – 0.45 μm. If larger suspended particles are taken into account, it can be assumed that the world ocean contains as much as some 30×10^9 tons of POC. This value disregards the colloidal OM which is contained

Table 24. The average content and total quantity of C-POM above the shelf, slope, and ocean floor in the world ocean

Ocean region	Area (Moïseyev 1969, Stepanov 1982)		Mean depth (km)	Water volume (10^6 km^3)	POC		
	10^6 km^2	%			g l^{-1}	g m^{-2}	10^9 tons
Shelf	26.7	7.4	0.08	2	200	8	0.4
Slope	57.4	15.6	1.6	91	40	64	3.6
Ocean floor	277.2	76.7	4.6	1243	10 – 15	46 – 69	13 – 19
World ocean	361.3	100.0	3.8	1336	22.5[a]	83[a]	30[a]

[a] Calculations made on the basis of the above given estimate of POC mass in the world ocean (30 × 10^9 tons POC)

in the surface and deep waters of the seas and oceans in significant quantities (see, for instance, Tables 14 – 16). The obtained value is close to that which can be derived from Riley's (1970) average POC content in the 0 – 4000 m layer (100 g m^{-1}) about 35 × 10^9 tons POC, as well as to Williams' estimates (1975) – 30 × 10^9 tons POC, but it is significantly higher than the amounts given by Menzel (1974) – 14 × 10^9 tons POC, Williams (1971) – 24 × 10^9 tons POC and Cauwet (1981) – about 20 × 10^9 tons POC.

In the ocean the bulk of POC (about 82%) occurs below the photic layer. About 50% of POC is present below 1000 m where bacterial activity is very low and OM oxidation proceeds slowly (Jannasch et al. 1971, 1973, Conover 1978, Sorokin 1978).

To solve the problem of OM accumulation in the sediments and assess the closeness of the general C_{org} cycle it is important to know what quantities of POC are contained above the different structural zones in the ocean. The figures below (Table 24) give a tentative idea of the POC quantities above the shelf, slope and ocean floor.

The dynamically balanced and interrelated masses of DOC, POC and living matter carbon (C_{living}) for the ocean as a whole relate as 1800 : 30 : 2.8 × 10^9 tons C_{org} or as 100 : 1.7 : 0.16. The mass of living organisms in the world ocean is given according to Bogorov's data (1967b, 1971) to which the slightly updated values (2.6 × 10^9 tons C) furnished by Conover (1978) are rather close. Although the separate estimates of the plant (1.74 × 10^9 tons C) and animal (0.45 × 10^9 tons C) biomass given by Woodwell et al. (1978) are strikingly different from Bogorov's (1967b, 1971) – 0.06 × 10^9 tons C for plant biomass and 2.7 × 10^9 tons C for animal biomass – the total mass of living matter in the world ocean according to the one and the other is similar.

Such a "pyramid of masses" of DOC, POC, and C_{living} is the result of a dynamic equilibrium of the conjugate processes of OM synthesis and decay. It reflects the different resistance of these OM forms to decomposition. If we apply to this succession (pyramid of masses) of carbon in the ocean (CO_2, DOC, POC, C_{living} – Degens 1979) the rule of inverse proportionality which links up the mass

Fig. 22. Distribution of POM calculated with respect to its maximum concentrations in the water column, %: *1* = <5; *2* = 5 – 10; *3* = 10 – 50; *4* = >50 (Bogdanov et al. 1971)

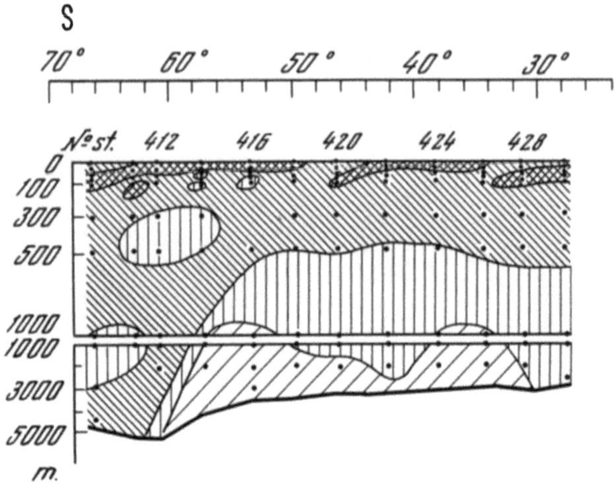

of each of the funds and its turnover rate in the system (Silen 1976; by analogy with the theory of compartment-funds in biochemistry and ecologic physiology of organisms) and assume the age of DOC to be 1500 years, the residence time in the ocean of the bulk of conservative POC will be about 25 years and that of living OM about 2 years. Some research workers, for instance Khailov et al. (1973), think that the turnover time of living OM in the photic layer, where a prevailing mass of living OM is concentrated, is equal to a single year. If we extend this inference to the entire mass of living OM in the ocean, then, according to the rule of inverse proportionality, the residence time of the bulk of POC in the ocean will amount to about 10 years. These limits (10 – 25 years) apparently characterize the average residence time in the world ocean of the bulk of POM. It is to be noted that with such an approach the age of the bulk of DOM will be 700 years.

The POC content in deep waters (several dozen metres from the bottom) in relation to the maximum content in the water column gives an idea of the relative quantity of OM reaching the seabed. A profile running south of the Aleutians as far as the Antarctic demonstrates that the maximum proportion (>10%) of POM produced in the surface layer attains the seabed in the moderate and high latitudes in the northern hemisphere (Fig. 22). A very insignificant proportion of OM (2 – 5%) reaches the seabed in tropical waters, north of the Marquises, and in the waters of the southern hemisphere's moderate latitudes, south of the Pasque il. A nearly similar decrease in the content of POM is observed in the Antarctic waters and only slightly more than 10% of OM produced in the surface ocean layer is observed to reach the deep-lying levels near the Antarctic coasts (Bogdanov et al. 1971).

According to other data, the lower reaches of the water mass are reached by the following quantity of POC in proportion to its content in the surface layer: in the West-Marianas and Caroline basins – 8 – 16%, in the Peruvian and equatorial part of the North-Eastern basins – 3 – 10%, on the Peruvian shelf – 11 – 35%, in the Andaman Sea – about 10%, in the northern part of the Bay of

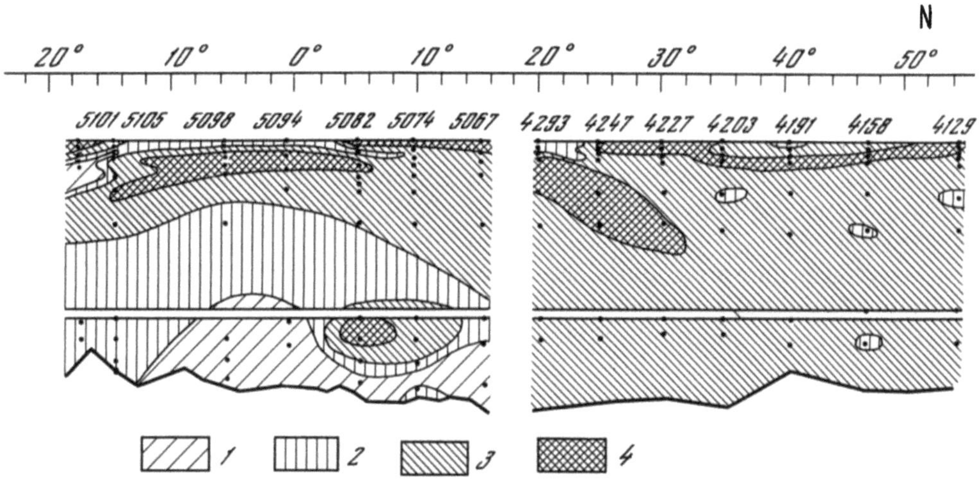

Fig. 22 (continued)

Bengal – 10–25%, on the Mosambique shelf – 10–30%, in the Red Sea – 5–10%. The available data on the quantity of POC reaching deep water layers and the bottom indicate that the annual average value for the world ocean amounts to 5–14%. Thus, as part of primary OM production and OM inputs from the land the ocean floor is reached by about $(1.2-3.7)\times10^9$ tons C_{org} annually. However, as will be shown later in this chapter, the larger part of this OM does not come to be deposited in the sediments but decays at the surface of the ocean floor.

3.7 Organic Matter Morphology

Particulate OM, as was pointed out earlier, comprises living and non-living components. The living matter in suspended material, including microplankton (200–20 μm), nannoplankton (2–20 μm) and ultraplankton (less than 2 μm), has long been the subject of study. The composition of this part of suspended material, representing a small proportion of OM below the photic layer, is not discussed here.

Non-living OM in ecology and hydrobiology starting with the work (Petersen 1918) is often called detritus. It can be subdivided into a coarse- and a finely-dispersed fraction with the particle size ranging from 200–150 to 1–0.4 μm and <1–0.4 μm, respectively. A light microscope can be successfully employed to study only the coarsly-dispersed part of suspended material, since calculating particles of less than 1 μm at the filters proves practically difficult due to their small size and the rough surface of the filters themselves. The finely-dispersed OM fraction is studied by means of electron microscopes with a resolution of 2–10 Å (0.0002–0.001 μm).

Coarsly-Dispersed Fraction of Suspended Material. Microscopic investigations with the use of various dyes have revealed that the usual and widespread types of organic and mineral-organic particles in suspended material are aggregates (60 – 80% of identified particles), fragments and flakes (Fig. 23), as well as OM confined inside skeletal remains of organisms, and unidentified particles (Bogdanov and Lisitzin 1968, Nemoto and Ishikawa 1969, Gordon 1970b, Riley 1970, Wiebe and Pomeroy 1972, Melnikov 1975, Degens 1976, Chesselet 1980, Cauwet 1981).

Aggregates are amorphous, heterogeneous formations. In the ocean they occur at all depths and are predominant over other forms of POM (Fig. 24). Largely, they do not represent identifiable formal remains of organisms but are rather fecal particles, pellets and their remains, collapsed masses consisting of small particles. They vary in size from those barely discernible under the microscope to 200 μm and larger. In the Pacific, as concerns the number of particles, the 5 – 25 μm fraction is predominant, whereas with respect to weight the prevailing fraction is under 5μg. In the Atlantic, the average diameter of such aggregates is about 25 – 50 μm, though particles of up to 1 mm and larger are observed to occur. Their shape may be most varied (irregular, laminar, fibrous), although it is mostly rounded. These are, as a rule, non-compact masses with numerous voids in which bacteria, algal remains and mineral inclusions have been found. In some of them are confined undamaged diatom and dinoflagellata cells, and vegetal fibres (Riley 1970, Melnikov 1975b).

The greater part of aggregates take colour from acid fuchsine with respect to carbohydrates, and the lesser part from acridine orange with respect to proteins and from Nile blue with respect to lipids. The larger part of aggregates, as revealed by the calculation of coloured particles, consists of carbohydrate-like compounds.

In the Pacific, two types of particles are prominent among the aggregates (Bogdanov and Lisitzin 1968; Fig. 25): (a) yellowish-grey, easily taking colour

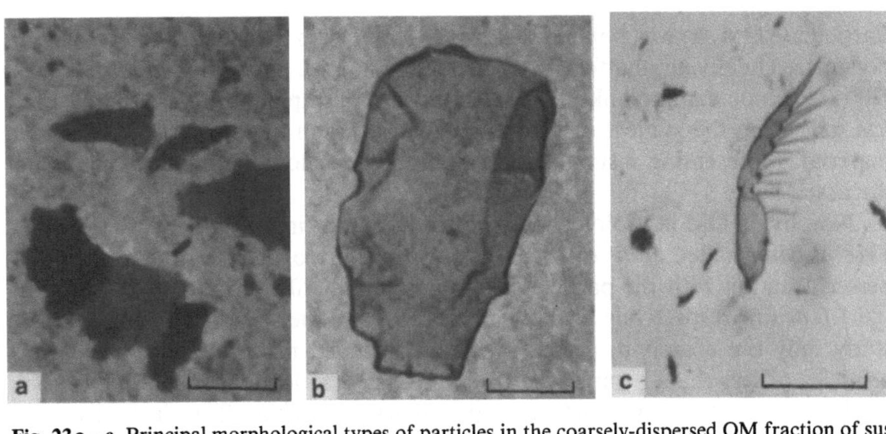

Fig. 23a – c. Principal morphological types of particles in the coarsely-dispersed OM fraction of suspended material (from about 200 to 1 μm): **a** aggregates; **b** flake-like particle; **c** a fragment (maxillipeda) of a copepod

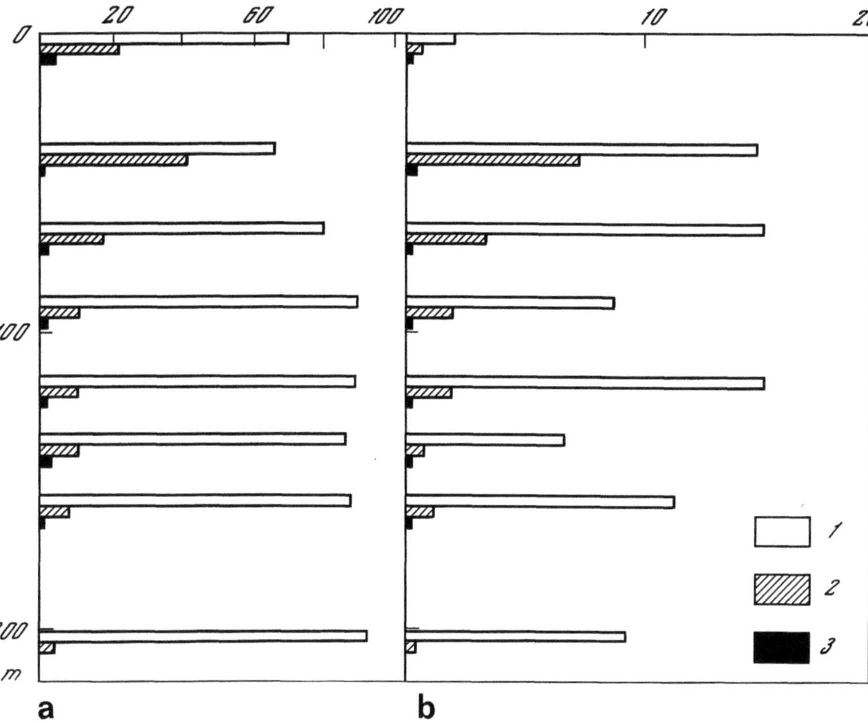

Fig. 24 a, b. The content of aggregates (*1*), fragments (*2*) and flakes (*3*) in suspended material of the Peruvian trough, st. M-558, 7°32′9 S, 102°43′2 Q: **a** % of OM; **b** in μg C · l^{-1} (Melnikov 1975)

from erythrosine and acid fuchsine; they are predominant in suspended material; (b) brown and yellowish-brown ones, which do not take colour from these reagents; their shape is more rounded; generally they are of a wax-like appearance and luminescent; as to size, they are significantly smaller than the former; particles less than 2.5 μm in size are predominant among them. The distribution of these particles at the vertical profile from the Aleutians to the Antarctic has shown that the number of the a-type particles decreases with depth, the decrease being more rapid in the low than in the high latitudes. As a result of a high vertical gradient about 10% of particles in the tropical zone reach the near-bottom layers, and in the boreal zone and off the Antarctic as much as 20% of their content in the upper layer. The quantity of the b-type particles declines insignificantly along the vertical (the decrease being greater in the low than in the high latitudes), therefore their relative proportion increases with depth. OM in the brown particles appears to be more humified and resistant to decay. Its principal source, judging from the character of meridional distribution, is plankton.

Organism fragments are mostly 50 – 200 μm in size. These are mainly crustacean crust and limbs and other parts of the bodies, as well as vegetal fibres and starch grains.

Shallow waters contain a significantly larger proportion of fragments of bottom algae, weeds, fecal lumps which are quantitatively far superior to zooplank-

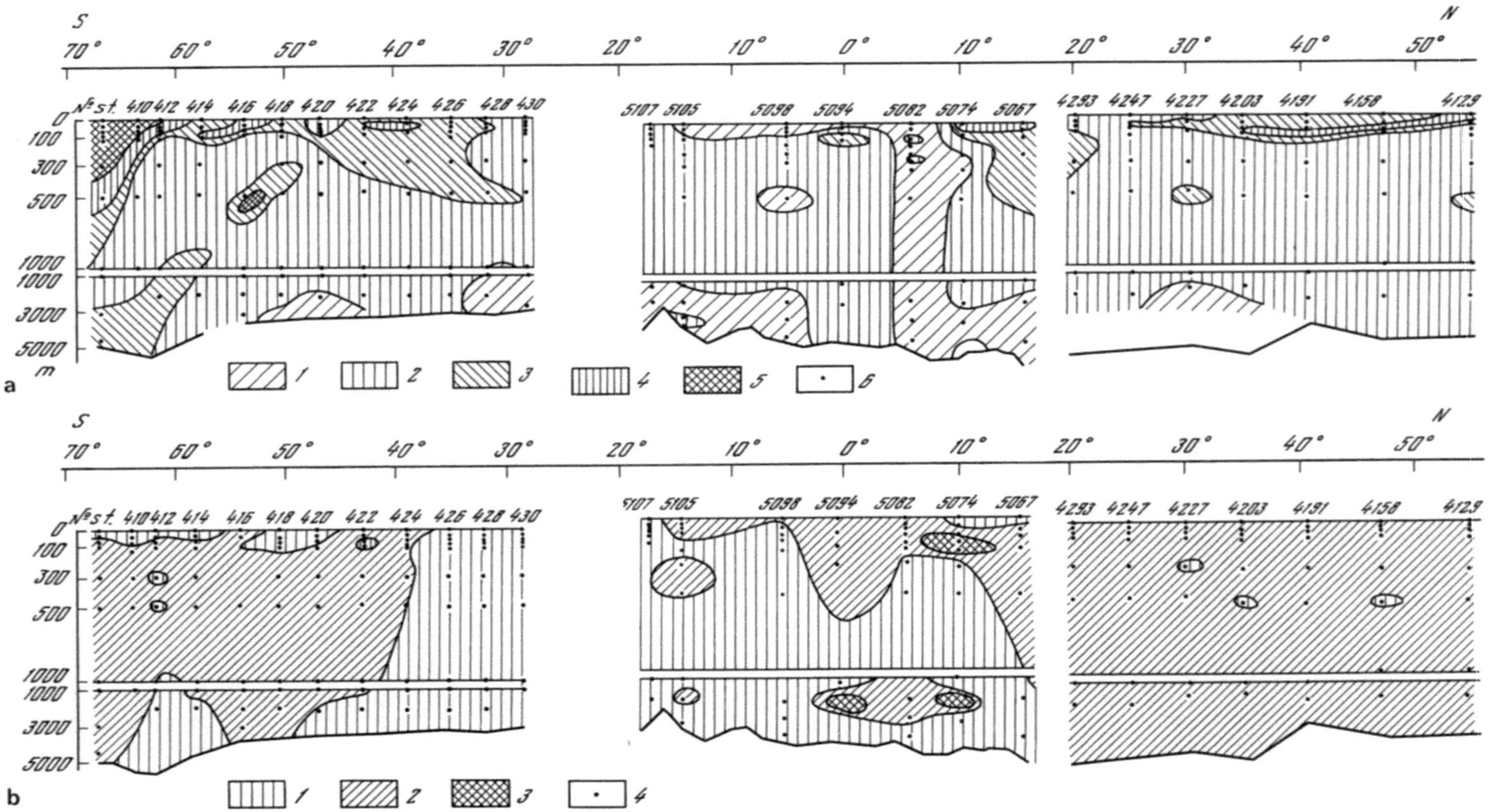

Fig. 25 a, b. Vertical distribution of organic particles on the profile running south of the Aleutians to the Antarctic (microscopy data calculated in terms of OM), µg OM l^{-1}: **a** yellowish-grey particles taking colour from erythrosine: *1* = <1; *2* = 1 – 10; *3* = 10 – 25; *4* = 25 – 50; *5* = >50; *6* = sampling sites; **b** brown particles taking no colour from erythrosine, luminescent: *1* = <1; *2* = 1 – 10; *3* = >10; *4* = sampling sites (Bogdanov and Lisitzin 1968)

ton remains. Near coral islands, organic aggregates are added, which form here from the slime liberated by corals. Fecal pellets are excreted by zooplankton and fishes – phytophages feeding on plankton and benthonic algae. These lumps contain labile OM partly deposited on the bottom and partly carried away by currents (author's observations, as well as, for instance, Odum and Cruz 1967, Johannes 1972). In deep-water areas the fragments of organisms mostly occur in the 0 – 200 m layer.

Flakes are thin, transparent, non-discrete, lamellar particles not infrequently having irregular and angular edges. As revealed by the study of suspended material in the eastern part of the Pacific and off Peru, these are mostly large particles of up to 100 μm, mainly 10 – 40 μm in size. The author happened to observe at some stations in the south-eastern part of the Pacific such kind of transparent particles – flat, lamellar (about 2 μm thick), mostly homogeneous but occasionally with mineral inclusions. They occurred in insignificant quantities in the upper part of the photic zone as often as in deep waters.

Flakes respond positively to proteins, carbohydrates and are externally not different from particles formed during air bubbling through sea water (Gordon 1970b). They possibly appear due to the flocculation of organic colloids. Their constituents carbohydrates, proteins and lipids give rise to geopolymers resistant to decomposition and bioutilization. Presumably this explains their significantly more homogeneous distribution in the mass of water as compared to aggregates.

The OM in suspended material may take the form of organic compounds confined in the valves of diatoms, silicoflagellates, coccoliths, as well as in foraminiferal shells occasionally occurring in suspended material. Clean (without plasma) diatom shells contain, according to our data, 0.13 – 0.63% (average about 0.30%) C_{org} and globigerine shells 0.15 – 0.42% (average 0.25%) C_{org}. In the composition of suspended material this OM is a minor component at all depths. However, in the mass of water it becomes selectively preserved during the decay of "pure" OM, since it is the most conservative form of OM and is little given to biochemical decomposition. Furthermore, OM is partly adsorbed by biogenic mineral and fragmentary material. The coating of organic compounds strongly affects the growth and solubility of crystals and their stability under aggressive conditions. These forms of OM do not make a great contribution even to the deep-water suspended material, but their importance sharply increases in bottom sediments.

The study of the distribution in the south-eastern part of the Pacific of particles taking colour from acid fuchsine (similarly to erythrosin, this reagent gives an idea of the total content of organic particles) revealed that among them at all depths the dominant group was made of particles 10 – 50 μm in size. The vertical distribution of particles was found to be uniform in coastal eutrophic waters off Peru and in mesotrophic equatorial-tropical waters and corresponded, as concerns the number, to the exponential law, and as regards concentration to the normal distribution (Melnikov 1975a). A decrease of the number of large particles with depth and the exponential character of the distribution of suspended OM particles were noted also in other ocean areas (Nemoto and Ishikawa 1969, Beardsley et al. 1970, Sheldon et al. 1972).

The biogenic genesis of the bulk of OM in the coarsely-dispersed fraction of suspended material cannot be doubted. The aggregates display red fluorescence, contain chlorophyll and certain chlorines which are indicators of OM of a phytoplankton genesis (Wiebe and Pomeroy 1972). The fragments are either of vegetal or animal origin. Some of flakes also appear to be organism remains, reminding in structure the transparent tissues of tantinides, infusoria, copepod remnants (Melnikov 1975a, b).

Finely-Dispersed Fraction of OM. Most of the works discussing the finely-dispersed OM fraction in sea water contain only information on the relationship between particles of different size (e.g., Sharp 1973, Ogura 1974, Baturina et al. 1975). Data on the morphological composition and systematics of particles relating to the finely-dispersed fraction of seawater OM are very scarce (Baturina and Mishustina 1976, Baturina et al. 1979).

In the south-eastern part of the Pacific the filtering technique was used to isolate OM fractions of different dispersion from sea water through ultrafilters XM-300, PM-10, UM-2, UM-0.5. From these fractions as well as from the filtrate (after UM-2 filter) and sea water passed only through a membrane filter with 0.4 μm pore size as many as 300 preparations were made and examined in the IEM-7 electron microscope (Baturina and Mishustina 1976).

Principal structural units of the OM fraction with a molecular mass superior to 10×10^3 are globular forms of 0.02 – 0.6 μm diameter and laminated forms of 0.11 – 0.66 μm diameter and thickness comparable with the resolving power of the electron microscope 2 – 10 Å (Fig. 26).

The varieties of globular forms are single round particles (0.02 – 0.6 μm); chains consisting of several globules; aggregates up to 1 μm in size; branched chain-filaments 4 – 10 μm in size consisting of globules of 0.01 – 0.16 μm diameter. Occasionally several globules have a common envelope and are reminiscent of similar forms known in artificial systems.

The varieties of laminated forms are single and usually large particles up to 0.5 μm in size, as well as tangled fibrous aggregates up to 3 – 5 – 10 μm in size consisting of fragments of about 0.2 μm diameter. Laminated forms are transparent to electron beams; when superimposed on the electron-microscope grid they form characteristic folds diverging raylike from the centre (cf. Fig. 26).

At present, the available data are too scanty to provide an insight into the differences in the distribution of globular and laminated forms or their varieties. Globular forms were observed in all the samples investigated, whether from the photic or from deeper layers. Chain-filaments composed of relatively large particles were observed in the photic layer alone. The rate of occurrence of laminated forms was higher in the Peruvian upwelling zone. A tentative ratio of occurrence rates for single globules, chain-filaments and laminated forms in the Peruvian upwelling waters can be given as 1 : 4 : 4, and in the waters of the open south-eastern part of the Pacific Ocean as 4 : 2 : 1.

Experimental research conducted during 6 – 8 months at 4 – 8 °C showed the collapse of some forms, the aggregation and neoformation of others. In the process, the globular particles disintegrated more rapidly than did the laminated ones, the number of the latter rising in time.

Fig. 26a – f. The electron-microscopic images of OM fractions in sea water with the molecular weight 10 – 300000 (**a** – **e**) and 500 – 1000 (**f**): **a** globular forms; **b** filmy forms; **c** round particles with a common envelope; **d** aggregates; **e** branched chains (filaments); **f** a variety of tangled-fibrous aggregates (Baturina and Mishustina 1975)

As for the genetic nature of the finely-dispersed OM fraction in sea water, it can be noted than no direct proofs of its biogenic origin are available, though, in the author's opinion, this is most likely. Comparing the size and morphology of structural units, as well as the manner of branching of chain-filaments with certain high-molecular organic compounds, and the microspectrofluorometric investigations of globular particles permit us to suppose that some of the discovered structural units happen to be the derivatives of nucleic acids and their compounds with phosphates of the type of nicotineamideadeninedinucleotide (NAD) and coenzyme A.

3.8 Supply of Organic Matter to Ocean Depths and Ocean Floor

The ways and means of OM supply to ocean depths and onto the ocean floor have been repeatedly discussed by research workers of different fields of study – biologists, geochemists, lithologists (for example, Riley 1970, Bogdanov et al. 1971, Romankevich 1977, Chesslet 1980, Suess 1980, Cauwet 1981, Gagosian and Lee 1981, Lisitzin and Vingradov 1982, Saliot et al. 1982, Lisitzin 1983, Lorenzen et al. 1983a, b, Suess and Thiede 1983, Degens and Ittekkot 1983).

At present we may speak of three methods of OM supply to ocean depths and the ocean floor: (a) passive or nearly passive settling of dead OM and organo-mineral particles of autochthonous and allochthonous origin, (b) neoformation of particulate OM from dissolved OM and its subsequent precipitation, (c) active OM transport by migrating animals.

3.8.1 Trophic System and the Passive Settling of Particles

The phytoplankton represents the first stage in the complex nutrient chain permeating the entire mass of waters. In the process of photosynthesis and metabolic exchange it expends about 30% of the generated gross OM itself. The energy confined in the remaining OM (net production) secures the existence of organisms at all the subsequent trophic levels.

In the open parts of water basins the grazing of phytoplankton by phytophages frequently exceeds 90% of daily production. It is important to note that in bays, gulfs and coastal marine waters a lesser part of phytoplankton is consumed, for example in the Sevastopol bay just a few percent (Petipa 1968). The rest of primary production here dissolves and mineralizes, supplements suspended material reserve, comes to be deposited on the seabed or carried away by sea currents. These areas are potent sources of DOC and POC in the ocean.

A decrease of grazing efficiency with the increase of food concentration stems from reverse dependence between animal filtration rate and food concentration, as well as by delayed development of phytophages compared with phytoplankton (Geinrikh 1971, Suschenia 1972, Petipa 1981). A lower degree of grazing is a favourable factor for OM accumulation in the shelf sediments.

Zooplankton, zoobenthos and practically all bacteria synthesize OM in much lesser quantity than they consume. The food eaten daily by zooplankton exceeds 50% of their bodily weight. However, a significant proportion of consumed food fails to be assimilated by zooplankton and is released into water in the form of liquid and solid excrements. Carbonaceous, siliceous, chitin skeletons, as well as mineral particles pass through the intestinal tract practically without being assimilated. The percentage of assimilated food depends on many factors (concentration of food, its ash content and biochemical composition, the size of the nutritive object, nutrition rhythm, food procuring mechanism, etc.) and varies within a very wide range, between 2 and 99% in copepods, for example (Petipa 1981). Liquid excrements being taken into account, the assimilated food amounts on average to 40–60% of the consumed food. Consequently, as much as

60 – 40% of consumed food returns into POM reserve from the first heterotrophic level. A similar situation exists at the subsequent trophic levels. As a result of the vital activity of consumers and bacteria, the bulk of primary production and OM resynthesized by heterotrophs is combusted. Thus, the more intricate the food web and the greater the number of trophic levels were OM is most actively utilized, the lower the quantity that is likely to reach the seabed. It would thus appear that zooplankton must unavoidably lead to a decline of OM quantity that could be supplied to ocean depths and deposited in the sediments. However, this is true for shallow waters alone, where part of OM due to primary producers can reach the bottom and be deposited. If other systems of OM transport into sea depths were unavailable, we would observe an even sharper boundary between OM content in the shallow-water and in the deep-water sediments which would be practically devoid of life. At the same time, about 5 – 15% C_{org} of primary production do reach the ocean floor at the depth of 3000 – 5000 m. An important role in the supply of this OM into ocean depths is played by zooplankton excretions, mostly compact pellets, as well as by the corpses of extinct organisms notable for high settling rates. The real supply of OM from these sources to ocean depths as well as its chemical composition, depends upon many factors but mainly on the relationship between the sinking and the decomposition rate of the settling particles as well as on the number of their consumers (e.g. Lorenzen et al. 1983a, b, Lee et al. 1983). Theoretical and experimental calculations made in recent years (e.g. Stepanov and Svetlichnyi 1975, Suess 1980, Gagosian and Lee 1981, Iseki 1981, Tseitlin 1981, Lisitzin and Vinogradov 1982, Lebedeva et al. 1982) have shown that corpses of mesoplanktonic animals (for instance, copepods) may reach the depths of some thousand metres, i.e. depths which are not usually attained by dead microplankton and phytoplankton organisms.

The bulk of OM from primary producers, among which prevail diatom algae with cells 1.3 – 1630 μm (average 50 μm, Semina 1974) in size, undergo 90% decay or more without reaching the 100 m depth. Microscopic studies of phytoplankton have revealed that diatoms with plasma do not occur, as a rule, below 200 – 350 m, though valves with and without plasma are often not easily distinguishable. Ohwada (1960) employed a direct microscoping of samples collected by means of water bottles east of the Tsugaru Straits and found that diatom cells with undamaged chromatophores occur down to the maximally explored depths (1000 m), but are mostly concentrated in the 0 – 200 m layer.

Observations made on Lake Baikal have demonstrated that the larger part of dead plankton is decayed in the upper 70 – 100 m. However, 80% of certain algae (melosira) remain alive at 500 m depth and even reach the bottom while still alive. Apparently, there is some mechanism which enables live algae to reach greater depths than the calculations based on non-living organisms suggest.

Plankton foraminiferal shells 50 – 250 μm in size and containing plasma reach 1500 – 2000 m depths, and pteropods with plasma (500 – 3000 μm size), 4000 m depth. However, in mass quantity both foraminifera and pteropods with plasma were observed in the ocean only at 0 – 200 (500) m depth (Beliayeva 1962, Vinogradov 1968, Parker 1973). At greater depth their quantity sharply declines. Little is known about the distribution, biomass and production of foraminifera in the ocean that may serves as OM source at great depths. Foraminifera are relatively large and at the end of their life cycle rapidly settle onto the seabed. Their contribution to the OM supply to ocean depths appears to be significant, but it is yet difficult to estimate it quantitatively.

Dead zooplankton undergo decomposition relatively quickly. Complete decomposition of copepods at 17 – 23 °C occurs within 30 – 40 days, small radiolarians – within 5 days, Sagitta – within 13 days (Pavlova 1968, cit by Tseitlin 1981). However, small crustaceans (0.5 – 1 mm) sink by some 50 – 100 m a day and within a day reach the low-temperature zone where the rate of decomposition decreases. Thus, the corpse of even a small copepod is likely to reach as low a depth as 1000 – 4000 m. Corpses of large zooplankton organisms have even larger sinking rates and may attain abyssal depths.

The sinking velocity of zooplankton fecal pellets is 3 – 5 times as high as that of dead organisms of a similar size, though the variability of velocity values is great (Tseitlin 1981). It greatly depends on the nature of food, on the content in it of mineral components, for instance, mineral particles of suspended material, diatom valves, etc. The more compact and heavy the fecal pellets, the higher their sinking velocity. The decomposition rate of OM in fecal pellets formed on natural food is much lower than that of dead organisms. In deep-water traps fecals decompose little over a period of many weeks. This is associated with a greater resistance to decay of OM in the pellets, as well as with the presence of tight envelopes in some of them. As a result, organism pellets prove to be the chief OM supplier to ocean depths and the ocean floor. They account for up to 80 – 90% of detrital flow. Corpses of organisms, equsions and suspended material generated from dissolved OM play a subordinate role as an OM source in ocean depths and at the ocean floor. The quantity of detrital OM in surface ocean waters is directly dependent upon primary production. On the evidence furnished by Lisitzin and Vinogradov (1982) in hypertrophic waters where primary production surpasses 10 gC m^{-2} day^{-1} there are formed on the average 9 g m^{-2} day^{-1} of detritus, in eutrophic waters with primary production equal to 6 – 2.5 gC m^{-2} day^{-1} – about 3 g m^{-2} day^{-1}, in oligotrophic waters with primary production below 1 g m^{-2} day^{-1} some tens mg m^{-2} day^{-1} of detritus are formed. A decrease of detritus quantity in the series hypertrophic – eutrophic – mesotrophic – oligotrophic waters is accompanied by an increase of OM mineralization in this series from 60 – 65% in the highest-productive waters to up to 90 – 95% and more in the lowest-productive ones. In the highly productive waters of coastal upwellings the composition of POM seems to be particularly strongly affected by the stage of development of the community. In the course of the plankton succession the composition of POM grows more conservative (e.g. Sellner et al. 1983). All this adds to the differences in magnitude of detrital flow reaching the ocean floor in the eutrophic and oligotrophic ocean areas.

It seemed interesting to estimate that part of OM contained in suspended material which is most strongly bonded with (being partly enclosed in) carbonaceous, siliceous and fragmentary terrigenous material and is little exposed to enzymic attack. Relying on indirect data and balance equilibria, we have attempted such an assessment (Table 25). The total amount of siliceous material leaving the cycle annually in the form of skeletal remains of diatoms, radiolarians and silicophlagellates is approximately equal to its quantity being annual-

Table 25. Supply to the seabed of organic matter carbon strongly bonded with terrigenous and biogenic material

Sources of C_{org}	Total quantity, 10^9 tons[a]	C_{org}	
		% of dry mass	10^9 tons
Terrigenous and volcanogenic material	25 – 26	0.1	0.025 – 0.026
Carbonaceous	1.36	0.3[b]	0.004
Siliceous	0.46	0.3[b]	0.001
Total	–	–	~0.03

[a] Lisitzin (1974)

[b] According to our determinations in foraminiferas and diatoms sampled respectively from strongly-calcerous foraminiferous (more than 95% $CaCO_3$) and strongly-siliceous diatom (more than 50% $SiO_{2(amorph.)}$) sediments. Prior to C_{org} determination skeletal formations had been treated with hydrogen peroxide under ultraviolet irradiation

ly supplied to the ocean by rivers – 0.462×10^9 tons (Livingstone 1963, Lizitsin 1974). The C_{org} content in this material being 0.3%, it can be computated that as much as 1.4×10^6 tons C_{org} year^{-1} arrives at the bottom. Similar calculations for the carbonaceous material yield 4.1×10^6 tons C_{org} year^{-1} and for terrigenous material, about 25×10^6 tons C_{org} year^{-1}. Summing up, we obtain 1 – 3% of the total C_{org} reaching the seabed and more than 30% of the total C_{org} buried in the sediments (cf. Chap. 4). Hence an important conclusion is that at the water – bottom interface, the bulk of OM undergoes decomposition while OM protected by mineral components and strongly bonded with them (mineral-organic form) escapes utilization. This gives a convincing proof of the fact that the production-destruction cycle in the ocean is generally well balanced.

3.8.2 Transition of Dissolved Organic Matter into the Particulate Form

The change-over of dissolved OM into the particulate form has been studied over quite a long period, but even at present numerous problems continue to be interpreted in dissimilar ways, the reason being mainly the differences at the methodological level of different investigations. The history of the problem and the principal published results were discussed in sufficient detail by Riley (1970). Our researches pursued the aim of determining whether particulate OM does form from dissolved OM under conditions similar to those prevailing in nature, and, if so, in what quantity.

Experiments were being conducted during the cruises of the research vessels "Vitiaz" (July 1971, 50th cruise) and "Dmitrii Mendeleyev" (July 1972, 8th cruise) in 20-litre black glass bottles washed with chromium mixture, distilled and bidistilled water. The samples were not hermetically sealed. On the 50th cruise, water was sampled in the western equatorial-tropical region of the Pacific

(St. V-6469; 0.6°49′8 N, 135°35′7 E, ocean-floor depth 4850 m, sampling depth 50 m), and on the 8th cruise – in the Peruvian upwelling area (St. M-552, ocean-floor depth 78 m, sampling depth 10 m). On each cruise we performed a 35–36-day experiment at two temperatures: 30° and 6 °C. Neither inhibitors nor stimulators of microflora growth were added to the water, and inoculation was not used. All the experiments were made under close to natural conditions. At the end of each term of exposition two water bottles (30° and 6 °C) were used in full. Water in them was gently stirred, most of it was filtered through glass fibre filters Whatman GF/C. The suspended material and non-filtered water ware analysed. Many analyses were performed with 2–4 parallel ones. Bacterial biomass was determined by Yu. I. Sorokin through the direct counting method and the potential production was assessed from a dark assimilation of radioactive carbonate over a period of four days. The results are given in Fig. 27.

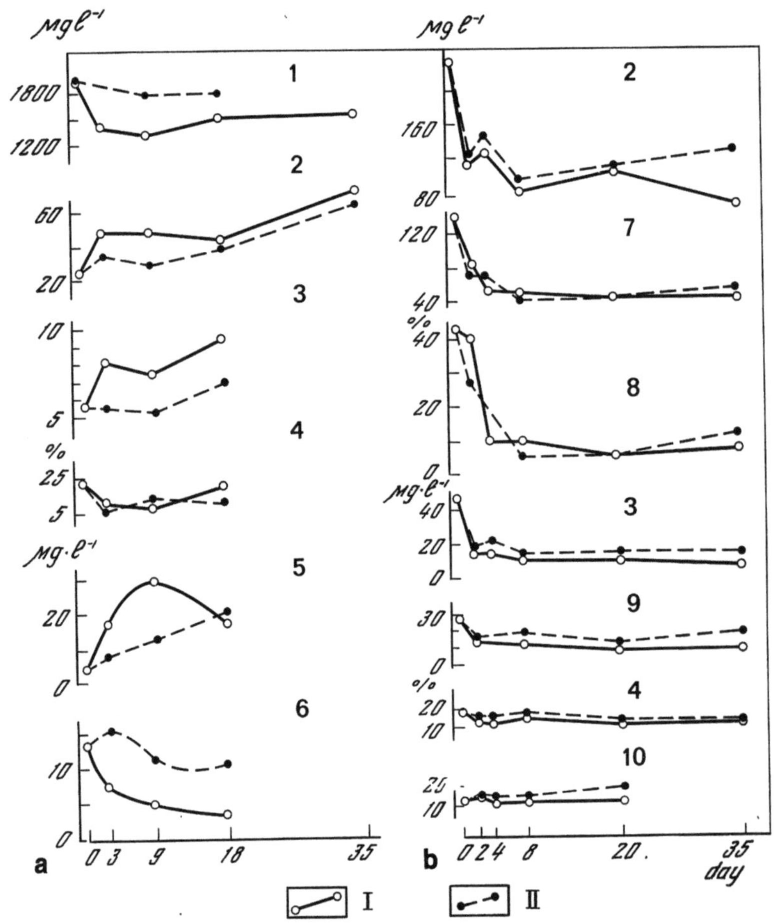

Fig. 27a, b. The kinetic of decay and neoforming of particulate OM from the dissolved one. **a** Western Equatorial-Tropical Pacific; **b** shelf off Peru. *I* Experiment carried out at 30 °C; *II* Experiment carried out at 6 °C. *1* DOC; *2* POC; *3* C_{CHC} in suspended matter; *4* part of C_{CHC} in POC; *5* bacterial biomass; *6* potential bacterial production; *7* C_{living}; *8* C_{living}/C-POM; *9* N-POM; *10* N-POM/C-POM

In the first series of experiments (cf. Fig. 27a) the DOC content at the beginning of the experiment was 1960 μg l^{-1} (98.8% of C_{org}). At 30 °C the DOC content declined by 28% within three days (to 1400 μg l^{-1}) and later changed but little (1330 – 1500 μg l^{-1}). At 6 °C the change was less great – by some 9% on the 18th day. The maximum number of bacteria was observed on the 9th day (30 °C) and thereafter declined sharply. The decomposition of dissolved OM was biochemical, judging from the character of the two curves. In the course of the experiment the POC quantity rose at 30 °C from 23.4 to 48 – 49 μg l^{-1} (on the third and ninth day) and up to 71.7 μg l^{-1} (on the 36th day). At 6 °C the increase proceeded more smoothly and the figure at the end of the experiment was 63.4 μg l^{-1}.

Thus, in the course of the experiment the growth of the bacterial population was accompanied by that of particulate OM which was found at the end to increase three-fold as compared to the initial water. The rate of POC increase in the first three days was 8 μg l^{-1} day^{-1} (30 °C) and 3.7 μg l^{-1} day^{-1} (6 °C) and for 36 days averaged 1.3 and 1.1 μg l^{-1} day^{-1} (correspondingly for 30° and 6 °C). Consequently, at 30 °C as much as 500 μg DOC · l^{-1} converted into CO_2 and 48.3 μg POC · l^{-1} were newly formed, at 6 °C the corresponding indices for 18 days were 160 – 180 μg DOC l^{-1} and 15.2 μg POC l^{-1}. Indisputably, the principal factor for the neoformation of POC was the bacterial destruction of DOC accompanied by aggregation of bacterioplankton and a transition of its remains into suspended material. In this case about 10% DOC converted into POC.

In the second series of experiments (cf. Fig. 27b) the DOC content was 1000 ± 100 μg DOC l^{-1} throughout the experiment. POC content by the end of the experiment declined from 236 to 62 μg l^{-1}. The decomposition rate of POC and C_{living} in the first two days was 63 and 28 μg C l^{-1} day^{-1}, and for the entire period 5 and 2.7 μg C l^{-1} day^{-1}, respectively. The relative quantity of living matter in the suspended material was invariable during the first two days and sharply decreased on the fourth day (from 43% to 10%). Further it was observed to vary between 5 and 10% and only at the end of the experiment in either case (30° and 6 °C) it rose slightly.

Consequently, dissolved OM sampled in the highly-productive area off Peru failed to mineralize during the experiment and, apparently, did not change over into the particulate form. During the experiment we observed the destruction of particulate OM, including living OM. The decrease of living OM in suspended material pointed to unfavourable conditions for bacterioplankton and microzooplankton. As for the causes of this phenomenon, we may only offer some suppositions. Most likely it was due to the presence in the water of organic inhibitors liberated during the phytoplankton blossoming. Furthermore, the labile dissolved part of OM could have been used up earlier by bacteria and Protozoa, the metabolic activity of which, judging from the ratio of primary production to respiration, is very high in the Peruvian upwelling.

The main inference ensuing from these investigations is that in the first series of experiments for the ratio DOC : POC = 100 ÷ 1.2 there was observed a bacterial decomposition of dissolved OM, an increase of bacterial biomass and POM at the expense of DOM (photosynthesis being excluded). In the second series of experiments for the ratio DOC : POC = 100 ÷ 21.4, which were conducted under similar conditions, there was only decomposition of POM and a decrease of the biomass of living matter (algae and bacteria); the content of DOM was in this case practically invariable (more accurately, it varied within the analytical error). Variations of the indices for 30° and 6 °C were unidirectional, while the rates of the processes at 6 °C were lower.

A similar conclusion about an important role of bacteria in the formation of POC from DOC on the basis of experiments with non-filtered and filtered water, with extracellular algal metabolites has been drawn by numerous researchers (see, for instance, Riley 1970). Thus our research, as well as the results obtained by other authors, confirm the possibility of neoformation of the POM from DOM with the participation of microorganisms. The data discussed here do not exclude the possibility of the formation of POM at interfaces by chemical means; on bubbles, salt crystals, mineral particles, at the water – air interface, particularly during bubbling (see, for instance, the reviews by Riley 1970, Cauwet 1981). These processes frequently occur in natural waters and happen to be an additional source of POC, though the scale of the process is yet unknown.

3.8.3 Active Transport of Organic Matter by Migrating Animals

It was in the late 1950's that the principles of the zooplankton "transport" of OM into ocean depths or the "migration ladders" were clearly formulated (Vinogradov 1968). The essence of the zooplankton "transport" of OM consists in the fact that migrating animals take food in the surface layers and thereafter descend into depths in the course of daily, seasonal and age-conditioned migrations, embracing practically the entire water column. As a result, OM is carried to depths that cannot be reached by phytoplankton or the remains of dead zooplankton. While performing their vital functions and after death the zooplankton organisms add to the DOC and POC reserve. The favourable effect of zooplankton on OM supply to ocean depths and the ocean floor, apart from speeding up its transport, is associated with the relatively large size of their remains and excretions in the composition of which, particularly in the case of excessive nutrition, the relatively little altered OM is likely to be preserved (Beklemishev 1954, Petipa 1981). In fact, this kind of OM brought by migrating animals has been discovered in medium-deep layers (Knauer and Martin 1980). Active OM transport is particularly effective in highly-productive waters of moderate latitudes. Even here under daily migrations the proportion of participating species is usually not great, while the range of vertical travel, as a rule, does not exceed 200 m.

Seasonal and age-conditioned migrations cover a water mass down to 1000 – 2000 m and participating in them in moderate latitudes are mass animal species accounting for 70 – 80% of mesoplankton biomass in the surface layers. In tropical waters the role of active OM transfer by animals into ocean depths appears to be insignificant due to a low number of migrating forms (Lisitsin and Vinogradov 1982).

Today it is still difficult to evaluate the role of animal migration in bringing OM to ocean depths where the probability of its preservation is much greater.

Zooplankton transport is noted for greater significance in OM supply to the ocean floor in highly-productive temperate and equatorial areas and is much less important in the oligotrophic tropical zones. This adds to the differences in OM supply to great depths and the ocean floor in both high- and low-productivity ocean areas.

Thus the quantity of OM reaching the ocean floor in one or other ocean area is dependent upon numerous indicators characterizing the balancing of the production-destruction system. Beside the magnitude of primary production, the role of which is particularly evident, other important factors are the number of trophic levels and the physiological characteristics of both zoo- and bacterioplankton. While in the pelagial zone the primary production transforms through five trophic levels with about 10% ecological efficiency at each of them, in the areas of continental submarine margins it goes through three levels with an efficiency of 15%, and in upwelling regions, such as the Peruvian and South-West African, it does so through two levels with 20% efficiency (Geinrikh 1971, Marti 1971, Schulz 1973).

The analysis of available data brings us to the conclusion that all three mechanisms of OM supply to great depths and the ocean floor operate simultaneously: (a) passive sinking of OM in the composition of fecal material, animal corpses and OM aggregates and a more active settling down of these particles in downwelling zones (the principal mechanism), (b) neoformation of POM from DOM, and (c) zooplankton transport.

In the series pelagial – continental submarine margin – upwelling zones, OM synthesis-decomposition becomes increasingly unbalanced and an ever greater quantity of primary OM escapes utilization in the metabolic furnace. In the same direction generally, decreases the depth and increases the quantity of little-altered OM arriving at the seabed from the autotrophic and the heterotrophic levels. Wherever the production-destruction cycle is poorly balanced, OM supply to ocean depths and the ocean floor is dependent in many respects also on the nature of the nutrition of the consumers, on the physiological possibilities of utilizing one or other food, on the efficiency of assimilating organic particles of different size, as well as on the character and size of fecaes.

Phytophages in the boreal and moderate ocean zones are larger than the tropical ones. Their fecaes are likewise larger and do not dissolve so much in the course of precipitation (Geinrikh 1971). This contributes to the fact that in the water mass in the north and south of the Pacific high concentrations of suspended material are maintained, as was noted above, down to great depths and a larger quantity of OM (up to 20% of primary output) reaches the ocean floor. As a result, these areas have a higher OM content in the bottom sediments.

It should be also taken into account that regional differences in OM supply to ocean depths and the ocean floor are associated with the local features of water circulation, water transfer and convective mixing. Thus, for instance, in moderate latitudes in the north Atlantic the active convective water mixing extends down to the bottom at a rate of about 1000 m a month (Stepanov et al. 1982). As a result, the deep waters here are notable for the ocean's highest O_2 concentration (up to 6 – 7 ml l^{-1}), high salinity (up to 35.1%), density (up to 27.8 conventional units), as well as an elevated temperature.

Under these conditions, increase of POM sinking rate fails to make up for the high rate of its accelerated biooxidation. As a result, with a similar or larger primary production, POC content in deep waters and the OM content in bottom sediments here are below those observed at the corresponding latitudes (40 – 60 °N) in the North Pacific.

Thus, the distribution of photosynthesis production in the ocean, the degree of its utilization in trophic networks, ecological and physiological characteristics of the consumers, the size, composition, precipitation and decomposition rates of their pellets and postmortem remains are factors which determine the varying degree of closeness of the production – destruction cycle and the OM quantity supplied to ocean depths and the ocean floor. These factors throw light on the meaning and causes behind the existence of the circumcontinental, latitudinal, and vertical zonalities in the distribution of POM in the ocean, which are inherited by bottom sediments.

4. Organic Carbon in Late Quarternary Sediments of Seas and Oceans

The state of inequilibrium in the synthesis-decomposition system of aquatic OM in the water mass of seas and oceans is reflected in the amount of suspended matter that sinks to the bottom, which in terms of C_{org} approximates $(1.2-3.7)\times 10^9$ tons yearly. The decomposition of the aquatic organic precipitation is practically the sole energy source of biogeochemical processes of authigenous mineral formation, that results in element dissemination and, in some instances, in the formation of ore concentrations. This implies that the organic content of sediments, which is expressed in terms of C_{org}, is the most important controlling factor determining the geochemical properties and type of sediments as a whole. Geochemical studies concerned with the principles governing the distribution and accumulation of OM in sediments are of high practical importance. For instance, the inferences drawn from such studies contribute to an understanding of the mode of deposition of large masses of OM, which serve as the source of petroleum and natural gas. This subject has long been a matter of lively concern, and the keen interest in the origin of petroleum has not lessened to date.

Investigations of OM in recent sediments date back to the end of the last century, when Verigo (1881, 1888) and Andrusov (1892, 1894) first conducted studies in the Black Sea. In later years Andrusov (1908) and Arkhangelsky (1927) stressed the importance of studies of OM in recent sediments for the elucidation of the processes and mechanism responsible for sedimentation and the origin of petroleum. The problem bearing on the accumulation of OM in contemporary and ancient water bodies Gubkin (1975) regarded as one of the fundamental problems of petroleum geology. Systematic analytical studies of C_{org} in bottom sediments were begun by Arkhangelsky (1927) and Trask (1932). During the 1930s and 1940s studies contributing to the knowledge of the content and composition of OM were advanced by Trask and Wu (1930), Waksman (1933), Ginsburg-Karagicheva, Pryanishnikov and Rodionova (1934), Gorshkova (1936), Anderson (1939), Wiseman and Bennett (1940) and Fox et al. (1944).

The years following 1950 witnessed a number of publications which led to a general recognition of the principles governing the distribution of C_{org} and (or) N_{org} in different environments; some papers contained information on the composition of OM. As a result of the extensive studies, over several thousand publications are now available on the composition, distribution and accumulation of OM in recent sediments.

Studies of aquatic bottom sediments were begun in the 1950s and continue to date in the P.P. Shirshov Institute of Oceanology of the USSR Academy of Sciences. During this period investigations were carried out in different areas of the Pacific, Indian and Atlantic Oceans as well as in the Arctic sea. A wealth of information contributing to the knowledge of the composition and gross levels of

Table 26. Organic carbon content of bottom sediments in seas and oceans (%, on a dry sediment basis)

Area investigated Type of sediments	Number of samples	Extreme values	Source
Pacific Ocean			
Bering Sea	152	0.08 – 2.56 (0.96)	Lisitzin (1966)
Okhotsk Sea	158	tr – 2.20 (1.02)	Bezrukov (1960)
Comandor-Kamchatka area	72	0.09 – 1.55 (0.58)	Romankevich (1970)
East China Sea	10	0.16 – 0.78 (0.38)	Bezrukov et al. (1958)
East China Sea	70	0.24 – 2.05	Gershanovich and Neyman (1964)
Sea of Japan	34	0.47 – 2.0 (0.90)	Gershanovich (1956)
	21	0.34 – 2.02 (1.02)	Skorniakova (1961 b)
Gulf of Alaska	–	0.3[a]; 1.0[b]	Gershanovich (1965)
		(0.70)[c]	Gorshkova (1973)
Kronotskii Bay	14	0.08 – 0.73	Romankevich (1959)
East of Simushire Island	9	0.15 – 1.20	Romankevich and Artemiev (1969)
North Pacific (down to 20°N)	360	0.06 – 2.5	Bezrukov et al. (1961)
Pacific ocean in all entirety (without seas)	928	0.01 – 7.44	Romankevich (1970)
Southeastern Pacific	57	0.2 – 10.6	Logvinenko and Romankevich (1973)
Between Tonga and Society Islands	17	0.10 – 0.29 (0.20)	Skorniakova and Petelin (1967)
Northwestern Sea of	44	0.25 – 2.0	Niino and Emery (1966)
Japan	20	0.37 – 1.83 (1.10)	Ohta (1953)
Yellow Sea	–	up to 1.8 (0.8)	Niino and Emery (1961)
Coastal waters off China	–	(0.3)	Niino and Emery (1961)
Antarctic, Anver Island	13	0.09 – 2.48 (0.75)	Warnke et al. (1973)
Southeast of Hokkaido	79	0.13 – 2.39 (0.89)	Kato (1956)
Near Hokkaido	140	0.21 – 3.04 (1.70)	Kato (1956)
Shallow Bay of Honshu	4	8.96 – 11.88 (10.9)	Koshimizu (1961)
Northeastern Pacific	57	0.14 – 2.30 (0.55)	Skorniakova (1961 a)
52 – 40°N	102	tr – 3.5	Gross (1967)
Peruvian-Chilean area	–	0.1 – 9.6	Trask (1961, quoted from Van Andel 1964)
Shelf and slope off Peru	69	0.2 – 11.6 (2.4)	Gershanovich (1981)
Gulf of California	111	0.4 – 7.4	Van Andel (1964)
Bousset Lagoon (Sakhalin Island)	–	0.14 – 4.29	Zadkova et al. (1975)
Posiet Bay (Sea of Japan)	14	0.49 – 3.67 (1.86)	Andreyeva and Agatova (1981)
Atlantic Ocean			
Atlantic Ocean	–	0.08 – 4.65[d]	Trask (1932)
North Atlantic	154	0.07 – 0.70	Vikhrenko (1964)
Central Atlantic	–	0.08 – 0.81	Vikhrenko (1966)
Pelagial areas of Ocean	10	0.21 – 0.49 (0.27)	El Wakeel and Riley (1965)
Near Greenland, Labrador and Newfoundland	87	0.30 – 3.60	Avilov (1965)

Table 26 (continued)

Area investigated Type of sediments	Number of samples	Extreme values	Source
Bay of Brest Roadsted	28	1.00 – 7.45	Guilcher and Pruleau (1962)
Mochima Bay	35	0.11 – 6.45 (1.31)	Kato (1961)
Shelf off Northwestern Africa	24	0.21 – 2.60	Senin (1970)
Shelf off Southwestern Africa	30	0.29 – 12.5	Emelianov and Senin (1969)
	63	up to 26[e]	Calvert and Price (1971)
Shelf and slope off Southwestern Africa	58	0.15 – 14.05 (2.34)	Gershanovich et al. (1972)
	11	1.33 – 11.79 (4.94)	Romankevich and Baturin (1972, 1974)
(0 – 25 cm layer)	47	0.18 – 16.06	Emelianov (1973 b)
Shelf of Southwestern Africa			
terrigenous sediments (<50% $CaCO_3$)	50	0.16 – 9.78 (3.25)	Emelianov and Romankevich (1979)
carbonate (>50% $CaCO_3$)	17	1.43 – 6.57 (2.86)	Emelianov and Romankevich (1979)
siliceous (>30% $SiO_{2(amorph)}$)	16	1.43 – 12.50 (6.20)	Emelianov and Romankevich (1979)
Atlantic Ocean without southwest-african shelf			
terrigenous sediments (<50% $CaCO_3$)	264	0.04 – 3.60 (0.69)	Emelianov and Romankevich (1979)
carbonate (>50% $CaCO_3$)	54	0.12 – 1.02 (0.71)	Emelianov and Romankevich (1979)
siliceous (>30% $SiO_{2(amorph.)}$)	6	0.18 – 0.60 (0.35)	Emelianov and Romankevich (1979)
Atlantic Ocean in all entirety	911	0.04 – 14.05 (0.92)	Emelianov and Romankevich (1979)
pelites	326	0.09 – 14.05 (1.16)	Emelianov and Romankevich (1979)
aleurites	323	0.04 – 9.78 (0.92)	Emelianov and Romankevich (1979)
sands	262	0.04 – 4.71 (0.62)	Emelianov and Romankevich (1979)
Shelf off South America	31	0.10 – 2.52 (0.72)	Emelianov (1973 b)
Greenland Sea	–	0.76 – 1.02	Belov and Lapina (1961)
Norwegian-Greenland Basin	130	0.14 – 1.80 (0.70)	Gorshkova (1962, 1973)
Norwegian Sea	–	0.5 – 2.1 (0.7)	Géochimie organique (1977)
Baltic Sea	250	0.11 – 5.72 (1.05)[c]	Gorshkova (1973)
	173	tr – 5.64 (2.07)	Blazhchishin (1976)
Kursch Bay	112	0.4 – 9.3 (2.2)	Pustelnikov (1983), this volume
Bay of Riga	49	0.07 – 2.96 (1.05)	Gorshkova (1970)
Gulf of Finland	24	0.33 – 4.98 (2.48)	Gorshkova (1970)
Azov Sea	45	0.44 – 2.96 (1.73)[c]	Gorshkova (1973)

Table 26 (continued)

Area investigated Type of sediments	Number of samples	Extreme values	Source
Aral Sea	–	0.07 – 2(0.65)	Brodskaya (1952, quoted from Gorshkova 1970)
Black Sea	–	0.24 – 4.70(2.15)[c]	Glagoleva (1961), Kochenov et al. (1965), Florovskaya and Gursky (1970), Gorshkova (1973)
0 – 1 m	–	(2.0)	Potapova (1980)
65 – 70 m	–	(1.2 – 1.5)	Potapova (1980)
Mediterranean Sea	215	0.20 – 1.60(0.60)	Emelyanov (1973a), Kochenov et al. (1965)
East of Brazil	–	0.4 – 2.0	Hobson and Menzel (1969)
Bay of Biscay shelf	21	0.07 – 0.82(0.28)	Blazhchishin (1972)
Barents Sea	–	0.15 – 3.12(1.28)[c]	Gorshkova (1975)
White Sea	119	0.09 – 2.92(1.14)[c]	Gorshkova (1975)
Kara Sea	–	0.25 – 2.00(0.83)[c]	Gorshkova (1975)
East-Siberian Sea	–	0.54 – 2.09(0.79)[c]	Gorshkova (1975)
Chukchee Sea	–	0.18 – 1.74(0.98)[c]	Gorshkova (1975)
Caspian Sea	101	0.20 – 3.67	Pakhomova (1961)
	56	0.02 – 3.49(1.02)[c]	Gorshkova (1973)
carbonate sediments	78	0.21 – 1.73(0.92)	Bordovsky (1969)
terrigenous pelites	114	(1.48)	Kuprin and Bagirov (1971)
South Caspian the western shelf			
Novocaspian layer (0 – 25 m)	18	0.37 – 5.09(1.24)	Kuprin et al. (1976), Kuprin and Potapova (1978), Potapova (1980)
Khvalyn (25 – 60 m)	8	0.38 – 2.99(0.80)	Kuprin et al. (1976), Kuprin and Potapova (1978), Potapova (1980)
Khazar (60 – 90 m)	9	0.33 – 3.76(1.17)	Kuprin et al. (1976), Kuprin and Potapova (1978), Potapova (1980)
Baku (90 – 280 m)	16	0.05 – 0.16(0.52)	Kuprin et al. (1976), Kuprin and Potapova (1978), Potapova (1980)
Apsheron (280 – 1120 m)	97	0.14 – 2.06(0.52)	Kuprin et al. (1976), Kuprin and Potapova (1978), Potapova (1980)
Akchagyl (1120 – 1280 m)	9	0.22 – 2.52(0.74)	Kuprin et al. (1976), Kuprin and Potapova (1978), Potapova (1980)
The productive mass (down to 1750 m)	51	0.05 – 1.31(0.23)	Kuprin et al. (1976), Kuprin and Potapova (1978), Potapova (1980)
Maine and Puget Sounds	45	0.24 – 4.74(1.61)	Bader (1955)
Southeast of Long Island	6	0.46 – 1.00(0.72)	Emery and Ross (1968)
North of Venezuela	–	<0.5 – 5.03	Okuda and Gomez (1964)

Table 26 (continued)

Area investigated Type of sediments	Number of samples	Extreme values	Source
Central Texas Bays	–	0.5 – 3.5	Oppenheimer (1960)
Gananeia Lagoons (Brazil)	95	2.24 – 6.37 (3.34)	Kutner (1964, quoted
Abidjan Lagoon (Western Africa)	–	4.5 – 12.8 (6.43)	from Van Andel 1964)
Gulf Paria	–	0.1 – 1.4 (0.71)	Van Andel and Postma (1954)
Mississippi Delta	–	0.1 – 1.3 (0.61)	Van Andel and Postma (1954)
Gulf of Mexico			
shelf	45	0.17 – 1.29 (0.53)	Trask (1953)
slope	13	0.31 – 1.21 (1.02)	Trask (1953)
Sigsby Basin	16	0.35 – 1.34 (0.63)	Trask (1953)
Southwestern Atlantic	126	tr – 5.17 (1.15)	Zakharov (1974)
Coastal sands of Georgia (USA)	–	~0.05 – 0.6 (<0.2)	Pinet and Frey (1977)
Area off Western Africa			
shelf and upper slope	31	0.05 – 1.66 (0.64)	Diester-Haass and Müller (1979)
slope between 500 – 2895 m	18	0.78 – 3.66 (2.22)	Diester-Haass and Müller (1979)
Bay of Benin	–	0.6 – 1.1	Klingebiel et al. (1976, quoted by Vigneaux 1980)
Cariaco trench	–	1 – 5	Daumas et al. (1978)
Cap Verde Ils.	–	0.3 – 2.5	Daumas et al. (1978)
Amazon detrital cone	–	0.5 – 2.9	Géochimie organique (1978)
Indian Ocean			
Arabian Sea	–	up to 6.74	Bezrukov (1964)
Arabian Basin	–	0.40 – 0.75	Bezrukov (1964)
Bay of Bengal	–	0.38 – 0.65	Bezrukov (1964)
Central part of Ocean	–	0.15 – 0.35	Bezrukov (1964)
Antarctic Region	13	0.13 – 0.73 (0.34)	Bordovky (1968)
Red Sea	28	(0.16)	Emery (1964)
Gulf of Aqaba (Red Sea)	5	0.06 – 0.15 (0.08)	Mohamed (1949)
Suez Bay	–	0.35 – 0.45	Mohamed (1949)
Persian Gulf	8	(0.8)	Emery (1956, 1964)
Tonkin Bay	–	up to 1.3 (0.6)	Niino and Emery (1961)
Gulf of Thailand	~92	0 – 1.3	Emery and Niino (1963)
Karachi-Cochin (shelf, slope)	30	0.32 – 8.93 (2.29)	Dietrich et al. (1966)
Bay of Bengal (shelf)	–	0.2 – 1.6 (0.77)	Rao (1960)
Aden Bay, Oman shelf and slope	9	1.3 – 6.9 (~3.0)	Pelet (1981)

Figures in parenthesis show mean values. Wherever sampling depth is not indicated, the figures refer to the 5 (10) cm layer

[a] Mean value for sands

[b] Mean value for muds

[c] Average value, relative to area occupied by sediments of variable C_{org} content

[d] Estimated from N × 8.3

[e] Values re-calculated on a salt-less basis

Table 27. C_{org} content in various environments, % of dry matter (Vigneaux et al. 1980)

Types of environments	Number of samples	Extreme values	Average	Sources (quoted by Vigneaux et al. 1980)
Lakes (England, Scotland, North America)	22	1 – 11.14	6.50	Brooks et al. (1976), Dungworth et al. (1977), Granwell (1977), Stuermer et al. (1978)
Fresh-water bogs (Everglades)	3	2.7 – 9.95	6.51	Lindberg et al. (1975), Hedges et al. (1976)
Soils	176	0.18 – 38	4.23	Duchaufour (1976)
Marshes (Bering Sea, Nova Scotia, Mexican Gulf, Porto Rico, Pacific coast of Mexico, Eilat bay, Mediterranean Sea)	16	1.3 – 13.3	5.07	Cauwet (1975), Brooks et al. (1976), Hedges et al. (1976), Rashid et al. (1976), Stuermer et al. (1978)
Mangroves (Gabon: Ikoy, Mombe; Senegal: la Casamance estuary; Australia: eastern Queensland; south of New Guinea)	6	1.1 – 3.35	2.27	Lafond (1967), Books et al. (1976), Duchaufour (1976)
Fiords (Norway, Scotland, British Columbia)	80	1.08 – 23.4	7.83	Strom (1955), Brown (1972), Krom et al. (1977)
Estuaries (Mississippi, St. Lawrence, Betsikoba, Escaut, Gironde, Agly Têt	150	0.7 – 5	1.84	Lindberg et al. (1975), Rapport CENTREAU (1975), Rapp Direct région des eaux intérieures (1967), Lafond (1967), Wollast (1972), Etcheber et al. (1977), Jouanneau (1979), Gadel (1974)
Shelf (summary data, Nova Scotia, Western Atlantic, Mexican and USA coast, Bay of Gascogne, Mediterranean Sea)	70	0.28 – 3.2	1.20	Latouche et al. (1970), Strakhov (1969), Rashid et al. (1970), Cauwet (1975), Farrington et al. (1977), Hedges et al. (1976)
Alluvial fans (Pacific Ocean, Amazon, Niger, summary data)	49	0.43 – 0.85	0.62	Romankevich (1969), Strakhov (1969), Klingebiel (1976), Géochimie organique des sédiments marins profonds (Orgon II) (1978), Wilde et al. (1978)
Canyons (Hudson, Capbreton)	15	0.67 – 5.39	2.87	Tauzin (1974), Farrington et al. (1977)
Upwellings (Walvis Bay)	10	2.5 – 22.2	10.53	Price et al. (1977), Gagosian et al. (1978), Stuermer et al. (1978)
Abyssal plains (summary data, Western Atlantic, Mexican Gulf, Cariaco trench, Argentine basin, Bermuda Rise, Blake Bahama basin, West of Antilles, Amazon river mouth, Demerara plain, Western coast of South America)	132	0.08 – 5.12	0.70	Hobson et al. (1969), Strakhov (1969), Koons (1970), Rashid et al. (1970), Stevenson et al. (1972), Aizenshtat et al. (1973), Farrington et al. (1977), Rapport Orgon II (1978), Stuermer et al. (1978)

OM in Late Quarternary sediments, was gained through investigations carried out by other research institutes, both in the Soviet Union and abroad. Some data on the content of C_{org} in various regions and environments are summarized in Tables 26 and 27.

4.1 Morphology of Non-Living Organic Matter in Recent Sediments

Organic substances of varied origin, together with the mineral base of the sediments, form intricate heterogeneous systems. An insight into the processes governing the latter can be gleaned from the distribution pattern and existing-forms of OM, which, in turn, are important factors controlling the microbiological activity, the trophic value of OM as well as degradation, redistribution and migration of organic and mineral compounds in the course of diagenesis. The study of OM is important fròm the point of view of the oil- and gas-formation, since the thermobaric conditions of OM transformation, oil- and gasogenetic potential of various form of OM may differ. Until recent years, only scant data were available on the forms of occurrence of OM in contemporary sediments (Andrusov 1892, Kriss 1959, Johnson 1974, Potapova et al. 1973, Suess and Müller 1980, Caratini et al. 1981, Zaslavsky 1981).

The evidence obtained from visual, microbiological and optical investigations carried out in various spectral ranges on Black Sea sediments, served to indicate that OM microaggregations are restricted chiefly to claystone layers. The organic matter entrapped in the latter consists of terrestrial plant scraps (elongated and acute-angled in shape), that at times preserve the structure of the initial tissue and are slightly translucent under visible and infra-red rays; numerous autochtonous algae remains with smoothed edges, that are more translucent under infra-red rays; phytoplanktonic remains of different rate of decayal. Colloidal hydrotroilite and small pyrite globules are sited on large-sized OM inclusion. Organic clots are often impregnated by hydrotroilite, which also forms infillings in the ostracoda and foraminifera shells. Observations in the Black and Caspian Seas (at the interchange from the suspension to the precipitate, at a depth of 20 – 30 m) revealed that the planktonic material is subjected to the utmost alterations; macroalgae remains are less altered, and terrestrial plant detritus – only slightly altered (Potapova and Gursky 1972, Potapova et al. 1973). The author of the present study observed similar phenomena in several highly productive areas of the Pacific and Atlantic Oceans (Logvinenko and Romankevich 1973, Logvinenko et al. 1973, Emelianov and Romankevich 1977).

Suess and Müller (1980) subdivided the OM of bottom sediments into living biomass, detrital and sorbed OM and offered a simple formula for the interrelationship of these forms: $C_{org.\,total} = C_{org.\,biom.} + C_{org.\,detr.} + C_{org.\,sorb.}$, which, apparently, is valid on the assumption that there exist no other forms of OM than these. The relationship of the forms was practically established for upper sedimentional layer in 8 areas, and it was shown that sorbed and biomass C_{org} are important OM components but in pelagic areas. In continental margin sediments the prevailing form is detrital OM. The absolute amount of biomass and detrital OM diminishes with depth.

Zaslavsky (1981) divided OM in bottom sediments into detrital OM (separate organic particles, fragment of organism tissues, organic aggregates) and molecular-dispersed OM (sorbed on mineral components and dissolved in interstitial waters). Flotation and dying of particles were employed for isolating and investigating detrital OM (Zaslavsky 1981). The study of the upper layer (0 – 5 cm) of bottom sediments in shelf and slope areas of the Chukcha Sea, Bering Sea, the Sea of Japan, Canari and Peruvian regions has revealed that detrital OM is represented by particles of up to 200 μm in the form of sharp-edged fragments, fibres, flakes and aggregates. Detrital content is the highest in sands and generally makes up 0.5 – 4% of OM in sediments, the rest 96 – 99.5% being accounted for by molecular-dispersed OM. Lipids, proteins and carbohydrates account, respectively, for 30 – 50%, 25 – 45% and 20 – 25% of total detrital OM.

On the basis of the investigations carried out on contemporary sediments, the following morphological (in part, genetical) types of OM can be distinguished: (1) organic detritus and aggregates (predominantly autochtonous OM component), (2) remains of land plants and macrophytes: pollen grains and spores (predominantly allochtonous OM component, and in off-shore areas – mixtures of auto- and allochtonous OM); (3) OM of biogenic calcareous and siliceous skeletal material; (4) OM of ichtyofaunal bone remains; (5) OM associated with pelitic (clay) materials; (6) OM enclosed in igneous rocks; (7) other forms of OM (Emelianov and Romankevich 1977).

4.1.1 Organic Detritus and Aggregates

The morphology of organic remains was best studied by the author in sediments off Western Africa and Peru. These investigations revealed, in the upper layer of bottom sediments, a profusion of organic particles of fibrous, oval and irregular shape, the particles varying in sizes from 0.01 to 10 mm. This material for the most part is made up of marine fossil remains. The organic particulate detritus is usually of grayish-green, gray and brownish-gray colors, but frequently even fresh green.

The fibrous and other relatively large particles envelop shells, their fragments and terrigenous material. They bind diatom valves, foraminiferal shells and terrigenous material into clumps (aggregates). In the sediments of the shelf and continental slope off southwestern Africa and Peru (between 8° – 12° S) of frequent occurrence are chiefly clay aggregates, that are bound by small particulate organic detritus, colloidal organic and phosphatic matter (Fig. 28). Many aggregates of the first and second type are, probably, fecal pellets. They often show a strong reaction for phosphorus. Gray or greenish-gray organic masses occasionally infill or enwrap foraminiferal shells. After bacterial degradation of the major part of OM, the residual mass left over within the shells consists of colloidal hydrotroilite (see Fig. 28).

The organic fibres and some finer particles of the detritus are highly resistant to decomposition. They are practically insoluble in 10%-solutions of hydrochloric acid, and undergo carbonization and partial dissolution when treated with ammonium molybdate solutions and nitric acid. The residue after these reactions

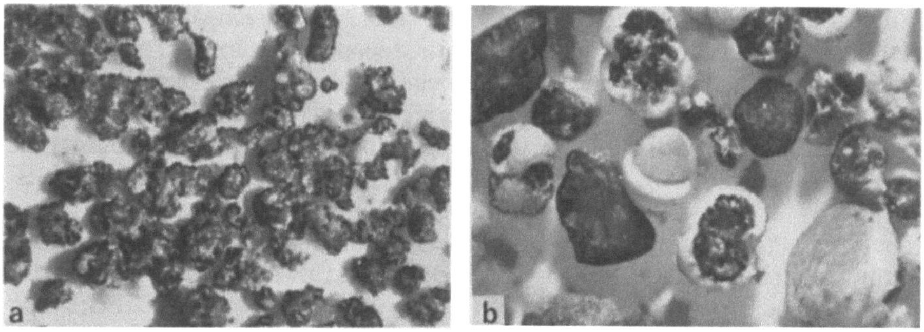

Fig. 28 a, b. Forms of OM occurrence in bottom sediments. **a** clay aggregates, agglutinated by OM (Southwestern Africa, Walvis Bay, 148 m in depth), size 100 – 250 μm; **b** foraminiferal shells, in which the OM is partly replaced by sulphide substance

consists of gray, brown and black fibres and sharp-edged particles. The organic fibres are only slightly soluble in hydrogen peroxide solutions and more completely dissoluble when oxidation (30% H_2O_2) is accompanied by ultraviolet radiation treatment.

The fecal pellets differ in their resistance to decay. The pellets of deposit-feeders are often insoluble in boiling sulphuric acid, strong alkalis, and are preserved even in ancient sediments. Zooplankton feces (calanuses and eufausiids) are of poor endurance and have been encountered only in the top-most layer of sediments (e.g., Morre 1933, Emelianov and Romankevich 1979).

Some samples, collected off southwestern Africa and on the Peruvian shelf, exhibit a strikingly high content of organic detritus and aggregates, making up as much as 43% (Walvis Bay) and even to 80% (Peruvian region, between 8 – 12° S) of the sediments. The C_{org} content of the aggregates varies from 3 – 15 to 30%, which corresponds approximately to 6 – 30, up to 60% of OM. This implies that an essential part of the aggregates is comprised of mineral substance. Investigations carried out near Woods Hole, also established that up to 69% of the mineral grains contain crustated particulate organic matter, and 13 – 71% of these are in the forms of organic-mineral aggregates (Johnson 1974). Organic detritus and aggregates are the principle forms of OM occurrence in the sediments in upwelling areas off southwestern Africa and Peru. The amount of organic particles decreases as the depth and the distance from the zones of high productivity increase, and consequently, the organic particles content is 1.5 – 2 times higher in the sands, that, as a rule, are restricted to the shallow-water zone.

In areas of low biological productivity organic detritus consists of small (0.1 – 0.001 mm), black and blackish-gray particles (the >0.01 mm size fraction was studied). In the Atlantic sediments (excluding the coastal area off western Africa), the amount of organic particles is in the range of 13 thous. to 180 thous. per gram of a dry sample weight (Emelianov and Romankevich 1977). This quantity is several times smaller than that estimated for the coastal sediments off South West Africa and Peru, where, besides, OM is more aggregated. Comparison of the amount of organic detrital and aggregate particles with the C_{org} content of the sediments of the shelf off South West Africa and the pelagial zone of the Atlantic Ocean, revealed, in the first case a direct relationship, notwithstanding the wide scatter in the values; in the second case, no such relationship has been established. The correlation between the values becomes apparent when the C_{org} content exceeds 1 – 3%. From this it is inferred, that the role of shaped OM particles in the gross balance of various types of OM in the sediments, is insignificant when the C_{org} content is less than 1 – 3%. Shaped organic particles are utterly lacking from the deep water foraminiferal ooze and red clays. Conversely, in the Azov, Baltic and Black Seas, also within the shelf areas off southwestern Africa and Peru, organic detritus and its aggregates settle in the bottom sediments, also are generated in situ by the benthos, and are better preserved under conditions of high sedimentation rates and hydrogen sulphide contamination.

4.1.2 Terrestrial Plant Remains and Macrophytes. Pollen Grains and Spores

In bottom sediments of seas and oceans a certain portion of OM is associated with terrestrial plant remains, their spores and pollen grains, and fragments of phytobenthos. Andrusov (1892) and Arkhangelsky (1927) reported the presence of numerous fragments of wood fibres, branches, bark and tree leaves in the Black Sea sediments. In later years with the aid of submarine television sets, land plants were observed in bottom sediments of many oceanic areas: in the Black, Mediterranean, Baltic Seas, offshore of Carolina. The author observed severely and slightly altered wood remains, occasionally scraps of macrophytes, that were recovered by dredging from Kurile-Kamchatka, Japan, Mariana, Phillipine, Bugenwille deep oceanic trenches. The plant material is brought into the ocean deeps from the land and shallow-water areas.

The evidence obtained from studies of the Atlantic sediments serves to indicate that pollen grains and spores are of much wider occurrence than plant remains. They are most abundant in shallow-water platform seas and coastal areas of large basinal water bodies. For instance, in the Baltic Sea the number of pollen grains amounts to 19,600 in one gram of sediments, averaging 7,900 grains per gram in the pelitic muds. As a whole, the pollen distribution pattern in various types of sediments and along the bottom of the Baltic Sea is similar to the C_{org} distribution pattern. This fact, as well as the high pollen content of the muds, strongly suggest that the pollen grains (chiefly pine types) are closely similar in hydraulic particle size to the major mass of OM. Pollen grains and spores comprise only a small fraction of OM in the upper sedimentary layer, even in the Baltic Sea (e.g. Kleimenova 1983). Assuming that the average weight of the pollen grains equals 20×10^{-9} g, the estimated pollen carbon in the Baltic pelitic muds, where the pollen content averages 8000, and the C_{org}-3.15%, amounts approximately to 0.3% of the C_{org} in the sediments.

During the later stages of diagenesis, the role of spore and pollen OM increases considerably, owing to their selective preservation during the degradation of OM. This is due to the fact that the spore vessel is a thick multilayered case, where the inner layer (intine) consists of cellulose, whereas the outer layer (exyne) – of sporepollenine (unsaturated lipoid-like polymer). The latter is of high resistance to destruction under pressure and high temperature. It is this outer layer that has a paramount role in the preservation of the spore and pollen constituents, that are mainly proteins (13 – 28%), carbohydrates (13 – 37%) and lipids (2 – 17%), from rapid biodegradation. Nevertheless, the spores and pollens can be utilized by many land animals, as well as marine invertebrates (Brooks and Shaw 1971, Koziar et al. 1974, Manskaya and Kodina 1975).

Spores and pollen grains are frequently completely absent in pelagic sediments of marginal seas of the basinal type and in deep open ocean areas. In those cases where they have been encountered, they are either of singular occurrence, or do not exceed a few hundred grains per gram of dry sediment, as, for example, in the Mediterranean – 3 grains, in the Pacific – up to 150 grains, in the Black Sea – from 9 to 381 grains (Emelianov and Romankevich 1979).

Data on the morphology of OM and excellent photos of pollen grains and spores, of lignified land plant remains, organic and organo-mineral flakes are contained in works by Combaz et al. 1977, Caratini et al. 1978, 1979, 1981, which consider particular regions of the Atlantic and Indian Oceans.

4.1.3 Organic Matter of Biogenic Calcareous and Siliceous Materials

Organic matter is an inalienable part of mollusc shells, foraminifera, calcareous algae, corals, bryozoan, diatom frustules. It is synthesized by living organisms together with the mineral substance.

Peptides and mucopolysaccharides, that are of natural occurrence in extrapallial fluids, undergo polymerization by enzymes and form insoluble structural proteins (scleroproteins). Morphologically they are in the form of filmy membranes, fibrulla, trabecula, and are incorporated in the carbonate skeleton. Scleroproteins of invertebrates are closely similar in composition to the tissue structural proteins, and identified as of the keratin – mizion – fibrion – elastin group (Barskov 1975).

Degens (1976) discussed the concepts of genesis, the composition and structure of the organic matrix of carbonate phosphate and siliceous skeletal formations, the molecular mechanisms of their development and the role of organic – inorganic interaction. The genesis, nature and functions of the mineralized tissues in calcification, silicification and biophosphate formation are only tentatively known.

The biochemical composition of skeletal OM has been studied mainly on the level of monomer protein and carbohydrate content. All 20 common amino-acids, chitozan, hexoze-amines and fatty acids have been identified in the hydrolysates of skeletal OM. A great portion of the skeletal OM, when subjected to hydrolysis and pyrolysis, is relatively stable; being tightly fixed in or in between the mineral grains, it survives microbial decomposition and can be of long-term preservation. Studies of the elemental composition showed that the C, H and N contents of mollusk shells undergo a progressive decrease from Recent sediments to the Quarternary and Jurassic. The shells in the process of aging lose a considerable amount of these elements. At the same time, the C/N ratio increases with age, which serves to indicate the more rapid loss of nitrogen.

C_{org} content in skeletal formations may vary from 0.01% in some mollusc shells to 20–30% in vertebrate bones and teeth (Degens 1976).

According to the data obtained by the author, calcareous shells, corals, and bryozoan contain 0.11–2.74% C_{org}. The highest concentrations occur characteristically in the coral-algae remains (0.51–2.74%). In the mollusc shell and bryozoan skeletal materials, which are the major constituents of biogenic calcareous sands and coarse silt in the Atlantic, the C_{org} content falls within the range of 0.12 to 1.27%, averaging 0.5–0.7%. From this it is inferred, that sediments comprised exclusively of shell rocks should contain about 0.5–1.0% of C_{org}. This is consistent with the author's observations. For instance, in the shell-rock deposits of the Ampère, Sen and Josephine shoalbanks and certain shelf areas off Africa and Argentine, where other biogenic materials, as well as terrigenous and volcanogenic admixtures are almost entirely absent, the C_{org} content was found to be in the range of 0.48 to 1.18%.

The C_{org} content of the foraminiferal and pteropod skeletal remains is 2–4 times less than that of many mollusc shells. This probably can be cited as one of the reasons why the C_{org} content of foraminiferal-, pteropod-, and coccolith-foraminiferal sediments is lower, as compared, for instance, to that of the shell-rock sands. However, the major reason that the foraminiferal, pteropod, and coccolith-foraminiferal sediments are poor in organic matter, is that they accumulate on the oceanic rises in the pelagial areas, where the planktonic productivity is, as a rule, very low, and the bottom currents prevent the precipitation of the fine suspended matter, with which OM is associated. Hence, the carbon contained in the shell materials is the major form of C_{org} occurrence in the highly calcareous abyssal sediments.

For the determination of C_{org} in siliceous materials special samples of diatom algae frustules were procured, that were treated with 30%-hydrogen peroxide solutions under ultraviolet radiation. The C_{org} content of the diatom siliceous matrix amounted to $0.3 \pm 0.11\%$. This form of OM has a paramount role in the highly siliceous diatom- and diatomaceous-radiolarian ooze in the Antarctic region and in several north-western Pacific areas. In the highly siliceous diatomaceous ooze of the African shelf and slightly siliceous muds of the Peruvian shelf, this form of OM does not play an essential role, owing to the abundant occurrence in the sediments of coprolites, of other aggregates, and of relatively little transformed identifiable fossil remains.

4.1.4 Ichtyofaunal Bone Remains

Singular bone remains of ichtyofaunas have been encountered in practically all types of sediments in the Pacific and Atlantic Oceans (Fig. 29). This form of OM is insignificant in the gross balance of various types of OM, considering that in most cases if makes up only a fraction of a percent of the total C_{org}. In areas of strong upwellings and high productivity, as, for instance, off South West Africa, Peru and Chile, bones, scales, fish teeth (bones and otoliths of marine mammals) comprise 1–6% of the total C_{org} content. The author observed outwash bone deposits in certain areas, for example, on the shelf margin, north of Callao,

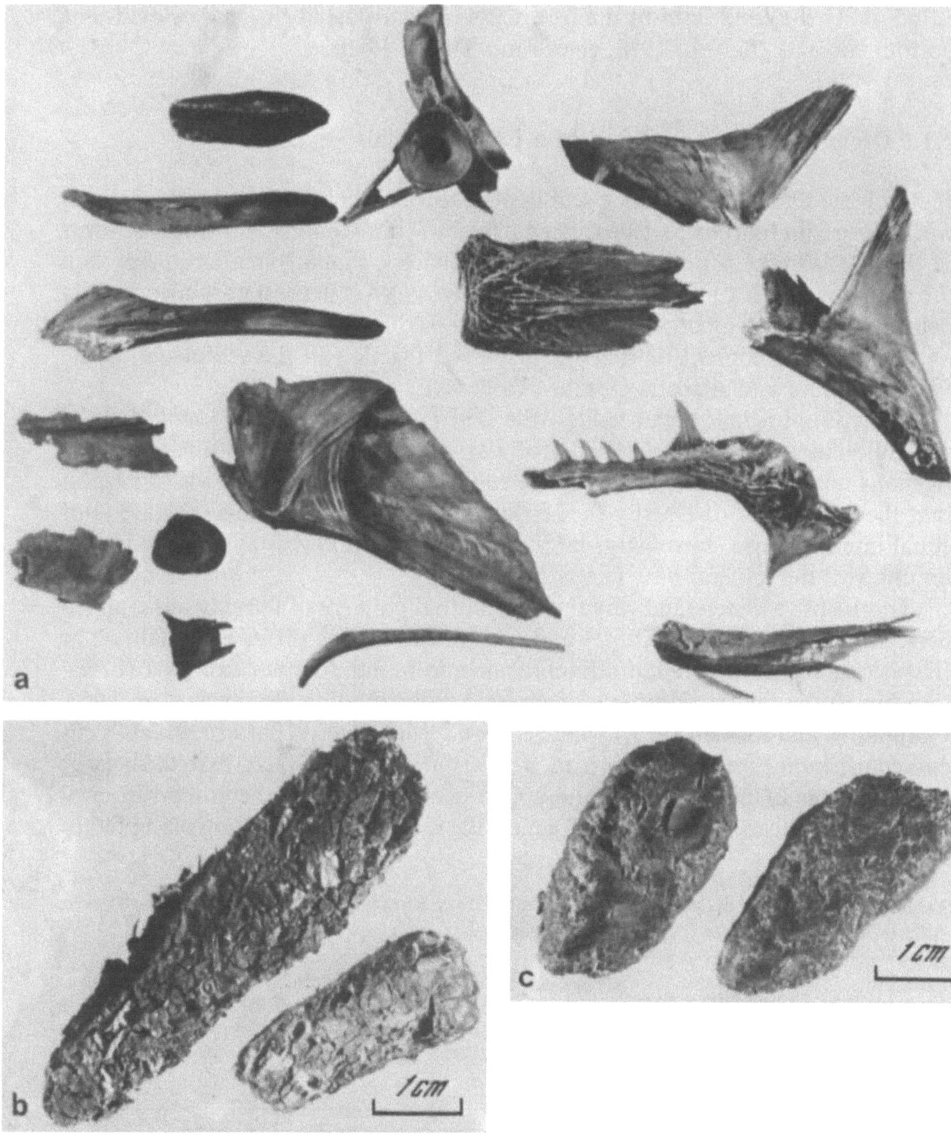

Fig. 29a – c. Ichtyofaunal bone remains and coprolites in sediments of the shelf off southwestern Africa. **a** bone remains; **b** gray, unconsolidated coprolites, containing scales, fish bones and phosphate detritus; **c** brown, firm fossilized coprolites

where such conditions prevail that prevent the precipitation of pelitic and aleuritic materials; the OM is practically entirely associated here with the ichtyofaunal remains.

The formation of bone deposits is linked with increased supply to the bottom of ichtyofaunal remains in the course of mass mortality phenomena (as during stronger southward propagation of El-Genio current off Peru) and their accumulation under the conditions of the fine material transport at elevated velocities of bottom currents (Baturin 1978, Suess and Thiede 1983).

4.1.5 Organic Matter Associated with Clay Materials

Most of the OM of deep water sediments shows no color change when treated with erythrozin, and cannot be distinguished in thin sections and smears. Some of the OM is in the form of fine-dispersed particles, that are smaller in size than the resolving power of the light microscope. This OM is in most part adsorbed on pelitic particles and incorporated within them. Indirect indication to this is found in the high OM content of the fine-dispersed fraction in the sediments of the Pacific, Indian and Atlantic Oceans (Table 28).

According to their mean values, the OM concentration coefficients for the sub-colloidal fraction of sediments of various facies types, for instance, terrigenous muds and eupelagic clays, are roughly the same. This serves to suggest that the adsorption of OM by the terrigenous muds and red abyssal clays is of equal intensity, and, seemingly, in carbonate deposits OM is even more tightly bound with the mineral base (see Sect. 4.3).

This helps to understand why the distribution patterns of the clay minerals (illites, montmorillonite, kaolinite) and C_{org} in the oceans differ essentially. The first is characterized by a latitudinal localization, and does not exhibit any relationship to the facies conditions of oceanic sedimentation (Rateyev et al. 1968, Gorbunova and Lisitzin 1979), whereas the C_{org} distribution pattern reveals a circumcontinental zonation. This can be attributed to the fact, that during the earliest stages of diagenesis (of the upper sedimental layer) the composition of the clay minerals is of negligible effect on the C_{org} content; the controlling factor

Table 28. Organic carbon content of the size fractions in the bottom sediments of the Pacific, Atlantic and Indian Oceans (%, on a dry weight basis)

Fraction, mm	Terrigenous sediments (n = 17)		Carbonate pelagic foraminiferal sediments (n = 14)		Red abyssal clays (n = 12)			
					miopelagic		eupelagic	
	I	II	I	II	I	II	I	II
Sandstone	0.14 – 0.30	0.23	0.18 – 0.28	0.24	–	–	0.30	–
(>0.1)	(0.3 – 0.7)	(0.4)	(0.7 – 1.7)	(1.1)	–	–	(1.7)	–
Aleuritic	0.11 – 0.57	0.31	0.22 – 0.31	0.26	0.23	–	0.19 – 0.22	0.21
(0.1 – 0.01)	(0.3 – 0.7)	(0.5)	(0.8 – 1.6)	(1.1)	(1.1)	–	(0.5 – 1.3)	(0.9)
Large and medium pelitic	0.27 – 2.29	1.02	0.27 – 0.42	0.34	0.38	–	0.20 – 0.45	0.30
(0.01 – 0.001)	(1.1 – 2.3)	(1.7)	(0.8 – 2.8)	(1.6)	(1.7)	–	(1.1 – 1.2)	(1.1)
Subcolloidal	1.22 – 1.90	1.50	0.72 – 1.54	1.13	1.01 – 1.40	1.20	0.41 – 0.63	0.52
(<0.001)	(1.3 – 3.2)	(2.2)	(2.2 – 3.3)	(2.8)	(2.1 – 2.2)	(2.2)	(1.4 – 3.2)	(2.3)

I = extreme values; II = mean values
Figures in parenthesis are concentration coefficients – C^{fr}_{org}/C_{org}, where C_{org} = 1

in the distribution of the latter is the complex of facial conditions of sedimentation including the whole of the pelitic fraction and the summary specific area of particles. It should be noted that montmorillonites have a much greater specific area as compared to illites, kaolinite and carbonate particles ($>150\ m^2\ g^{-1}$) and are distinguished by a high sorption capacity in relation to numerous organic substances. Nevertheless the distribution pattern of montmorillonites in the ocean does not appear to be the controlling factor for the global distribution pattern of C_{org} in bottom sediments. Only in some ocean areas the content of montmorillonites in recent sediments seems to affect the content and composition of OM (e.g. Diester-Haass and Müller 1979). During the late stages of diagenesis, the influence exercised by the clay minerals on the conservation and composition of OM increases. Many authors have, for example, observed a direct relationship between the montmorillonite content and the bitumoid concentration, and also with the hydrocarbon constituent of the latter (Weber and Gorbunova 1961, Degens 1967, Klubova 1971, Galwey 1972).

Experimental studies of the adsorption of planktonic degradation products, OM from soils and ancient rocks, served to establish that the amount of OM that can be adsorbed by clay minerals is 2–6 times greater than their OM content. Under mild desorption conditions the major part of this OM, and under rigid conditions ~25–75% of the OM is firmly retained on the clay minerals, and does not pass into solution (Khan 1951, Bader et al. 1960).

The condensed arrangement of the layers and basal sheet in the tetrahedral lattice favors ionic substitution in various positions, and results in more firm bonding of the organic ions with the mineral part of sediments, as compared to that in the octahedral lattice, and causes the better preservation of organic compounds than of individual free molecules. This factor favors the preservation of OM from microbial break-down during the early stages of diagenesis (see e.g. Klubova 1971).

The probe data from two columns of the Arabian Sea, comprised to terrigenous muds with a predominance of illites, revealed an increase in the C_{org} content of the subcolloidal fraction, probably resulting from the adsorption of OM, during its breakdown, from the interstitial waters. The OM concentration coefficients undergo, at first, a sharp decrease (from 2.8 to 1.8, at a depth about 2 m from the ocean floor), following which they remain practically unchanged (1.7–1.9, to a depth of 7.4 m from the bottom surface), which evidences the stabilization of the OM decomposition and sorption processes. At the same time, the sulphate reduction process, formation of hydrogen sulphide and sulphides, undergoes gradual extinction.

4.1.6 Organic Matter in Igneous Rocks

Investigations of igneous rocks, carried out by the author in the Indian Ocean, revealed C_{org} values ranging from 0.01 to 0.23%, the average being 0.09% (22 specimens). The highest C_{org} values were determined for the ellipsoidal and pillow lavas, taken from the Central Indian Ridge (0.18–0.23%, with a mean value of 0.20%). The basalts from the Arabian-Indian and Western Australian Ridge contained 0.03 – 0.04% C_{org}, the ultrabasic rocks – traces to 0.12% (averaging 0.03%, in 11 specimens). The C_{org} content of the igneous rocks in the Atlantic Ocean was found to be in the range of 0.03 – 0.42%. Hence, it becomes apparent that the oceanic igneous rocks are characteristically of very low C_{org} content. This carbon is brought into the sediments with the physical and chemical degradation products. It plays only a meager role in the gross balance of OM accumulation in the sediments. However, in some areas (as in the rift valleys of the mid-Atlantic ridge, in the Ross sea) its content is up to 90% of the total C_{org} in the bottom sediments.

4.1.7 Other Forms of Organic Matter

Organic matter in the form of clots and gelatinous substances the author encountered occasionally in shallow water sediments. The origin of these substances still remains a puzzle. One suggestion is that these clots are remnants of certain types of "jellous" organisms. Probably, some of them were formed as a result of OM coagulation, either from sea (?), or interstitial waters.

The C_{org} content of the interstitial waters in the anoxic pelitic sediments of the submarine continental margins, is $\sim 10 - 20\ mg \cdot l^{-1}$, the moisture content – 50 – 75%. Calculations of the OM in the interstitial waters show that it amounts to 1 – 8% of the total OM in the sediments, whereby it is inferred that this form of OM does not play an appreciable role. During the process of OM decay the relative amount of the dissolved organic matter in the sediments increases.

During the two last decades there has arisen a general awareness of the increasing pollution of the world ocean and particularly its marginal seas, resulting from the deleterious effects of industrial waste, including solid organic substances. This pertains chiefly to crude oils, various petroleum products, effluents of sewage sludge, domestic terrestrial outfalls and ship waste. Black and blackish-brown bituminous substances, in forms of lumps, 10 mm in diameter, were encountered in the Sargasso Sea. In the area of the Azores, 50 lumps of this kind were collected with a pleuston trawline in the course of a few hours, which estimates 10 tons per km^2. Such bituminous substances can be found in many oceanic areas. Numerous publications have drawn attention to this fact. These substances undergo gradual compaction and are ultimately incorporated into the bottom sediments. They are as yet only of minor contribution to the gross balance of OM in sediments.

4.2 Principles Governing the Distribution of Organic Carbon in Bottom Sediments

4.2.1 Regularities of Regional Distribution in the Upper Sedimentary Layer

A new map showing the distribution of C_{org} in the world ocean sediments was developed (Fig. 30). It offers an insight into the principles governing the areal dis-

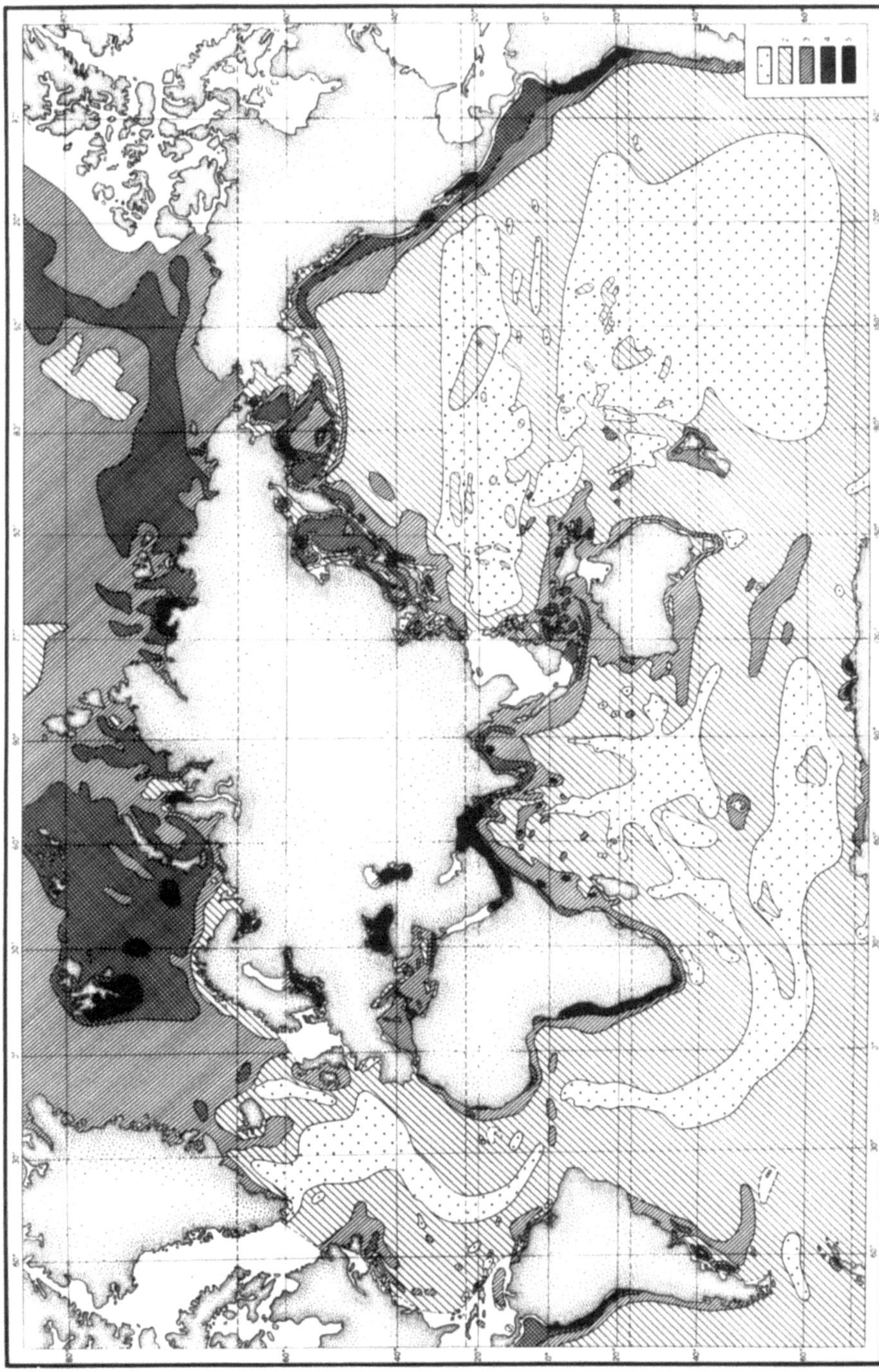

Fig. 30. Distribution pattern of C_{org} in the upper sedimentary layer (0 – 5 cm) of the world ocean, in %, on a dry weight basis: *1* = <0.25; *2* = 0.25 – 0.50; *3* = 0.51 – 1.00; *4* = 1.01 – 2.00; *5* = >2.00

tribution of OM in bottom sediments and helps elucidate its relation to the facial conditions of sedimentation and diagenesis. The basis for the map were direct measurements of C_{org} values (in terms of $CO_{2\ org}$), made in about 5000 bottom sediment samples, most of which were collected and analysed in the P. P. Shirshov Institute of Oceanology, of the Academy of Sciences of the USSR. The principle source materials were earlier maps of the distribution of C_{org} in the bottom sediments of the Pacific (Romankevich 1970), Atlantic (Emelianov 1975) and Indian (Bezrukov and Lisitzin 1975) Oceans, in which certain ammendments and additions were made by the author. The data on the C_{org} content of the offshore sediments in some oceanic regions, and the marginal and inland seas, are plotted on Fig. 31.

The regularities in the C_{org} distribution in the world ocean sediments are reflected in the circumcontinental and latitudinal zonation. According to the distribution of particulate OM, best expressed is the circumcontinental zonation, which is the most characteristic feature of the OM distribution pattern in oceanic bottom sediments. This zonality is revealed in the general trend of the high OM concentrations around the peripheral zone, including the marginal seas, oceanic shelf, continental slope and marginal basin areas. In the peripheral zone two OM maxima can be recognized – a shallow water and relatively deep water – which are divided by a zone of low C_{org} values.

The shallow water maximum is restricted to the relatively fine-grained sediments, occurring in gulfs, lagoons, bays, inland shallow seas, and within the upper part of the open ocean shelf. This is exemplified in the Pacific Ocean, in the muds of the northern shallow part of the Okhotsk Sea, the Shelekhov and Anadira Bays, where the C_{org} content is over 2%; in the muds of many inlets of Hokkaido and Honshu (up to 3.6 – 11.9% C_{org}); Korea and Primorye (up to 2% C_{org}); the Gulf of California (up to 7.4% C_{org}); the internal shelf of Ecuador and Peru (up to 10.6% C_{org}); in the Atlantic Ocean – in the muds in the Gulf of Finland, Riga Bay and other areas of the Baltic Sea (up to 5.72% C_{org}), the eastern continental margin of the USA (Maine and Puget Sounds – up to 4.7% C_{org}), Central Texas Bay (up to 3.5% C_{org}), Venezuela (up to 5.0% C_{org}), Mochima Bay (up to 6.45% C_{org}), lagoons of Brazil (up to 6.4% C_{org}), the bays of western (up to 12.8% C_{org}) and southwestern Africa (up to 16 – 26% C_{org}), and many other coastal areas of the world ocean (see Tables 26 and 27).

The oxidized sedimentary layer is of least thickness here, owing to the decayal of large masses of OM, and is completely absent from areas of stagnant waters. The negative Eh values, which are characteristic of the sediments that are deposited in such environments, favor the conservation of the OM proper, due to the fact that in the absence of free oxygen biochemical oxidation reactions are shifted towards the formation of intermediate products resulting from partial degradation of OM. Unfavorable living conditions for macrobenthos susceptible for periodic mass mortality, lead to OM undergoing a lesser degree of destruction. This is best expressed in the sapropel muds of the inner shelf of Peru and southwestern Africa (Emery 1960, Romankevich 1971, Romankevich and Baturin 1974).

The OM in this zone, from a genetical and morphological viewpoint, is most heterogeneous in composition, and reveals a spotty and mozaic distribution pattern, which, to a large extent, results from the irregular distribution of POM and the benthic biomass. As was shown in the previous section, an essential part of the OM occurs here in a relatively unaltered form. It consists of planktonic decomposition products, and besides, is largely comprised of benthos algae (uni- and multicellular), occasionally of flower plant types, of excretions of various OM consumers and their organic remains. In water bodies with a high river runoff, the role of allochtonous OM increases conspicuously.

Near the headwaters of large rivers, such as La-Plata, Congo, Cunene, Nile, Po, the muds that accumulate in the shallow waters usually reveal high C_{org} contents. This can be attributed to its high content in the aleuritic-pelitic portion of the river suspended matter the high planktonic productivity and rapid burial of OM particles and sorbated OM. In such areas the terrigenous influx does not cause a decrease in C_{org} content, whereas in zones of sand deposition it usually exerts a strong influence on the C_{org} content, although certain exceptions can be cited (Zakharov 1974). The decrease in the sands of the percentage concentrations of C_{org}, frequently, is not conducive from the decrease in the absolute amount of C_{org}, which in some places, owing to the large input by rivers and high sedimentation rates in the zones of river influence, accumulates in very large masses. Notwithstanding the fact, that in most areas allochtonous matter, as a whole, and planktonic in particular, predominate, in certain discrete areas autochtonous OM plays a rather significant role. For instance, the low C_{org} content of the muds in the Amazon river mouth, results not only from the low planktonic productivity, but also from the low activity of the sapropel-humus and predominantly allochtonous (terrigenous) OM. This type of OM is, seemingly, incapable of producing a reducing environment in the muds, which are brown-colored here. The predominantly terrigenous composition of OM at the brown-coal (lignite B) diagenetic stage in the sediments of the Amazon fan does not create the reducing environment, even at high C_{org} content (up to 2%), due to the less active sulphate reduction process.

The OM in the zone of the shallow water maximum has not as yet been sufficiently studied. At the same time, these shallow water facies deserve special attention, being of high interest in regard to the origin of petroleum, formation of authigenous phosphates and the balneological properties of the sediments, which in some areas are enriched in organo-mineral compounds of high radioactivity. Sediments with high OM concentrations are also of interest as a source of natural fertilizers.

The deep water OM maximum is restricted to the muds on the continental slopes and adjacent marginal parts of oceanic basins. In some inland water basins, as, for instance, the Black Sea, this maximum occupies the central part of the basin. In the Pacific and Indian Oceans it extends through the sediments of the continental slopes and deep oceanic trenches, located near the continents.

In the open ocean areas, marginal seas and many inland water bodies, the shallow- and deep-water maxima are commonly divided by a zone of low C_{org} values. This can feasibly be explained by the rock-fall of sedimentary materials, including OM, from the external shelf, and its subsequent deposition within the lower part of the slope. A certain portion of this OM reaches to great depths and settles in the central halistatic areas of inland seas, or at the bottom of deep

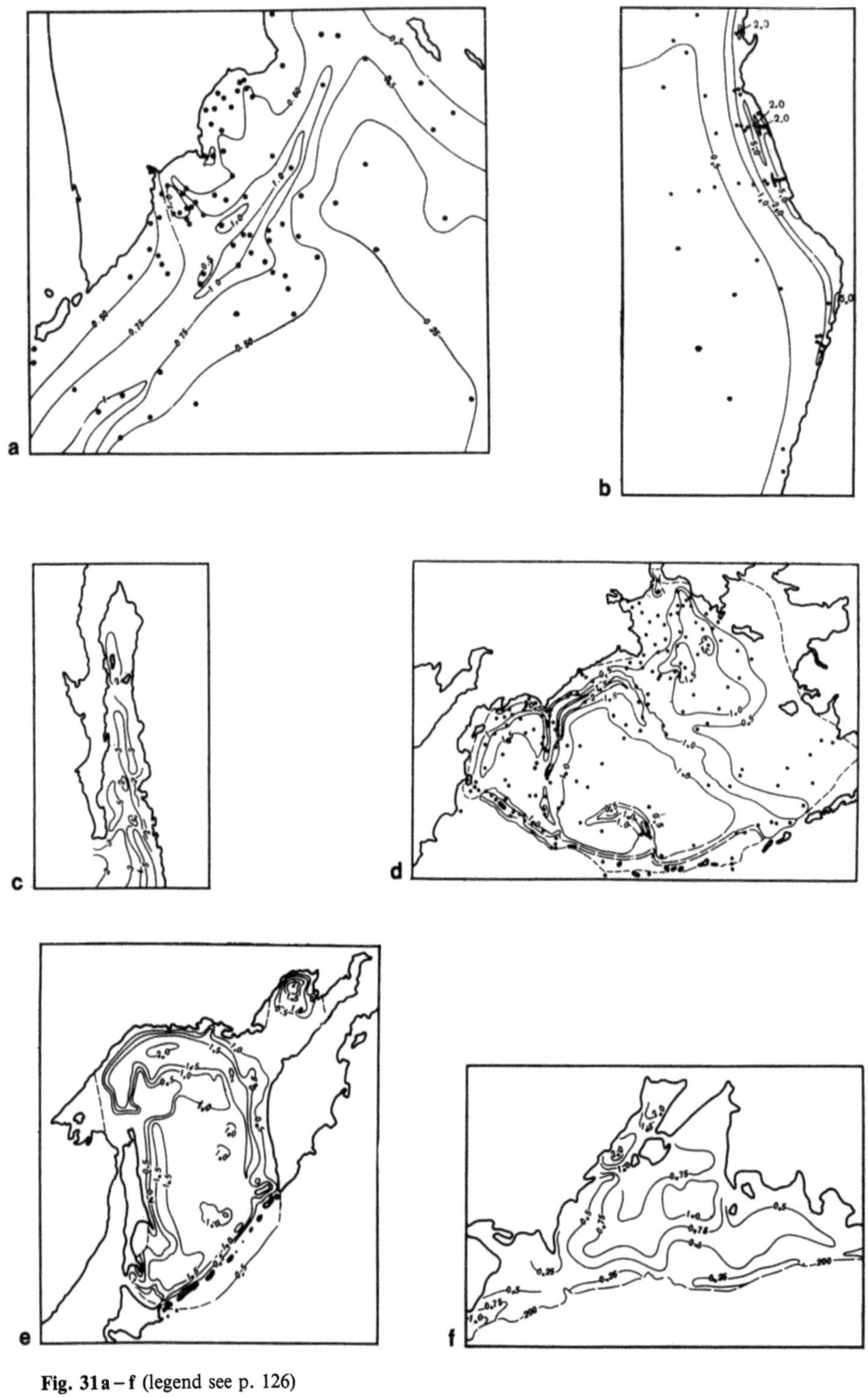

Fig. 31a–f (legend see p. 126)

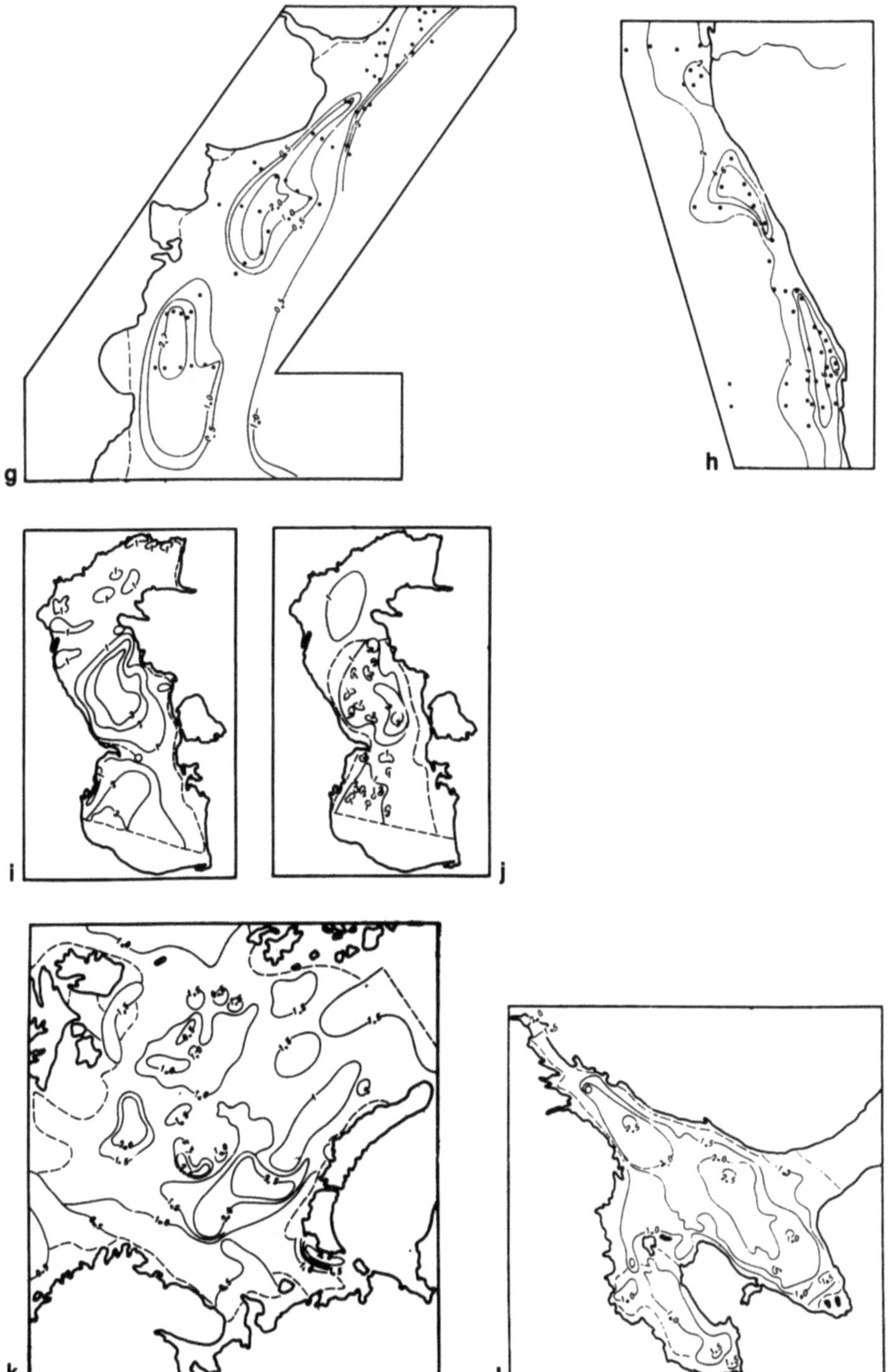

Fig. 31g–l (legend see p. 126)

m

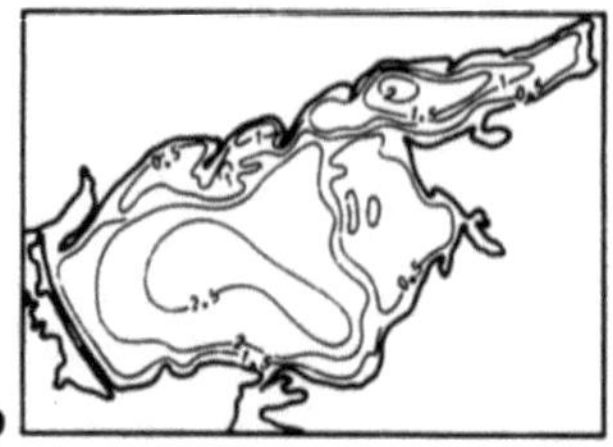

o

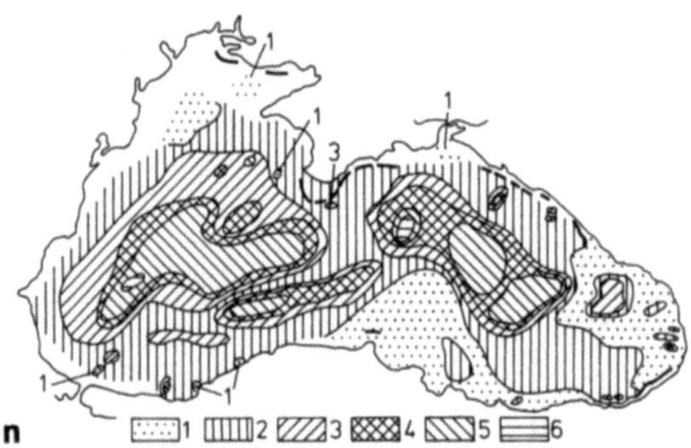

n

Fig. 31a – p. C_{org} distribution pattern of the upper sedimentary layer in the near-coast areas of oceans, marginal and inland seas. **a** Kuril-Kamchatka area of the Pacific Ocean (Romankevich 1970); **b** Southwestern Pacific (Logvinenko and Romankevich 1973); **c** Gulf of California (Van-Andel 1964); **d** Bering Sea (Lisitzin 1966); **e** Okhotsk Sea (Bezrukov 1960_1); **f** Southern Primorye (Koniukhov 1975); **g** Southwestern Atlantic (Zakharov 1974); **h** Southeastern Atlantic (Emelianov and Romankevich 1977); **i** Caspian Sea (Bordovsky 1974); **j** Caspian Sea (Bordovsky 1974); **k** Barents Sea (Gorshkova 1975); **l** White Sea (Gorshkova 1975); **m** Baltic Sea (Blazhchishin 1976); **n** Black Sea (Shimkus and Trimonis 1974); **o** Azov Sea (Gorshkova 1970); **p** Mediterranean Sea (Emelianov and Romankevich 1979). **a – i, k – m,** in % of dry sediment; **j** C_{org} in g cm^{-2} per 1000 years

oceanic trenches. A good example is the deep Peru – Chile Trench, where the bottom sediments contain up to 2 – 4.5% of C_{org} (Table 29). The OM which is retained in the sediments of the external shelf, under the influence of constant water currents, rich in oxygen, undergoes rapid dissolution (Vallentyne 1962) and decomposition; a small portion experiences burial and fossilization.

Thus, the major factors responsible for the separation of the upper and lower OM maxima by a zone of relatively low OM concentrations in the lower part of the shelf, are as follows: (a) the rapid rate of bottom water flow, that results in the redeposition of sedimentary materials, increase of its fraction size, and re-

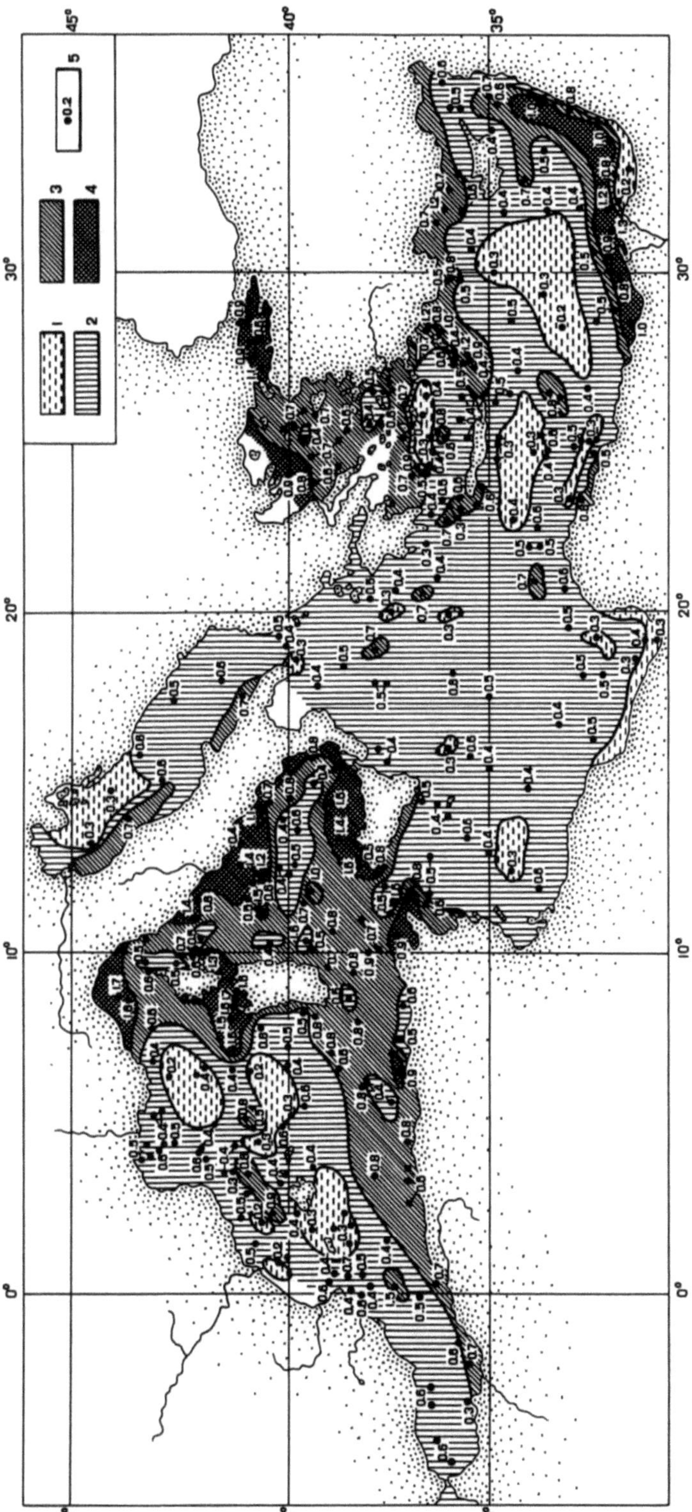

Fig. 31 p (legend see p. 126)

Table 29. Biogeochemical composition of the sediments in the Pacific Ocean Deep Trenches (in %, on a dry sediment basis)

Deep trench, sampling depth	C_{org}	$CaCO_3$	$SiO_{2(amorph)}$	Fraction <0.01 mm
Peru – Chile (6100 – 7320 m)	*0.26 – 4.47* 2.26(4)	*0.07 – 8.23* 2.23(4)	*3.53 – 9.27* 7.27(3)	*62.52 – 83.82* 75.50(3)
Japan (7360 m)	*1.34* – (1)	*0.16* – (1)	*18.30* – (1)	*67.43* – (1)
Kuril – Kamchatka (6008 – 9530 m)	*0.36 – 1.30* 0.79(20)	*0.0 – 1.32* 0.23(20)	*3.90 – 24.12* 13.10(18)	*21.40 – 73.09* 60.11(17)
Aleutian (6323 – 7266 m)	*0.57 – 0.76* 0.65(4)	*0.0 – 0.18* 0.08(4)	*4.15 – 4.73* 4.39(3)	*65.05 – 80.8* 70.52(4)
Bougenville (76 – 9022 m)	*0.61 – 0.63* 0.62(2)	*1.04 – 3.78* 2.41(2)	*1.53 – 1.68* 1.60(2)	*55.05 – 57.13* 56.09(2)
"Vityaz" Trench (6135 m)	*0.44* – (1)	*0.91* – (1)	*1.25* – (1)	*76.46* – (1)
Jap (7988 m)	*0.42* – (1)	*0.14* – (1)	*41.21* – (1)	*79.13* – (1)
Marianas (7990 – 10710 m)	*0.14 – 0.95* 0.42(3)	*0.09 – 0.59* 0.32(3)	*1.44 – 7.10* 4.27(2)	*57.34 – 80.79* 69.06(2)
Idzu – Boninian (7315 m)	*0.35* – (1)	*0.18* – (1)	*2.52* – (1)	*65.86* – (1)
Kermadec (6000 – 9223 m)	*0.21 – 0.39* 0.30(3)	*0.05 – 0.93* 0.36(3)	*1.15 – 3.87* 2.51(2)	*31.48 – 68.70* 50.09(2)
Ryukyu (6810 m)	*0.22* – (1)	*0.09* – (1)	*0.96* – (1)	*55.66* – (1)
Volcano (7132 m)	*0.20* – (1)	*0.52* – (1)	–	*68.50* – (1)
New Hebrides	*0.20* – (1)	*0.36* – (1)	*1.05* – (1)	*53.37* – (1)
Tonga (6340 – 10680 m)	*0.07 – 0.30* 0.19(6)	*0.16 – 0.70* 0.38(6)	*1.60 – 3.12* 2.40(6)	*23.03 – 65.41* 47.29(4)

Italics = extreme values; Roman = mean values; () = number of samples

moval of OM; (b) low rate of accumulation and even bottom erosion; the presence of ancient sedimentary relics, which in certain places occupy up to 70% of the ocean floor (Emery 1970, Blazhchishin 1972); (c) good aeration of the sediments and enhanced dissolution and mineralization of the OM.

With the increase of the distance towards the central part of the ocean, the C_{org} content of the sediments decreases to minimum values (<0.25%). Areas of such low C_{org} values are widespread in the central parts of the Pacific. In the Atlantic and Indian Oceans the areal coverage of sediments with such low C_{org} values is much less. Sediments containing ~0.3 – 0.4% of C_{org} are confined chiefly to areas of anticyclonic circulation, small rate of supply of biogenic elements to the euphotic zone, and low rate of reproduction of POC. In the Pacific and Indian Oceans, ferro-manganese concretions are formed in areas of low C_{org} content.

In the oceanic basins OM is seemingly of uniform distribution and equally low content. However, the actual distribution pattern is more intricate. Many features have not as yet been detected. Some result from the differences in the age of the sedimentary cover. Pelagic clays of the Peruvian and Carolina basins, also of the Boreal and Equatorial Provinces of the Northwestern basin, reveal characteristically high C_{org} values (miopelagic red clays and diatomaceous-radiolarian ooze are of chief occurrence here). Minimal values are typical of the Southern, Central and Phillipine basins, where eupelagic red clays are of wide occurrence. These differences in the quantity of OM are not large, but they serve as an important controlling ecological factor and reliable biogeochemical indicator of the environmental conditions of the deposition of eupelagic muds and formation of ferro-manganese nodules.

The regional variations in the C_{org} content of the basinal bottom sediments can be attributed to many factors, but, considering that they show a good correlation with the productivity of the euphotic zone, it may be inferred that the law of the inherited distribution of OM in the sediments from the distribution of POC is manifested here, although in a vague form.

The relation between the C_{org} content of the sediments and the POC persists to the maximum ocean deeps and is reflected in the different content of OM in the bottom sediments of the deep water trenches of the Pacific Ocean (see Table 29). The maximal C_{org} values are characteristic of the sediments in the Peru – Chile Trench, whereas the minimal values – of the trenches further distant from the continents and areas of input of POC (Ryukyu, Marianas, New Hebrides, Tonga, Kermadec). The high OM concentration in the sediments of certain trenches is reflected in the abundance of their benthos fauna (Beliayev 1974).

This implies that the facial conditions of sedimentation are the controlling factors in the areal distribution of C_{org} in the bottom sediments. In the submarine continental margins the major factors that exert an influence on the OM content of the sediments are as follows:

1. The biological productivity and supply of autochtonous OM to the bottom; in some areas, the biological productivity of the land and allochtonous POC supply are of similar influence (a direct correlation).
2. The granulometric composition of the sediments and OM burial conditions, which depend upon the hydrodynamics, bottom relief, aggregation of the sedimentary material by benthos, the rate of input by rivers and particle size of the suspended load, and abrasion products (the relationship is intricate).
3. The living conditions of the benthic population, which depend upon the amount of requisite (labile) OM, the gas regime of the habitat and physiological functions of the heterotrophic organisms.
4. The relation between the rate of OM supply and terrigenous influx, which may cause the decrease, as well as increase of OM in the sediments.

Maps showing the distribution of the C_{org} content in the sediments, as calculated on a non-carbonate and non-siliceous basis, are presented in Figs. 32 and 33. Superficially, it may appear that such maps are meaningless, considering that the carbonate and siliceous materials, same as the terrigenous materials, are the

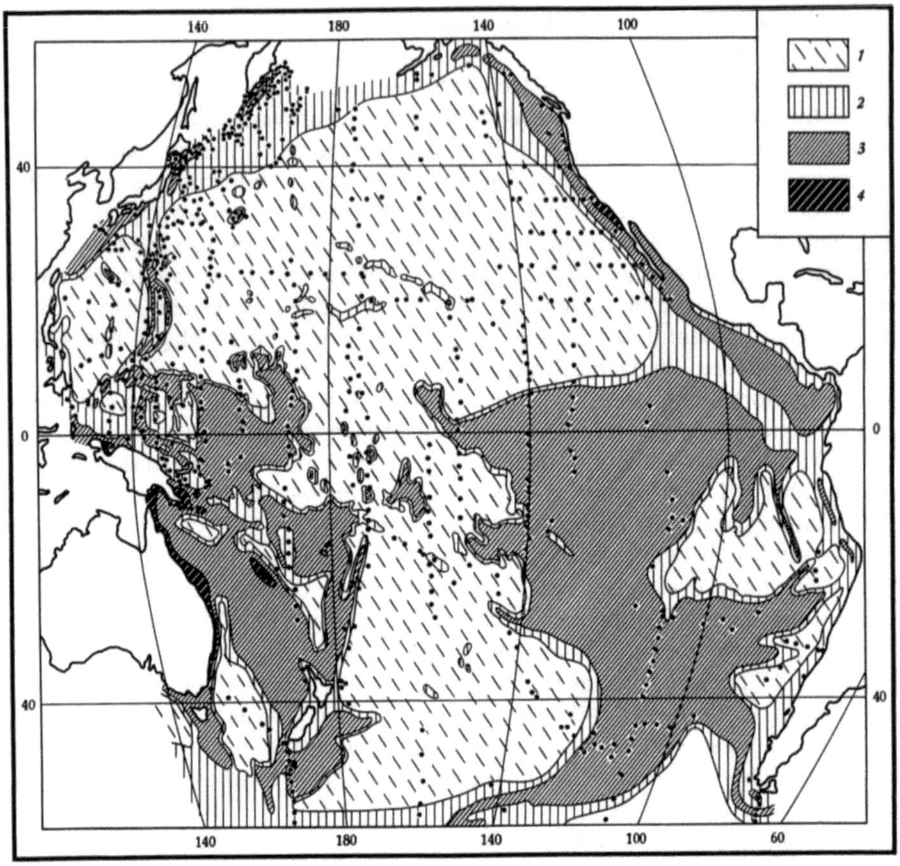

Fig. 32. C_{org} distribution pattern for the upper sedimentary layer in the Pacific Ocean, re-calculated on a non-carbonate and non-siliceous basis, in % of dry sediment. *1* = <0.5; *2* = 0.5 – 1.0; *3* = 1.0 – 3.0; *4* = >3.0

genetical carriers of OM. Nevertheless they contribute essentially to the understanding of the distribution and fate of OM in the ocean.

The areal distribution in the bottom sediments of the C_{org} values, as calculated on a non-carbonate and non-siliceous basis, differs conspicuously from the distribution pattern of C_{org} in natural sediments. However, the trend in the decrease of the C_{org} values with the distance from the coast persists. The main differences are obvious in the central parts of the ocean, where even all the elevated parts of the floor that are covered by carbonate sediments are reflected as areas of high C_{org} content (1 – 3%). Consequently, the differences between the red clays and pelagic carbonate sediments are more pronounced. An important implication from this is that in the areas of the red abyssal clays the foraminiferal shell materials undergo nearly complete dissolution, mainly at the bottom surface, and their OM is contributed to the deep waters. In the area of the carbonate and siliceous sediments, the OM, being of tight bonding with the mineral base, escapes dissolution, utilization, and undergoes fossilization.

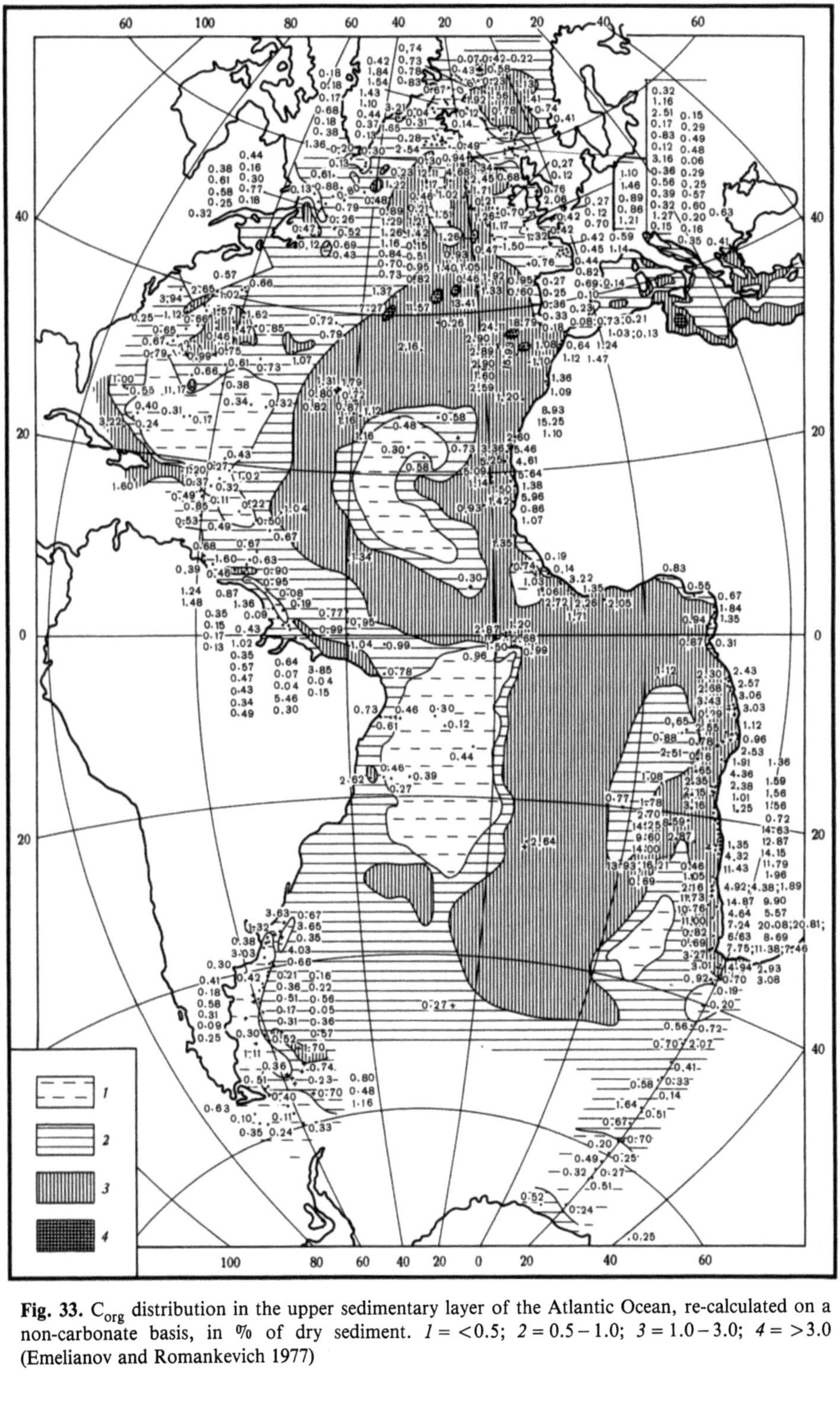

Fig. 33. C_{org} distribution in the upper sedimentary layer of the Atlantic Ocean, re-calculated on a non-carbonate basis, in % of dry sediment. *1* = <0.5; *2* = 0.5 – 1.0; *3* = 1.0 – 3.0; *4* = >3.0 (Emelianov and Romankevich 1977)

The C_{org} distribution pattern of the sediments exhibits, besides a circumcontinental zonality, a latitudinal zonality, which is typical of the planktonic, benthal and particulate OM (e.g., Filatova 1977, Vinogradov and Lisitzin 1981). This zonality is best expressed in the Pacific sediments, where it is reflected in the occurrence of two high-latitudinal zones (Boreal and Antarctic) and one Equatorial zone of enhanced C_{org} values, that are divided by extensive tropical zones of low C_{org} values. The latitudinal zonality is less distinct in the Atlantic and Indian Oceans. In some places the C_{org} content increases here towards the southern high-latitudinal areas of enhanced bioproductivity. However, the area of the Antarctic divergence is only slightly reflected in the C_{org} distribution pattern of the sediments. This can be attributed to the relatively low total annual biological productivity in the pelagic areas of the southern ocean.

In summarizing the foregoing, once again emphasis is laid on the fact that the circumcontinental zonality is an inherent feature of the distribution pattern of OM, which is most markedly depicted in the distribution of POM in the ocean water column and the OM in the upper layer of the bottom sediments. The inference is therefore drawn, that it is inherited by the sediments from the suspended matter.

During the cainozoic-mesozoic period the circumcontinental zonality also was the most prominent feature in C_{org} distribution in the sedimental mantle of the ocean (Bogdanov and Romankevich 1976, Bogdanov and Levitan 1980). The black clay formation most widely spread in the Atlantic Ocean (thickness 45 – 270 m, sedimentation rate 15 – 70 mm per 1000 yrs, C_{org} up to 16.8%) and dating back mainly to Barrenian and Cenomanian time usually concentrates in the near-continental ocean areas (Bogdanov and Levitan 1980). In the Paleogene, high C_{org} concentrations in sediments of near-continental regions occurred in the form of discrete patches. In the Neogene, along the western coasts of the continents there had developed the upwellings in the trade-wind areas, resembling the ones that exist at present, and a mass of sediments of specifically high C_{org} content had accumulated. Similar conditions took place in the Neogene along the eastern coast of the continents in the moderate latitudinal zones where persistent eastern winds prevailed.

In the pelagic oceanic area the distribution of C_{org} in the mesozoic-cainozoic period followed the latitudinal zonality pattern similar to that which takes place at present. In the pelagic arid zones the content of C_{org} was minimal throughout the mass uncovered by deep drilling. The C_{org} content of sediments was higher in the pelagic humid areas of enhanced bioproductivity. The equatorial zone has a typically elevated C_{org} content in sediments beginning at least from lower Paleogene. In contrast to the present time, the near-Antarctic zone of high C_{org} concentrations in bottom sediments was non-existent during the Late Cretaceous and Paleogene, whereas in the northern high-latitude areas high C_{org} concentrations evolved as discrete patches.

4.2.2 Organic Carbon in Various Genetic Types of Sediments in the Pacific and Atlantic Oceans

In the upper sedimentary layer (0 – 5 – 10 cm) in seas and oceans the C_{org} content, according to the data obtained from our studies, is in the range 0.01 – 14.05%, and according to the data available from other publications, reaches to 26% (see Table 26). The Pacific and Atlantic Ocean sediments reveal the same wide range of C_{org} values: 0.03 – 10.59% (Pacific Ocean), 0.04 – 14.05% (Atlantic Ocean) (Tables 30 and 31).

Both low, as well as high extreme C_{org} values have been reported for sediments of practically all size fractions. Through investigations carried out earlier in the Pacific Ocean, the fact was established that the lower C_{org} level (0.03 – 0.08%) is roughly the same in all granulometrical types of sediments, whereas the upper level shows a trend towards the shallow-water pelitic muds, and the highest $C_{org}^{max}/C_{org}^{min}$ ratios (up to 220) occur characteristically in the shallow water pelitic sediments (Romankevich 1974). However, sand and aleurite facies enriched in OM, have been encountered in many oceanic areas, as for instance, in the shelf areas off Kamchatka, southwestern Africa and South America (Romankevich 1974a, Zakharov 1974, Diester-Haass and Müller 1979). Consequently, the $C_{org}^{max}/C_{org}^{min}$ ratios in the sands are almost as high as in the pelitic muds (up to 70 in the Pacific, and up to 120 – in the Atlantic). All this suggests that the environmental conditions of sedimentation are of primary influence on the C_{org} content in the sediments.

Large fluctuations in the C_{org} content have been observed not only in all size fractions and most types of sediments (terrigenous, siliceous, various carbonates), but also in all litho-facies types within the submarine margins of continents and marginal areas of the ocean floor. The range of the C_{org} values, as expressed in terms of percentage of the weight of sediments, and in the $C_{org}^{max}/C_{org}^{min}$ ratios, is usually narrower in the central parts of the deep ocean floor. This implies that the supply and burial conditions of OM are more stabile in these areas. The large fluctuations of the C_{org} content in many types of litho-facies is an alternative explanation for the high intensity of the diagenetic processes in local areas, however, the wide areal development of these processes is determined by the average level of the OM content (Romankevich 1974b).

Table 30. Organic carbon content of the upper sedimentary layer (0 – 5 cm) in the structural-geomorphological zones of the Pacific Ocean (%, on a dry sediment basis)

Litho-facies types	Sands	Aleurites	Pelites
Sediments in shelf areas (33 – 260 m in depth)			
Terrigenous[b]	*0.08 – 5.27* 0.66(18)	*0.26 – 3.35* 1.24(9)	*3.65 – 10.59*[a] 7.34(4)
Sediments on continental slopes (330 – 9334 m in depth)			
Terrigenous, within upper slopes[b]	*0.15 – 1.45* 0.47(20)	*0.13 – 1.60* 0.71(38)	*0.41 – 3.48* 1.48(25)
Glauconites, within upper parts of slopes	*0.25 – 2.66* 1.59(7)	*0.61 – 3.92* 2.31(5)	– –

Table 30 (continued)

Litho-facies types	Sands	Aleurites	Pelites
Carbonates, within upper slopes[c]	*0.32 – 3.51* 1.19(5)	*0.67 – 1.42* 1.05(2)	*0.41 – 6.80* 2.66(3)
Siliceous[d]	– –	*1.33* – (1)	*0.81 – 2.03* 1.28(5)
Terrigenous, of the lower slopes[b,e]	*0.62* – (1)	*0.21 – 1.46* 0.63(10)	*0.22 – 2.75* 0.82(70)
Sediments in the marginal parts of the ocean floor (2880 – 7315 m in depth)			
Terrigenous hemipelagic[b]	– –	*0.15 – 0.96* 0.31(19)	*0.13 – 2.13* 0.47(136)
Sediments in the central parts of the ocean floor (26 – 10680 m in depth)			
Siliceous[d]			
Antarctic regions	– –	– –	*0.14 – 0.23* 0.20(5)
Tropical Ethmodiscus	– –	– –	*0.19 – 0.42* 0.30(5)
Carbonate foraminiferal and coccolith-foraminiferal			
30 – 50% $CaCO_3$	*0.54* – (1)	*0.17 – 0.44* 0.29(8)	*0.17 – 0.96* 0.34(29)
30 – 70% $CaCO_3$	*0.18 – 0.34* 0.25(4)	*0.21 – 0.48* 0.34(4)	*0.14 – 0.89* 0.37(28)
70% $CaCO_3$	*0.08 – 0.36* 0.23(15)	*0.09 – 0.56* 0.28(35)	*0.06 – 0.68* 0.29(64)
Carbonate[f]	*0.12 – 0.59* 0.29(12)	– –	– –
Diatomaceous-radiolarian and radiolarian-foraminiferal	– –	– –	*0.11 – 0.48* 0.28(48)
Red abyssal clays			
miopelagic	– –	– –	*0.09 – 0.56* 0.29(113)
eupelagic	– –	– –	*0.04 – 0.48* 0.22(158)
Volcanic clastics[b]	*0.20 – 0.22* 0.21(2)	*0.03 – 0.55* 0.19(28)	*0.11 – 0.39* 0.24(15)
Metalliferous	– –	– –	*0.23 – 0.31* 0.29(7)
Ocean total	*0.08 – 5.27* 0.56(85)	*0.03 – 3.92* 0.53(159)	*0.04 – 10.59* 0.46(708)

Italics = extreme values; Roman = mean values; () = number of samples

[a] Values, characteristic of sediments deposited in the Peruvian-Chilean area of high productivity

[b] <30% $CaCO_3$ and <30% $SiO_{2(amorph)}$

[c] Chiefly benthogenic-foraminiferal

[d] Diatomaceous, >30% $SiO_{2(amorph)}$

[e] Include sediments of the inner slopes and bottom sediments of the near-continental deep trenches

[f] Algal-coral, coral-shell and bryozoan-skeletal materials

Table 31. Organic carbon content of the upper sedimentary layer (0 – 5 cm) in the structural-geomorphological zones of the Atlantic Ocean, %, on a dry sediments basis (Emelianov and Romankevich 1979)

Litho-facies types	Sands	Aleurites	Pelites
Sediments in shelf areas (25 – 360 m in depth)			
Terrigenous[a]	*0.04 – 3.87*	*0.19 – 9.78*	*0.16 – 12.60*
	0.53(90)	1.23(52)	2.06(29)
Western Africa	*0.05 – 3.87*	*0.29 – 9.78*	*0.16 – 12.60*
	1.27(28)	1.95(15)	3.29(12)
Other areas	*0.04 – 1.20*	*0.19 – 2.85*	*0.55 – 3.30*
	0.28(68)	0.95(37)	1.18(17)
Carbonate[b]	*0.10 – 4.71*	*0.36 – 6.57*	*0.27 – 12.29*
	0.93(64)	1.43(18)	1.76(16)
30 – 50% $CaCO_3$	*0.10 – 2.69*	*0.36 – 2.79*	*0.42 – 12.29*
	0.62(19)	0.86(14)	3.66(4)
Western Africa	*2.66 – 2.69*	*0.99 – 1.13*	*0.42 – 12.29*
	2.67(2)	1.06(2)	4.56(3)
Other areas	*0.10 – 0.75*	*0.36 – 2.79*	*0.93*
	0.37(17)	0.83(12)	– (1)
>50% $CaCO_3$	*0.12 – 4.71*	*1.74 – 6.57*	*0.27 – 1.43*
	1.04(45)	3.43(4)	0.60(10)
Western Africa	*1.15 – 4.71*	*1.74 – 6.57*	*0.27 – 1.43*
	2.90(6)	3.96(3)	0.60(10)
Other areas	*0.12 – 1.82*	*1.82*	–
	0.75(39)	– (1)	–
Siliceous[c]	–	–	*3.66 – 14.05*
	–	–	6.44(13)
Glauconite and hydrogo-ethitechamozite	*0.15 – 3.42*	*1.02 – 4.71*	*0.98 – 3.02*
	1.33(11)	2.04(7)	2.27(10)
Volcanic clastics[a, b]	*0.07 – 0.42*	–	–
	0.27(5)	–	–
Sediments of continental slopes (300 – 4500 m in depth)			
Terrigenous[a]	*0.05 – 1.25*	*0.10 – 4.16*	*0.10 – 3.88*
	0.42(46)	0.81(144)	0.98(92)
Western Africa	–	*0.67 – 1.95*	*1.20 – 3.88*
	–	1.24(6)	2.38(14)
Other areas	*0.05 – 1.25*	*0.10 – 4.16*	*0.10 – 3.64*
	0.42(46)	0.80(138)	0.74(78)
Carbonate[b]	*0.10 – 2.91*	*0.14 – 6.57*	*0.21 – 4.06*
	0.50(30)	0.90(64)	0.70(75)
30 – 50% $CaCO_3$	*0.10 – 1.26*	*0.15 – 3.86*	*0.40 – 1.14*
	0.43(9)	0.92(25)	0.73(27)
>50% $CaCO_3$	*0.17 – 2.91*	*0.14 – 6.57*	*0.21 – 4.06*
	0.52(21)	0.88(39)	0.68(48)
Western Africa	*0.22 – 2.91*	*2.52*	*0.29 – 4.06*
	1.16(3)	– (1)	1.84(3)
Other areas	*0.17 – 0.95*	*0.14 – 6.57*	*0.21 – 1.22*
	0.42(18)	0.84(38)	0.61(45)
Volcanic clastics[a]	*0.11 – 1.43*	*0.04 – 1.71*	*0.49 – 1.36*
	0.40(9)	0.67(17)	0.78(7)
Sediments in the marginal parts of the ocean floor (4500 – 5000 m in depth, down to 5500 m in the North American Oceanic Basin)			
Terrigenous[a]	*0.27*	*0.11 – 0.59*	*0.21 – 1.25*
	– (1)	0.35(4)	0.47(21)

Table 31 (continued)

Litho-facies types	Sands	Aleurites	Pelites
Carbonate[e]	–	*0.20 – 0.71*	*0.17 – 0.56*
	–	0.40(11)	0.35(16)
Sediments of the mid-Atlantic Ridge and submarine shoals			
Carbonate[f]	*0.12 – 0.50*	–	*0.09 – 1.39*
	(11)	–	0.30(16)
Sediments in rift valleys	*0.10*	*0.09 – 0.16*	*0.29 – 0.99*
	– (1)	0.12(6)	0.64(2)
Siliceous[c, g]	–	–	*0.18 – 0.60*
	–	–	0.34(4)
Sediments in the central areas of the ocean floor (pelagial, at depths over 5000 – 5500 m)			
Terrigenous[a]	–	–	*0.19 – 0.50*
	–	–	0.33(8)
Red abyssal clays	–	–	*0.26 – 0.83*
	–	–	0.40(9)
Carbonate[h]	–	–	*0.21 – 0.50*
	–	–	0.36(8)
30 – 50% $CaCO_3$	–	–	*0.32 – 0.50*
	–	–	0.43(4)
>50% $CaCO_3$	–	–	*0.21 – 0.45*
	–	–	0.31(4)
Total for Ocean	*0.04 – 4.71*	*0.04 – 9.78*	*0.09 – 14.05*
	0.62(262)	0.92(323)	1.16(326)

Italics = extreme values; Roman = mean values; () = number of samples
[a] <30% $CaCO_3$ and <30% $SiO_{2(amorph)}$
[b] Shell materials, coral-algal, bryozoan, foraminiferal, coccolith-foraminiferal (>30% $CaCO_3$)
[c] Diatomaceous, >30% $SiO_{2(amorph)}$
[d] Area of Iceland
[e] Shell materials, foraminiferal, coccolith-foraminiferal (>30% $CaCO_3$)
[f] Shell materials, bryozoan, foraminiferal, coccolith-foraminiferal, pteropod-coccolith-foraminiferal (>50% $CaCO_3$)
[g] Antarctic region
[h] Foraminiferal and coccolith-foraminiferal

Table 32 shows the average C_{org} content of the upper sedimentary layer within the major geomorphological zones of the Pacific and Atlantic Oceans[3]. In the

3 In the Pacific Ocean to the sediments of the submarine continental margin are assigned the deposits within the shelf areas, continental slopes and the deep trenches (including the external slopes), that are directly contiguous with the latter, as, for example, the Kuril-Kamchatka, Japan, Aleutian, Peru-Chile Trenches, and are not separated from them by a transitional zone. In those areas where the slope passes more or less gradually into the floor of the Ocean, the boundary is taken at the 4000 m isobathic curve.

In the Atlantic Ocean the sediments of the submarine continental margins include those that occur within the shelf areas, continental slope and rise, to 4500 m in depth. The ocean floor sediments include the basinal sediments, those of the mid-Atlantic Ridge and other submarine rises located beyond the limits of the continental margin.

The boundary between the marginal and central parts of the ocean floor is provisional. Various genetic types of sediments (predominantly terrigenous and sparsely siliceous in the northern part) occurring around the periphery of deep oceanic basins, are referred to the marginal part of the ocean floor

Table 32. Average organic carbon content in the upper sedimentary layer of geomorphological zones in the Pacific and Atlantic Oceans (%, on a dry sediment basis)

Structural-geomorphological zone	Pacific Ocean			Atlantic Ocean		
	sands	aleurites	pelites	sands	aleurites	pelites
Submarine continental margin	0.76(51)	0.91(65)	1.29(107)	0.63(248)	0.96(302)	1.42(242)
Including						
shelf	0.66(18)	1.24(9)	7.34(4)[a]	0.73(163)	1.35(77)	2.86(68)
slope	0.81(33)	0.86(56)	1.05(103)	0.44(85)	0.83(225)	0.85(174)
Ocean floor	0.26(34)	0.26(94)	0.31(601)	0.27(14)	0.30(21)	0.40(84)
Including						
marginal part	–	0.31(19)	0.47(136)	0.27(1)	0.37(15)	0.42(37)
central part	0.26(34)	0.25(75)	0.27(465)	0.27(13)	0.12(6)	0.39(47)
Average	0.56(85)	0.53(159)	0.46(708)	0.62(262)	0.92(323)	1.16(326)
Subabundance ratios[b]		0.40			0.83	

Figures in parenthesis show the number of samples

[a] The value is given for sediments deposited in the Peruvian-Chilean area of high productivity

[b] Mean values, relative to the areas occupied by various types of sediments

Pacific Ocean the decrease in the OM content of the sediments in the series: submarine continental margin – marginal parts of the ocean floor – central oceanic areas is in the proportion of 100: – : 34 (for sands), 100:34:27 (for aleurites), 100:36:27 (for pelites); the summary average proportion may be expressed as 100:35:27. The sediments in the marginal parts of the ocean floor differ only slightly in their OM content from the pelagic sediments proper and, according to the mean values, their C_{org} content is about three times less than that in the sediments of the submarine margin of continents. In the Pacific Ocean sediments the highest C_{org} gradients were recorded in the areas of the near-continental deep trenches and marginal parts of the ocean floor.

Similar calculations made for the Atlantic Ocean revealed a decrease in the C_{org} content from the sediments of the submarine continental margin towards the ocean floor sediments, including those in the marginal and central parts, the relative proportions being 100:43 (for sands), 100:31 (for aleurites) and 100:28 (for pelites), and averaging about 100:30. Thus the sediments of the Atlantic Ocean floor, like those of the Pacific, contain approximately three times less C_{org} than the sediments of the submarine continental margins.

Of importance and demanding special attention, is the fact that the mean value of the C_{org} content is higher in the sediments of the Atlantic Ocean than that in the Pacific; this is reflected in the estimates for the dry natural sediments (1.10 and 0.49%, respectively), as well as on a non-carbonate and non-siliceous basis (1.84 and 0.87%). Comparison of the sub-Clarke ratios (that is, the mean values relative to the areas occupied by sediments of various genetic types) shows that in the Atlantic they are about twice as high as in the Pacific (0.83% and 0.40%). This is typical of all types of sediments and most of their size fractions. Worthy of note is the fact that the differences in the OM content of the Pacific and Atlantic Ocean sediments decrease progressively in the sequence: shelf facies (terrigenous, slightly siliceous, glauconite) – carbonate facies – red abyssal clays, which is in accordance with the circumcontinental zonality of OM distribution.

This phenomenon can be explained by the following: (1) the greater (about 50% higher) annual phytoplanktonic productivity of the Atlantic Ocean; (2) the higher rate of accumulation of bottom sediments here; (3) the larger area of the shelf (10.3%), continental slope (8.5%) and rise (13.8%) in the Atlantic, as compared to the Pacific – 5.5%, 3.0% and 1.7%, respectively (Leontiev et al. 1975); (4) artifacts associated chiefly with the irregular collection of samples, variable in their C_{org} content. Analyses of the above-cited data strongly suggest that the first two factors, the relative higher productivity of the Atlantic Ocean and higher rate of OM burial, are the most important.

The fact that the rate of deposition of mineral matter exerts a strong influence on the content and accumulation of OM, was repeatedly stressed in earlier publications (e.g., Trask 1953, Strakhov 1960, Romankevich 1968, 1970, Bogdanov et al. 1979, Nazarkin 1979, Müller and Mangini 1980, Suess and Müller 1980). The influence of the rate of sedimentation is most obvious in the dilution of the OM by the mineral matter, which results in the decrease in the percentage content of OM in the sediments. At the same time, the sedimentary mass causes the rapid removal of OM from the sphere of active destruction by the benthos, and, thereby, its ultimate preservation (Romankevich 1968, 1971).

Comparison of the C_{org} mean values in different genetic types of sediments in the Pacific and Atlantic Oceans, revealed that the highest values occur characteristically in the terrigenous and slightly siliceous shallow-water diatomaceous muds of different climatic zones: boreal humid of the Pacific Ocean and tropical arid southeastern parts of both oceans. Exceptionally high C_{org} values were recorded in the areas of near-coastal upwellings, as the Peruvian, Californian, certain areas of the Northwestern Pacific, and also in the area of the upwelling off the coast of Western Africa. The OM content of many different sediments (terrigenous, siliceous, diatomaceous, calcareous) is here several times higher than that of similar facies in other areas of less productivity. This again points to the important role of the biological productivity, which is largely dependent upon hydrodynamic conditions (Gershanovich and Muromtsev 1982).

In contrast to the diatomaceous muds that accumulate in areas of general cyclonic circulation and upwelling of waters, including coastal upwellings, the pelagic diatomaceous (Ethmodiscus) silt in the tropical zones of the Pacific Ocean and the diatomaceous silt of the Antarctic Belt are characteristically of low C_{org} content (averaging 0.30% and 0.20%, respectively). They accumulate at great depths in areas of low productivity, far from the coast and under conditions of a low rate of sedimentation, all of which is unfavorable for the preservation of OM. As a consequence, their OM content is 2 – 4 times less than even the diatomaceous ooze of the Boreal Belt in the Pacific Ocean, and ten times less than the OM content of the shallow-water siliceous and slightly siliceous muds. The relatively shallow water carbonate sediments (coral, foraminiferal-coral, coral-algal types) contain more OM than the pelagial foraminiferal muds and terrigenous sediments. A very high C_{org} content (3.5 – 6.8%) was recorded in a benthogenic-foraminiferal facies encountered on the continental slope off Peru. A specific feature of this facies is the occurrence of OM in forms of aggregates, flakes, clots and plasm remains inside of shells. The presence of OM in such forms (and not within the carbonate matrix) invoked intensive sulphate-reduction and the occurrence of hydrogen sulphide and iron sulphides at the surface of the sediments. According to the intensity of the reduction processes, these sediments are similar in type to the benthogenic bituminous shales.

All the pelagial sediments in the central regions of the Pacific and Atlantic Oceans are poor in OM. The red abyssal clays, particularly the eupelagic types in the Pacific Ocean, characteristically have the lowest OM content compared with other sediments of similar granulometrical composition. This can be attributed to the low amount of POC that sinks to the bottom, and more complete utilization of the OM at the bottom surface as a result of slow rate of sedimentation.

Metalliferous sediments, that were encountered in the regions of deep fractures and spreading of the ocean floor in the southeastern Pacific, contain only 0.23 – 0.31% of C_{org} and up to 20 – 40% of ferric hydroxides.

The volcanic clastics, which are widespread in the pelagial areas of the western Pacific, are of low C_{org} content (0.03 – 0.39%, averaging 0.23%). The C_{org} content of the volcanic clastics within the submarine continental margin of the Atlantic Ocean, is in general, higher than those in the Pacific (0.04 – 1.7%, averaging 0.57%). These discrepancies are probably due to the fact that in the

Table 33. C_{org} and nitrogen contents in the ferro-manganese nodules of the Pacific and Indian Oceans[a]

Object studied	C_{org}	N_{org}	N/C weight ratios in %
	%, on a dry substance basis		
Outer crust	*0.040 – 0.120*	*0.001 – 0.007*	*1.7 – 11.6*
	0.074 (7)	0.003 (10)	3.1 (6)
Ore body	*0.060 – 0.130*	*traces – 0.004*	*1.0 – 3.8*
	0.100 (4)	0.002 (8)	2.1 (4)
Core	*0.050 – 0.270*	*0.005 – 0.021*	*6.3 – 30.3*
	0.128 (4)	0.012 (4)	8.7 (4)
Host sediment	*0.130 – 0.350*	*0.035 – 0.092*	*18.9 – 37.0*
	0.234 (9)	0.061 (12)	24.4 (9)

Italics = extreme values; Roman = mean values; () = number of probes

[a] Concretions were collected from the Pacific Ocean in the Tropical north-eastern Basin, Southern Basin, Central Basin; from the Indian Ocean in the West-Australian Basin

Atlantic Ocean these deposits occurs in areas where the POC that settles at the bottom is significantly higher.

In the ferrous-manganese nodules collected from the Pacific and Indian Oceans, the C_{org} content was found to be between 0.04 – 0.13%, that is, less than in the enclosing red abyssal clays (Table 33). The outer crust of the concretions does not differ in its C_{org} content from the rest of the ore-body (excluding the core). The cores of the nodules are comprised of altered sedimentary materials of the red clay type and contain less C_{org}, than the host rocks. Thus is is conceivable that the C_{org} content decreases in the order: host sediments > core > ore-body plus outer crust.

According to Gorshkova (1970), the ferrous-manganese nodules from the Riga and Bothnian Bays have a much higher C_{org} content (about 1 – 2%) than the oceanic concretions, whereas their C/N ratios are much lower. This implies that the OM in the oceanic concretions, being much older than those from the Riga and Bothnian Bays, experienced more profound alterations.

The phosphorite authigenous formations and concretions that are formed in shallow water facies in the zone of high biological productivity, that is, on the facies profile opposite the ferrous-manganese nodules, characteristically have a high C_{org} content (0.8 – 3.4%). During the process of diagenesis and lithification of the phosphate material, the OM content decreases, and its composition undergoes a substantial change (Fig. 34). The carbonic acid, which during the degradation of OM is released into the interestitial waters of the muds, undergoes partial bonding into phosphate-carbonate compounds of the carbonate-fluoro-apatite type and forms metasomatic replacements of opal in the diatom tests. Up to 18 – 39% of the carbonic acid, that is releated during the degradation of OM entrapped in the sediments, can be bound in the formation of phosphates (Shadsky et al. 1980, Romankevich and Baturin 1972).

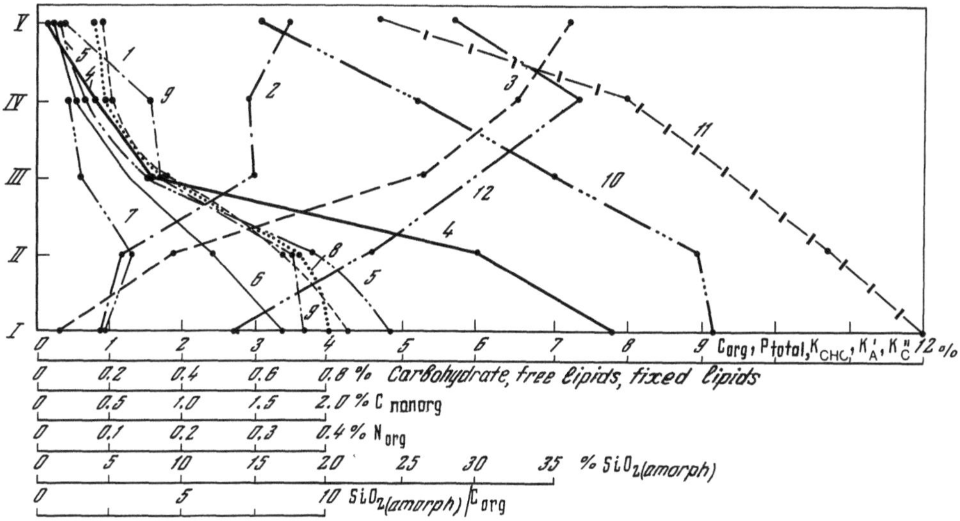

Fig. 34. Biochemical composition of the phosphorite concretions from the shelf off southwestern Africa. *I* siliceous diatomaceous mud enclosing phosphorite concretions; *II* phosphated diatomaceous mud *III* friable phosphorite nodule; *IV* globular phosphorite nodule; *V* firm phosphorite concretion. *1* C_{org}; *2* C_{nonorg}; *3* P_{total}; *4* $SiO_{2(amorph)}$; *5* carbohydrates; *6* lipids "A" (mixed chloroform-methanol (2:1) Soxhlet extraction); *7* lipids "C" (mixed alcohol-benzene (1:1) Soxhlet extraction, after removal of lipids "A" and 10% HCl treatment of the precipitate); *8* N_{org}; *9* $SiO_{2(amorph)}/C_{org}$; *10* C_{CHC}/C_{org} 100 (K_{CHC}); *11* lipids "A" in OM (K_A); *12* lipids "C" in the OM K(K_C). (After Romankevich and Baturin 1972, with complementary data)

4.2.3 Organic Carbon Distribution in the Late Quarternary Sediments of the Pacific Ocean

Measurements of the C_{org} content were made by uniform analytical techniques (the Knopp method based on separate weight determinations of the $CO_{2\ carb}$ and $CO_{2\ org}$ contents) in 120 cores of bottom sediments that were collected (the total approximates 2300) during the cruises of the research vessels "Vityaz", "Academik Kurchatov", "Dmitrii Mendeleyev" in the Pacific (Fig. 35). The lithology and chemical composition of the sediments were studied in most of the cores; stratigraphic studies and age-determinations were made in many cores by means of various methods (biostratigraphic, geochemical, ionium-thorium, paleomagnetic, protactinium) and their combination. Similar studies, though on a lesser scale, were conducted in the Atlantic and Indian Oceans. The differences in the age-datings of the sediments, also in the amount of supplied and buried OM, allowed for the distinction of four major types of vertical OM distribution and corresponding types of sediments (I – IV) in which OM is entrapped, that gradually replace each other from the land towards the central halistatic areas of the Ocean.

I-Type C_{org} Distribution. The sediments within shelf areas and the upper parts of the continental slopes exhibit the most intricate C_{org} distribution pattern, charac-

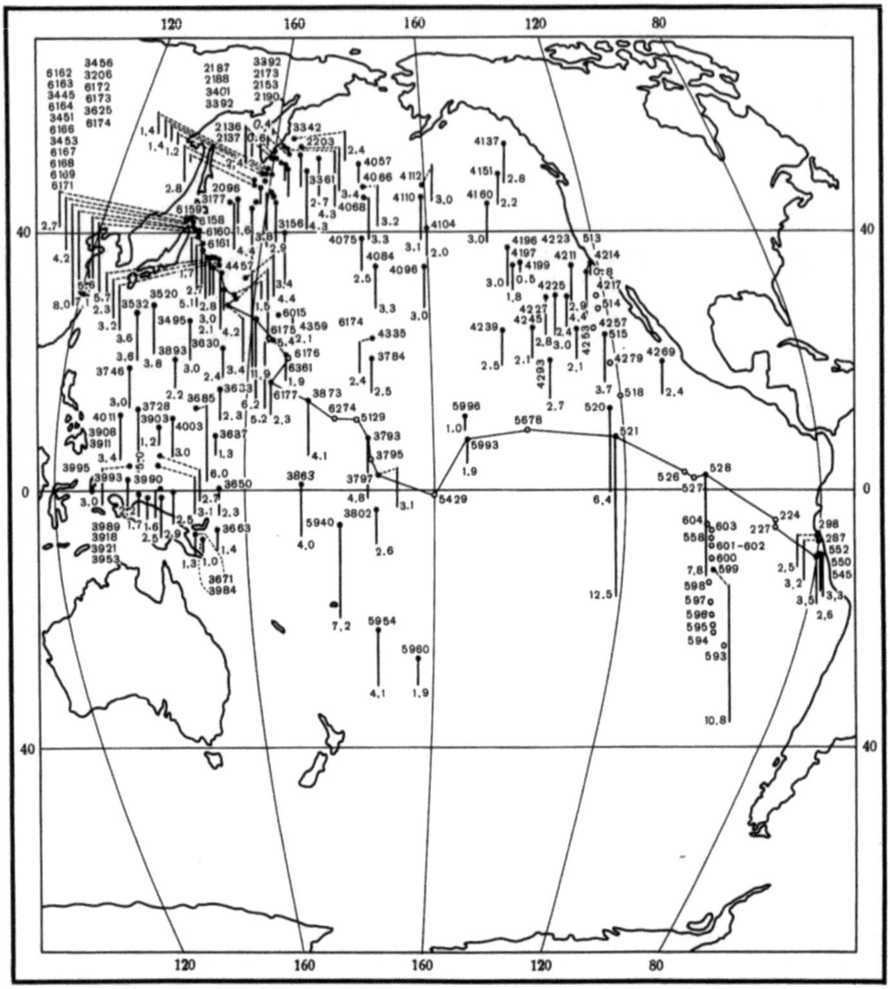

Fig. 35. Sampling sites of bottom sediment columns, in which measurements were made of the C_{org} content and, in some, the OM composition was studied, during the cruises of the Research Vessels Vityaz, Academik Kurchatov, Dmitrii Mendeleyev. *Figures* show station numbers and column length. *Line connecting columns*: cross-section location

terized by repeated alternations of layers with low and high C_{org} values, large lateral and vertical variations in the OM content of the layers, and the presence of layers almost entirely lacking shaped organic remains and OM aggregates, their deposition being strongly inhibited by the hydrodynamic conditions. As a consequence, besides high C_{org} values, the distribution pattern reveals background values, that are associated with the sorbated forms of occurrence of the major part of C_{org}.

This pronounced heterogeneity in the vertical distribution of the C_{org} content is exemplified in the sediments off the coasts of Peru and South West Africa. The cores from the Peruvian shelf penetrated 2 – 6 m layers of terrigenous Holocene

cene and Wurmian slightly siliceous, phosphate-terrigenous and terrigenous-glauconite sediments of various size-fractions, which are characterized by marked fluctuations in their C_{org} content. Similar variations were observed in the terrigenous muds and diatomaceous ooze of the southwestern shelf of Africa. The general similarity in the C_{org} distribution pattern of both areas can be ascribed to the variations of the content and the rate of supply onto the bottom of sedimentational material, the stirring up of sediments and their redeposition, the change in the fraction-size and substance composition of the deposits, that determined the burial of OM. Within the 2–4-m layer of sediments no substantial changes are apparent in the productivity of the areas. The transgressive overlapping of the sediments which is distinctly apparent in the columns, is evidently associated with the ocean level rise during the Post-Glacial period. Consequently, the sediments in each layer reveal from bottom to top a decrease in the particle sizes and $CaCO_3$ content, and an increase of the C_{org} content.

The decrease of C_{org} content from the top to the bottom layers in all the cases considered resulted chiefly from lower OM supply to the bottom during sedimentogenesis (the bioproductivity at that time could well remain the same), and not from OM expenditure during biomixing and diagenetic decomposition. In spite of the large fluctuations in the C_{org} values, the Upper Quarternary sediments that accumulated in oceanic zones of high productivity, inherited the high C_{org} concentrations that are typical of the surface layer. The cores procured from the shelf off southwestern Africa suggest that intensive reduction conditions (Eh from −43 to −293 mV) persisted in the sediments, at least from the Wurmian, and from this time onward there began an intensive upwelling of water and accelerated C_{org} accumulation in the sediments in the near-shore areas.

II-Type C_{org} Distribution. This type can be observed in the cores from the continental slopes, the bottom of deep trenches, that are not separated from the former by a transitional zone, and also in the cores from the continental rise. In high-latitudinal areas these cores are comprised chiefly of aleurites and aleurite-pelite, terrigenous and siliceous-terrigenous muds, whereas in the tropical areas they contain also carbonate foraminiferal and foraminiferal-coccolithic sediments. Characteristic of the II-type sedimentation are the intermittently varying conditions of deposition, particle by particle, and in non-uniform rate of supply of sedimentational material onto the bottom as it slides down in slope, brought about by turbidity currents. They may also be deposited in river fans.

The sediments in the cores of the II-type are variable in composition in vertical profile, so that layers enriched in C_{org}, $CaCO_3$, $SiO_{2\,(amorph)}$ can be distinguished. The changes in the composition of the sediments can be attributed chiefly to the repeated changes in the climate and productivity of the ocean waters during the Quarternary. The relatively high rate of deposition of the sedimentary material and high OM content invoked sharp changes in the cores from oxidizing to reducing conditions. The decrease in the sulphate supply caused the extinction of the sulphate-reduction process linked with the OM transformation process, that added to the better preservation of the OM and its fossilization.

The subsequent expenditure of OM on bacterial methane formation which is carried out in bottom sediments principally by reducing CO_2 with hydrogen, is

1 – 2 times less, in the Baltic sea 20 times less (Lein et al. 1981), than its consumption for sulphate reduction (Ivanov 1979, Germanov et al. 1981, Geochemical activity of microorganisms ... 1981).

III-Type C_{org} Distribution. This type was observed in cores procured from the marginal parts of the ocean floor. The cores comprised hemipelagial terrigenous muds, which in temperate latitudes are very abundant in biogenic siliceous materials and in tropical latitudes in carbonate materials. The sediments of the III-type are characteristically of lower C_{org} content than those of the II-type. The C_{org} content varies considerably in the vertical profile, and in some individual layers it is higher than in the upper layer.

The impoverishment and enrichment of individual sedimentary layers in biogenic components, can be attributed to the different productivity of the ocean waters during the Glacial and Interglacial periods. The warming of the surface waters in the Boreal regions of the Pacific during the Post-Glacial and Interglacial periods caused the enhancement of planktonic productivity and of the settling rate of the skeletal remains of diatoms, foraminifers and OM to the bottom. In the equatorial regions of the Pacific Ocean the productivity increased during the Glacial periods, when, owing to the increase in the circulation and inflow of cold deep waters, large quantities of biogenic elements were brought into the photic zone. The sediments of the III-type were deposited, in most part, particle by particle, and reflect the facies changes in the sedimentation environment. Thus within this type of sediment key levels can be recognized, which being relatively enriched in OM, $SiO_{2(amorph)}$, $CaCO_3$, microfloral and microfaunal remains, provide the means for correlations and age-datings of the layers over wide areas of the Pacific Ocean floor, despite the relatively low sedimentation rates and widespread bioturbation processes involving the upper sediment layer.

Some cores of the III-type did not reveal a systematic decrease in the C_{org} content within the depth profile of the penetrated sediments. They contained the same amount of OM, at times even higher, in their lower layers as in the upper layers. An example of this is the core taken at st. V-3163 from the marginal part of the western basin in the Pacific Ocean (12 m in depth; the lowermost layers approximate 0.7 million years in age). In most of the III-type of cores the C_{org} content decreases with depth.

A characteristic feature of the sediments in the marginal part of the ocean floor compared with II-type sediments is the change in the vertical profile from oxidizing (Eh = 300 to 450 mV) to slightly reducing conditions (Eh = +70 to –70 mV), and the much lower rate of sedimentation (about 5 – 30 mm per 1000 years, seldom more). The coloration of the sediments and its change in the vertical profile is a good indicator of the intensity of the oxidation-reduction processes. The upper sediments of the core, up to 10 – 60 cm in depth, are brown owing to the presence of Fe and Mn oxides. Lower in the section the color of the sediments undergoes a sharp or gradual change to gray or greenish-gray, which marks the transition to reducing conditions for Fe and Mn, while the OM, on the other hand, is subjected to further oxidation.

IV-Type C_{org} Distribution. This type has been observed in the ocean floor cores that were taken at a considerable distance from the land and consist ubiquitously of eupelagic and miopelagic red clays. The metalliferous sediments, that in their C_{org} content and distribution pattern do not differ from the red abyssal clays, are also assigned to this type. In the cores of the IV-type the C_{org} content undergoes a conspicuous decrease within the uppermost layer during the first 40 – 100 thousand years. Further below the C_{org} content changes insignificantly, revealing only a tendency to its reduction. Marked oxidation conditions prevail throughout the core, with Eh = 400 – 600 mV (the age of some of the lower layers exceeds 10 million years, as, for instance, in the core at st. M-599, which penetrated down to the Miocene).

The pelagic sediments from the equatorial zone and regions of the tropical divergence are transitional between the III- and IV-types. These areas are characterized by enhanced rates of sedimentation and supply of POC onto the bottom. The diatomaceous-radiolarian ooze (st. M-520, M-521), that accumulated in areas of relatively low productivity, differ only slightly in their C_{org} content from the miopelagic muds, whereas the diatomaceous-radiolarian-foraminiferal and coccolith-foraminiferal ooze of the equatorial region (st. M-528), according to their C_{org} content, more closely resemble the hemipelagic terrigenous sediments.

4.3 Organic Carbon Distribution in the Size-Fractions of Sediments: Relation Between Organic Carbon and $CaCO_3$ and $SiO_{2\,(amorph)}$

For a better understanding of the principles governing the accumulation and distribution of OM, investigations were performed on various size-fractions of terrigenous, carbonate-foraminiferal sediments and red abyssal clays, procured from the Pacific, Atlantic and Indian Oceans. Analytical studies of the samples served to show that in the terrigenous, carbonate sediments and miopelagic clays the C_{org} content increases from the sand fraction (>0.1 mm) to the subcolloidal (<0.001 mm) (see Table 27 and Fig. 36).

In the eupelagic clays, the increase in the C_{org} content is obvious only in the medium-pelitic (0.005 – 0.001 mm) to the subcolloidal. The sand and aleuritic fractions of the eupelagic clays contain relatively large-sized bone remains that under natural conditions are resistant to decay. The preferential preservation of such OM, resistant to degradation, results in the relative OM enrichment of the large-size fractions of the eupelagic clays. Another major constituent of the eupelagic clays is the tightly sorbated OM. The latter is also one of the most conservative OM components. Even in the surface layer (0 – 3 cm) of the sediments, which ages 50 – 200 thousand years, the metamorphic state of this OM is the brown coal (lignite B_I) diagenetic stage. Owing to this, various inorganic compounds are practically not involved in reducing processes, which require the expenditure of OM. The "saturation" factor of the pelitic fraction shows a gradual decrease from the shelf towards the ocean floor.

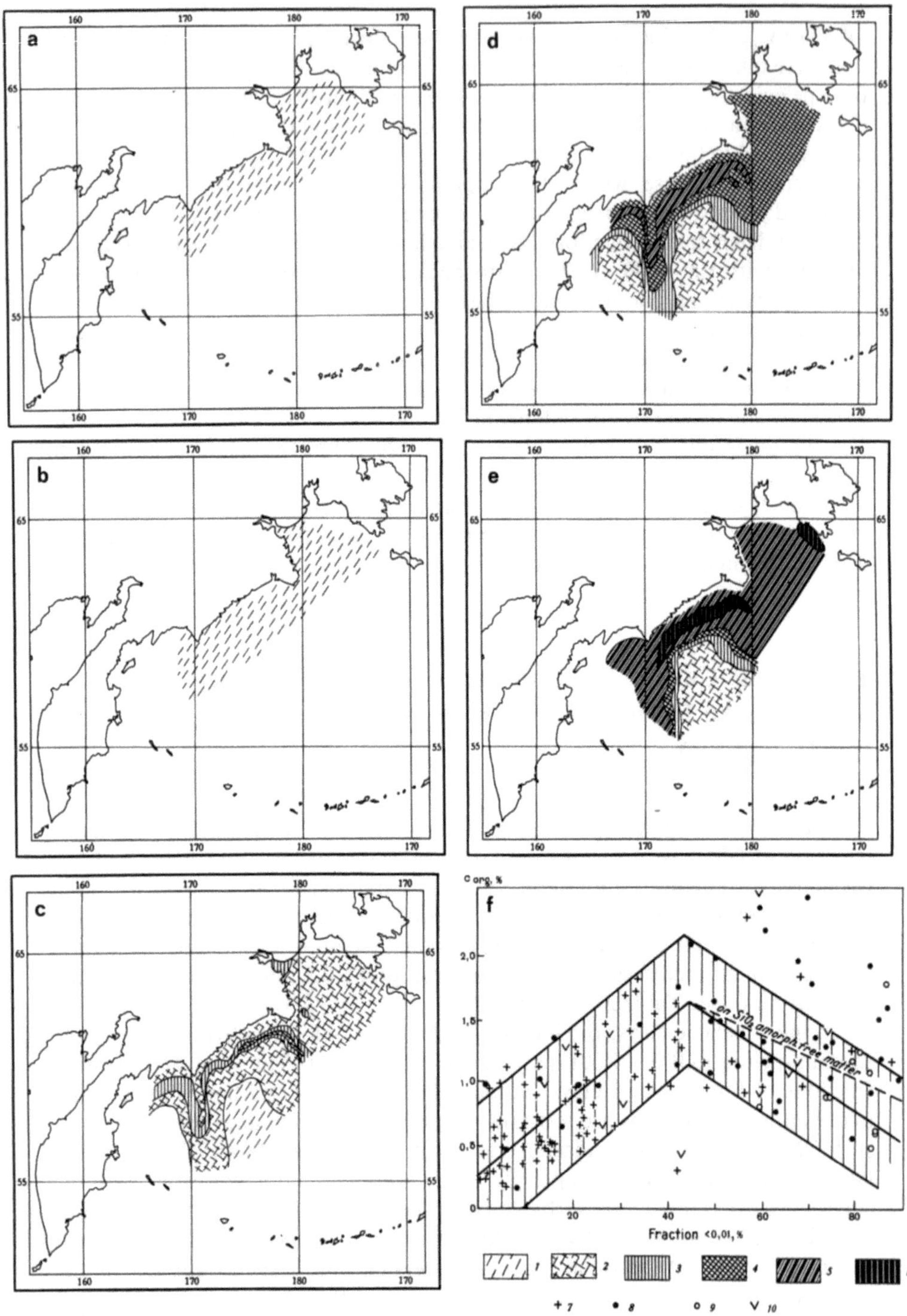
a
b
c
d
e
f
160
170
180
170
65
55
C org, %
2,0
1,5
1,0
0,5
0
20
40
60
80
on SiO_2 amorph free matter
Fraction <0,01, %
1
2
3
4
5
6
7
8
9
10

Fig. 36 a – f

The percentage concentrations of C_{org} in the subcolloidal fraction serve as indicators of the factor of OM "saturation", whereas, by contrast, the concentration coefficients reflect mainly the intensity of the bond between the OM and inorganic moiety.

In the terrigenous sediments the sand and aleuritic fractions are 2.5 – 2 times poorer in C_{org} than the natural sediments, whereas the large pelitic plus the medium-pelitic and subcolloidal fractions are 1.7 – 2.2 times richer (respectively). The latter two fractions in the carbonate sediments and miopelagic clays are also richer in OM. In the eupelagic clays the C_{org} concentration coefficients were found to be characteristically high in the sand (1.7) and subcolloidal (2.3) fractions. The fact that the OM concentration coefficients in the subcolloidal fraction of the terrigenous sediments and red clays are roughly the same, serves to suggest that the bond between this fraction and the mineral base is practically of the same intensity in both types of sediments. If this indeed proves to be the case, then the OM retained in these fractions during the process of diagenesis, is practically of the same degree of preservation. The higher concentration coefficients in the subcolloidal fraction of the carbonate sediments (the $CaCO_3$ content is 85 – 90%) is evidence for the existence of a strong bond between the OM and the carbonate matrix, and explains the stability of this form of organic compounds.

The analytical techniques applied to carbonate sediments for evaluation of the actual amount of OM involved in the process of diagenesis, should differ from those applied to silicate rock types. The effect of HCl treatment on sediments composed of $CaCO_3$, is similar to that of hydrofluoric acid on silicates in that it causes the breakdown of the mineral carbonate skeleton. This results in the artificial liberation of that part of the OM which is tightly bonded with inorganic matter and is inactive during the early stages of diagenesis (prior to recrystallization). In natural environments this OM is little involved in reduction processes, and, according to its chemical-analytical properties, is traditionally defined as readily hydrolysable OM.

In general, no correlation is obvious between the C_{org} content and the <0.01 mm fraction content in the sediments of the Pacific and Atlantic Oceans. Despite the large scatter in the values, a direct relationship is observed in the terrigenous sediments of the shelf areas and upper parts of the continental slopes (Fig. 37).

Investigations on various size fractions of the Bering Sea sediments showed that the C_{org} content increases in direct proportion to the size of the <0.01 fraction, but only to a certain level (see Fig. 36). The increase in the pelitic content

Fig. 36a – f. The C_{org} content in the size-fractions of sediments from the western part of the Bering Sea (**a** – **e**, in % of dry sediment weight) and the relationship between the C_{org} and <0.01 mm fraction content (**f**). **a** sand fraction (1.0 – 0.1 mm); **b** coarse aleurite fraction (0.1 – 0.05 mm); **c** thin aleurite fraction, re-calculated on a non-siliceous basis (0.05 – 0.01 mm); **d** large- and medium-pelitic fraction (0.01 – 0.001 mm); **e** pelitic fraction, recalculated on a non-siliceous basis (0.01 – 0.001 mm); **f** relationship between the C_{org} and <0.01 fraction contents.
C_{org} content, in %: *1* = <0.5; *2* = 0.5 – 1; *3* = 1 – 1.5; *4* = 1.5 – 2; *5* = 2 – 3; *6* = >3. Types of bottom sediments: *7* terrigenous (<10% $SiO_{2(amorph)}$; *8* slightly siliceous (10% – 30% $SiO_{2(amorph)}$; *9* siliceous (over 30% $SiO_{2(amorph)}$); *10* sediments in which the $SiO_{2(amorph)}$ was not determined (Lisitzin 1966)

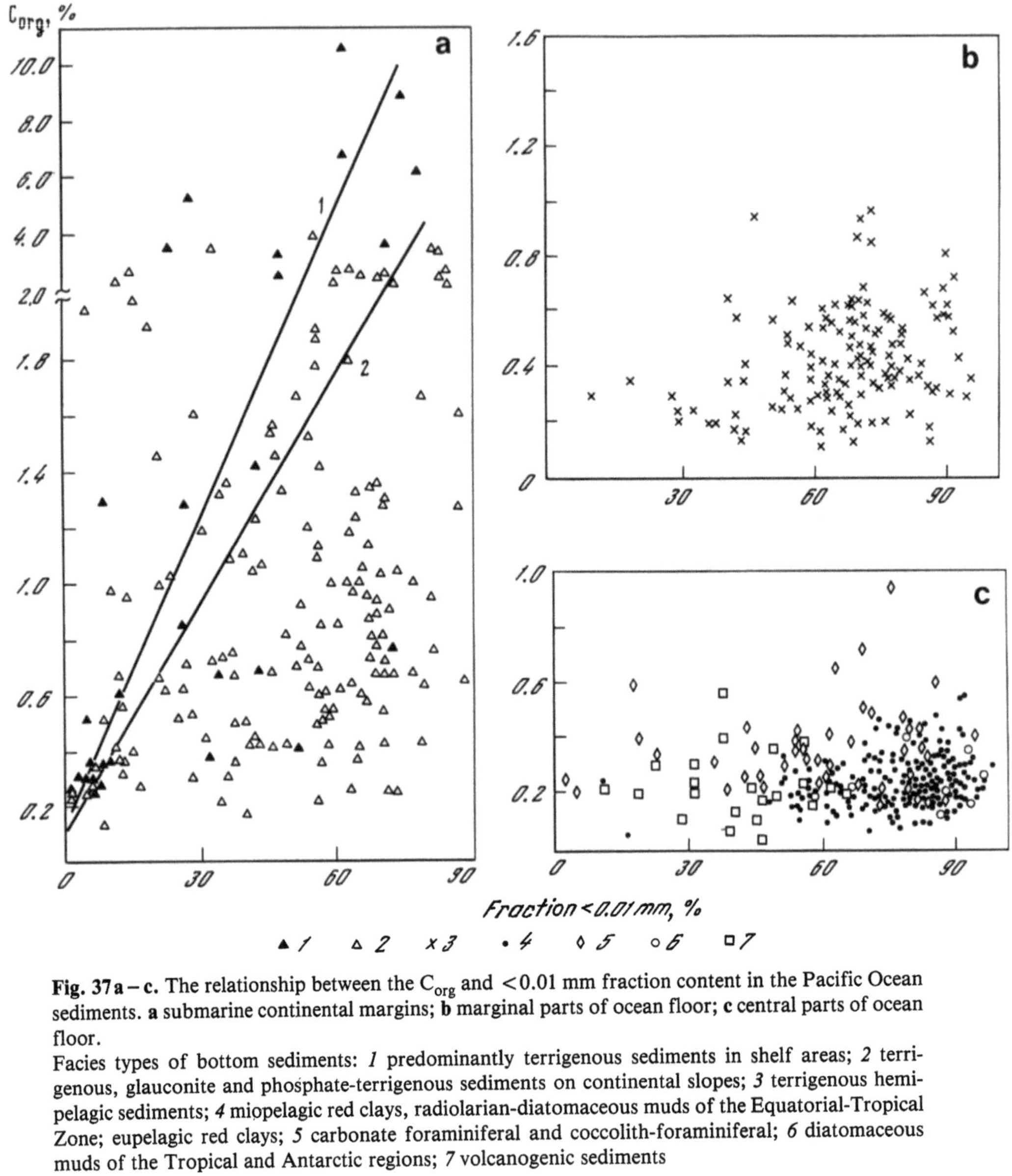

Fig. 37 a–c. The relationship between the C_{org} and <0.01 mm fraction content in the Pacific Ocean sediments. **a** submarine continental margins; **b** marginal parts of ocean floor; **c** central parts of ocean floor.
Facies types of bottom sediments: *1* predominantly terrigenous sediments in shelf areas; *2* terrigenous, glauconite and phosphate-terrigenous sediments on continental slopes; *3* terrigenous hemipelagic sediments; *4* miopelagic red clays, radiolarian-diatomaceous muds of the Equatorial-Tropical Zone; eupelagic red clays; *5* carbonate foraminiferal and coccolith-foraminiferal; *6* diatomaceous muds of the Tropical and Antarctic regions; *7* volcanogenic sediments

above 50% is concomitant with the increase in the concentration of C_{org} in the shallow water areas of high biological productivity, whereas in the deep water areas the C_{org} concentrations are of concurrent decrease. This is consistent with the above-cited generalizations: the existence of two "pelitic" shallow-water and deep water C_{org} maxima, and the different OM "saturation" of the pelitic material. It also explains why the C_{org} content is enhanced in the fine aleuritic fraction and not in the pelitic muds in the area of the deep ocean OM maxima. The decrease in the OM saturation of the pelitic fraction in the sediments of the ocean floor, as compared to those of the ocean shelf, feasibly explains why no relationship is obvious between the C_{org} content and the content of the <0.01 mm fraction in the sediments located beyond the limits of the submarine margin, and, in most instances, beyond the upper part of the continental slope. Thus it is conceivable that the above mentioned relationship may serve as an indicator of relatively shallow water and near-continental terrigenous facies in the Earth's history.

A direct correlation between the C_{org} content and the size of the <0.001 mm fraction has been observed only in discrete areas. A tendency towards such a correlation is apparent in the sediments within the lower parts of the continental slopes in the Pacific Ocean, where the environmental conditions are, seemingly, relatively homogeneous, which is most favorable for its manifestation.

In the carbonate sediments, both shallow- and deep-water, as a rule no relationship is obvious between the C_{org} content and the <0.01 mm fraction, although in certain local areas it may be distinctly reflected, as, for instance, in the shelf area off the Amazon River. In the carbonate sediments both maximum and minimum C_{org} values have been recorded in the sands, as well as in the pelitic muds. In the terrigenous sediments extreme values are of rare occurrence, being encountered only in the shallow water areas. There are a number of reasons for this. First of all, the biogenic carbonate material is of highly variable C_{org} content. For example, in the Atlantic Ocean, the C_{org} content of some sand and aleuritic facies, that consist predominantly of shell fragments of *Mytilus, Modiola* and *Mesodesma*, is higher than in the sands comprised of *Cardium, Pecten* and *Ostrea* fossil remains. The C_{org} content of sediments, consisting of corals, bryozoan, lithothamnium, is much higher than in sediments of similar size fractions comprised of mollusc detritus. Secondly, the carbonate matrix is in part comprised of terrigenous and more seldom, authigenous materials. The terrigenous materials in shallow water environments are more enriched in various forms of OM; in the pelagial areas, in comparison to the carbonate materials, they are poor in C_{org}, owing to the tighter bonding of the OM with the mineral base. As a consequence, the C_{org} content can vary considerably in equidimensional size-fractions of carbonate sediments, relative to the depositional environment of the sediments and the admixture of terrigenous OM. The C_{org} content in the carbonate sands of the Pacific and Atlantic Oceans usually shows an increase from the calciferous (30–50% $CaCO_3$) to calcareous (50–70% $CaCO_3$) and highly calcareous sediments (>70% $CaCO_3$). In the more fine-grained sediments an inverse relationship has been observed (Emelyanov and Romankevich 1977). The reason for this was stated above. In general, no correlation is obvious between the C_{org} content and the $CaCO_3$ content.

The biogenic siliceous material, 70% of which is produced by diatom algae, is associated with the OM in organo-mineral complexes. The plasms of the diatoms and radiolarians are of spontaneous destruction and undergo rapid utilization; only in shallow water areas, particularly in zones of strong upwellings, may they be retained within the shells that settle at the bottom. Organic-siliceous skeletons are more enduring to decomposition, and accumulate in areas where the influx of terrigenous and carbonate matter is low, whereas the silicon bioproductivity is

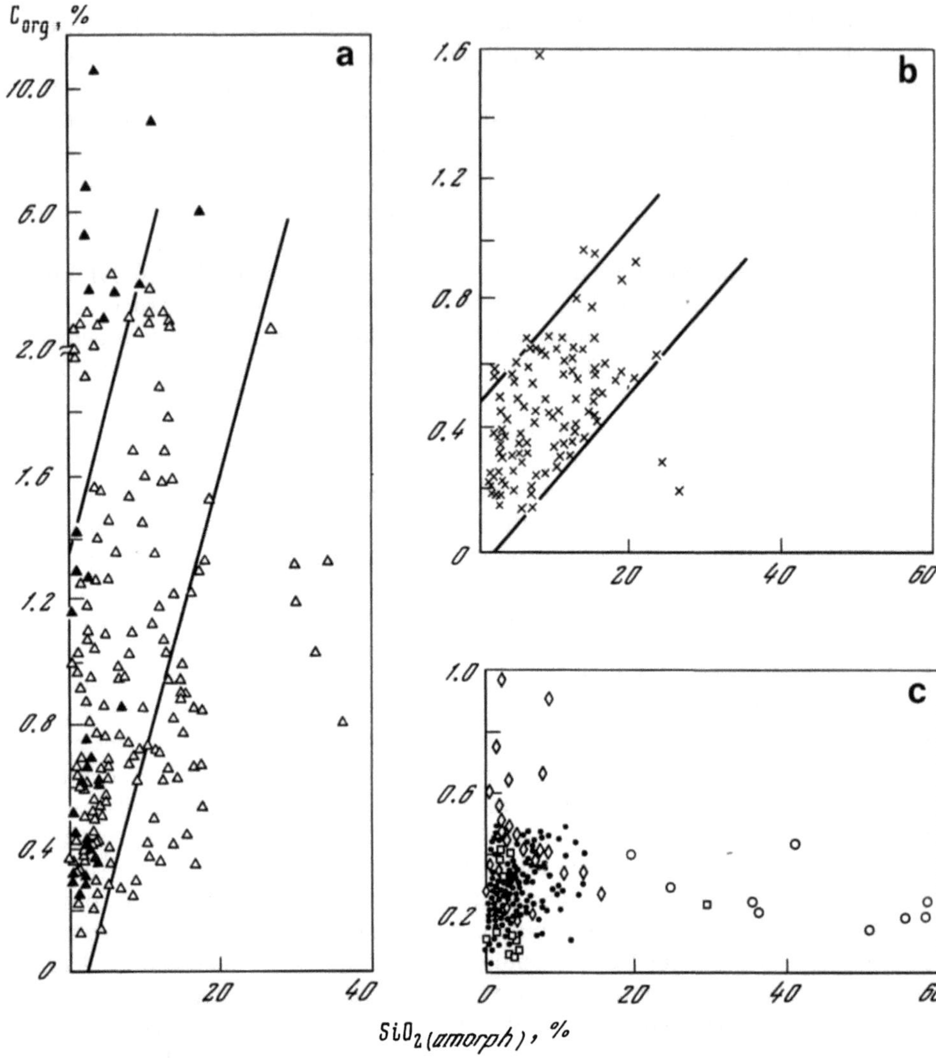

Fig. 38 a – c. The relationship between the C_{org} and $SiO_{2(amorph)}$ contents in the Pacific Ocean sediments. For explanation of symbols see Fig. 37

significant. This fact is reflected in the existence of Boreal, Antarctic and Equatorial belts of silicon-accumulation (Strakhov 1966, Lisitzin and Vinogradov 1982). The first two belts are restricted to the lower parts of the slopes and adjacent areas of the ocean floor. According to the regularities in the OM distribution in sediments, a direct relationship would be expected between the C_{org} and $SiO_{2(amorph)}$ in the sediments within the lower parts of the slopes and adjacent basin zones, i.e., in the area of the second C_{org} maximum. As a matter of fact, such a correlation is reflected precisely in this area (Fig. 38). It is more pronounced in the marginal parts of the ocean, where terrigenous hemipelagic slightly siliceous sediments occur (diatomaceous muds of the Boreal belt), and less distinctly expressed within the lower parts of the continental slopes. In the latter areas only a portion of the OM is directly bound with the siliceous material, whereas most of it occurs in complexes with the terrigenous material (plus aggregates and zooplanktonic fecal pellets).

Among the numerous factors that exercise an influence on the relationship between the C_{org} and $SiO_{2(amorph)}$, most predominant (besides the OM source) are hydrodynamic and trophical factors. The plotted C_{org} values against the $SiO_{2(amorph)}$ (see Fig. 38) reveal a steeper gradient of the curves in sediments of the submarine continental margin, than in the hemipelagic sediments. Hence, the amount of C_{org} for each $SiO_{2(amorph)}$ units is higher in the first case than in the second. This apparently reflects the general trend towards the pelagial realm in the more enhanced rate of degradation of OM, as compared to that of $SiO_{2(amorph)}$.

4.4 The Absolute Masses of Organic Carbon in the Sediments

4.4.1 A Concise History of the Problem

Theoretically, estimations of the amount of OM that accumulates in various parts of the ocean during a specific time-span, should not be difficult. The necessary parameters are the mean C_{org} values and average absolute masses of sedimentary materials that were deposited during a given period. Calculations of the absolute masses of sediments are based on the average rate of sedimentation, the moisture content and volumetric weight of the sediments, so that these parameters are also of essential importance in such estimations. As a result of the many variables (both syn- and diachronic factors) that are involved in these calculations, difficulties arise in obtaining reliable estimates of the absolute masses of sediments. For this reason until the 1970s only few valid estimates were made of the absolute masses of C_{org} in the seas and oceans.

Quantitative evaluation of the amount of OM that was deposited per unit time, was first made for the Black Sea gray clays (3.3 tons of C_{org} km^{-2} per year) by Arkhangelsky (1927), who compared these estimates with those obtained for the North Caucasus Maikopian clays (4.2 tons of C_{org} km^{-2} per year). The absolute mass-method, later proposed by Strakhov (1947), made possible the development of the first map showing the accumulation of OM in the Black Sea, and established the fact that the amount of OM that accumulated in these deposits over the last 2500 years varies in the range between less than 0.6 to over 3.0 g cm^{-2} (from less than 0.1 to over 0.6 g of C_{org} cm^{-2} per thousand years). The distribution pattern of the absolute masses of C_{org} and the sedimentary materials revealed a similarity in their localization: their restriction to the near-shore areas of the seas, the shelf and uppermost parts of continental slopes; whereas the maximum C_{org} percentage concentrations were confined to the halistatic regions of the sea (Strakhov 1960).

In the Caspian Sea the distribution pattern of the absolute C_{org} masses likewise differs from that of the C_{org} percentage concentrations (Bordovsky 1969). The absolute masses of C_{org} show an increase from the peripheral areas towards

Table 34. Average content and total amount of organic carbon deposited during the Holocene (10 thousand years) in various structural-geomorphological zones of the Ocean (Gershanovich et al. 1974)

Indices	Shelf	Continental slope and rise	Ocean floor
Area, km^2	26.7×10^6	26.7×10^6	257.0×10^6
Average thickness of Holocene layer, in m	0.8	1.5	0.05
Average C_{org} content, %	0.7	1.3	0.3
Amount of C_{org}, in tons	22.4×10^{10}	195.0×10^{10}	5.4×10^{10}
Total amount of C_{org}	222.8×10^{10} tons of C_{org} during 10 thousand years		

the lower parts of the slope and adjacent basinal areas, and thereafter a decline. The percentage concentrations, on the other hand, reveal a general trend to increase towards the central parts of the basin, where the C_{org} reaches the highest level.

In the Okhotsk Sea, Bezrukov (1960a) observed that the areas of enhanced C_{org} percentage concentrations coincide with those of high absolute masses of C_{org} in the sediments, and are restricted to the peripheral parts of the Sea.

Despite the differences in the localization of the absolute masses and percentage concentrations of C_{org}, a common feature of all three water bodies is the accumulation of the maximum quantities of C_{org} in those areas where the highest amount of sedimentary material was deposited. This implies that the percentage concentration is not the sole controlling factor in the formation of the absolute masses of C_{org} in the sediments.

Interesting data on the mean values and gross content of C_{org} in the Holocene sediments of the shelf, continental slope and floor of the world ocean were published in 1974 and are shown in Table 34 (Gershanovich et al. 1974). Calculations of the average thickness of the Holocene sedimentary layer were based on available data on the stratigraphy of the Upper Quarternary deposits in the Pacific and Atlantic Oceans, and were extrapolated for entire ocean. As the authors noted, the figures cited can be regarded only as approximate values. Nevertheless, they distinctly illustrate the fact that the amount of OM that accumulated in Recent sediments within the continental margin is tenfold greater than that in the sediments of the ocean floor.

Owing to the difficulties in making reliable estimations of the rate of oceanic sedimentation, the first valid data on the absolute masses of C_{org} in the pelagic areas of the ocean were obtained only at the end of the 1960s and the beginning of the 1970s (e.g., Romankevich 1968, 1970, Strakhov et al. 1973). As a result of the application of new nuclear geochronological techniques, biostratigraphical, paleomagnetic, as well as lithological and geochemical methods, several thousand age-datings are now available on world ocean sediments. Critical evaluation of these data was made by Lisitzin (1974). Lisitzin (1974) and Blazhchishin and Emelianov (1981) have published schematic maps showing the rate of sedimentation and distribution of absolute masses of sediments in the Oceans.

4.4.2 The Rate of Organic Carbon Accumulation in Bottom Sediments

On the basis of the data obtained in the P. P. Shirshov Institute of Oceanology of the USSR Academy of Sciences, as well as others from publications (Romankevich et al. 1966, Lisitzin 1974, 1980, Blazhchishin and Emelianov 1981), calculations were made of the average rate of accumulation of various types of Late Quarternary sediments in the Pacific and Atlantic Oceans, and also of the volumetric weight, moisture content and areal coverage of these sediments (Table 35). These data furnished the basis for the estimations of the accumulation of the absolute masses and gross content of C_{org} in the sediments of the Pacific and Atlantic Oceans, which were calculated from the following equations (Table 34).

$$V_{sed} = \frac{\bar{v}\bar{\Delta}(100 - \bar{w})}{100},$$

where V_{sed} is the rate of accumulation of the absolute masses of sediments (in terms of dry sediment weight, in g cm^{-2} per 1000 years); $\bar{v}$ = average rate of sedimentation for Recent and Late Quarternary sediments that accumulated in the course of 1000 years, as stated in centimeters of sediments of natural moisture content; $\bar{\Delta}$ = average volumetric weight of sediments of natural moisture, in g cm^{-3}; $\bar{w}$ = average moisture content, in % of sediment of natural moisture.

$$V_y = V_{sed}\bar{C}_{org}\,10^{-2},$$

where V_y is the rate of accumulation of the absolute masses of C_{org}, in g cm^{-3} per 1000 years; $\bar{C}_{org}$ = the mean value of C_{org}, in % of dry sediment.

$$A_y = V_y S,$$

where A_y is the total amount of C_{org} (in g), that accumulated in the sediments of the water body in the course of 1000 years; S = the area of the water body, in cm^2.

Comparison was made of the values obtained for different genetic types of sediments with similar values calculated independently for sediments of the continental margin and the ocean floor (the initial data are the same). The areal coverage of various structural-geomorphological zones, for which calculations were made of the average rate of sedimentation, C_{org} content, moisture content and volumetric weight of the upper sedimentary layer, in the Pacific Ocean were 20% (continental margin), 23% (marginal areas of the ocean floor), 57% (central part of the ocean floor), and in the Atlantic – 33% (continental margin) and 67% (ocean floor), relative to the total area of each Ocean.

It should be noted, that due to the irregular distribution of sediments in all structural-geomorphological zones of the ocean, an essential difficulty arose in the evaluation of the average rate of C_{org} accumulation (Bezrukov 1964a, 1973).

Table 35. Sedimentation rates and absolute masses of sedimentary material and C_{org} in the Late Quartenary Pacific and Atlantic sediments

Genetic type of sediments	Average rate of sedimentation, cm per 1000 years[a]	Average volumetric weight of sediments of natural moisture[b] g cm^{-3}	Average moisture content, %	Average C_{org} content, %	Rate of accumulation of absolute masses		Areal coverage of sediments, 10^3 km^2	Total amount of deposited matter, 10^9 tons per 1000 years	
					dry sedimentary materials, g cm^{-2} per 1000 years	C_{org}, mg · cm^{-2} per 1000 years		sedimentary	C_{org}
Pacific Ocean									
Terrigenous and iceberg sediments, including slightly siliceous and calcareous (<30% $CaCO_3$)	6.0	1.50	50	0.80	4.500	36.0	36164	1627.4	13.02
Terrigenous-calcareous and carbonate (>30% $CaCO_3$)	3.0	1.50	45	0.30	2.475	7.42	50301	1244.9	3.72
Siliceous diatomaceous of the Antarctic region (>30% $SiO_{2(amorph)}$)[c]	1.0	1.20	68	0.25	0.384	0.96	3041	11.7	0.03
Diatomaceous-radiolarian and radiolarian-foraminiferal	0.7	1.30	60	0.28	0.364	1.02	17096	62.2	0.17
Miopelagic (red) clays	0.4	1.30	65	0.29	0.182	0.53	28890[d]	52.6	0.15
Eupelagic (red) clays	0.1	1.50	55	0.22	0.068	0.15	28890[d]	19.6	0.04

Sediments of the									
– submarine continental margin	6	1.50	50	1.05	4.500	47.25	32876	1479.4	15.53
– marginal part of ocean floor	2	1.30	60	0.45	1.040	4.68	37808	393.2	1.77
– central part of ocean floor	0.3	1.50	50	0.27	0.225	0.61	93697	210.8	0.57
Total (for ocean excluding seas)	–	–	–	–	1.2 – 1.8[f]	10.65[e]	164381	~2000 – 3000	17.5
Atlantic Ocean									
Terrigenous and iceberg	10	1.50	50	0.7	7.50	52.50	27563	2067.2	14.46
Transitional calcareous-terrigenous (30 – 50% $CaCO_3$)	4	1.50	45	0.5	3.30	16.50	15705	518.3	2.59
Biogenic carbonate (>50% $CaCO_3$)	3	1.50	45	0.6	3.48	14.88	33883	840.3	5.04
Biogenic siliceous (>30% $SiO_{2(amorph)}$)	1	1.35	60	0.4	0.54	2.16	7281	39.3	0.16
Volcanoclastics (only the Iceland region)	6	1.60	50	0.9	4.80	43.20	250	12.0	0.11
Red clays	1	1.35	45	0.3	0.72	2.16	7284	52.4	0.16
Sediments of the									
– submarine continental margin	10	1.50	50	1.00	7.5	75.00	30384	2278.8	22.79
– ocean floor	2	1.50	50	0.37	1.5	5.55	61688	925.3	3.39
Total (for ocean excluding seas)	–	–	–	–	3.7[f]	26.5[e]	92072	3367	24.4

[a] Data abstracted from various sources on Recent and Late Quarternary sediments (amended from Lisitzin, 1974)
[b] Data based on volumetric weight measurements and derived from the relationship between the size-fraction and volumetric weight of the sediments
[c] Includes Tropical diatomaceous (Ethmodiscusids) ooze
[d] For miopelagic and eupelagic clays the estimates are based on the assumption of their equal areal coverage (totalling 35.11%)
[e] Calculated on the basis of general C_{org} mean value
[f] Calculated from the total mass of precipitated material and the summary area covered by sediments

The rate of accumulation of the upper sedimentary layer varies largely. Considerable bottom areas are wholly devoid of sediments (zero sedimentation rates), and in some places the deposited sediments are washed away (negative sedimentation rates). The latter include some submarine mountain tops and slopes of bottom rises, such as Shadsky rise in the Pacific. In the halistatic areas of the Tropical Pacific Ocean, where eupelagic clays accumulated, the sedimentation rates were found to be 0.01 – 0.02 cm per 1000 years, whereas in the sediments of the Okhotsk Sea and Californian Borderland the rate of sedimentation at times exceeds 50 – 180 cm per 1000 years (Bezrukov 1960a, Bezrukov and Romankevich 1970, Lisitzin 1974, 1980). Similar fluctuations in the rate of sedimentation were noted in the Atlantic Ocean, where it ranged between zero, within the vast areas of the North American shelf (Emery 1970), to 500 cm in the Gulf Paria, and even to 1000 cm per 1000 years in the avant-delta of the Mississippi (Lisitzin 1974). However, the rates of sedimentation in the major areas of the continental slope (central and lower parts) and continental rise of the Atlantic Ocean are characteristically of the order of 5 – 20 cm per 1000 years. The rate of sedimentation in the central part of the Atlantic Ocean is about 3 – 10 times greater than in the central part of the Pacific, which can be explained by the greater input of sedimentary materials from the land to the Atlantic Ocean.

The average rates of sediment accumulation in the Atlantic, as calculated by the author for the Late Quarternary period (about 3.7 g cm^{-2} per 1000 years), are 2 – 3 times as high as those in the Pacific (1.2 – 1.8 g cm^{-2} per 1000 years). Therefore the average absolute amount of sedimentary material annually deposited in the Atlantic in the Late Quarternary (around 3.4×10^9 t) is greater than in the Pacific ($2 - 3 \times 10^9$ t), despite the former ocean's lesser dimensions. Similar relative differences are characteristic for the average thickness of the whole sedimentary mass, excluding shelf areas and marginal seas (Lisitzin 1980): in the Atlantic (860 m) it is three times that of the Pacific (280 m), as well as for the overall volume of accumulated sediments (61.86×10^6 km^3 and 44.98×10^6 km^3, respectively).

The figures presented in Tables 34 and 35 should be regarded only as approximate values, owing to the marked changes in the thickness of the sediments even of similar genetic types and size-fractions, and the variations in the volumetric weight and moisture content of the sediments (the latter factors are of minor effect). Nevertheless, in the author's opinion, these figures duly reflect the general order of magnitude of the values.

The average rate of accumulation of absolute masses of C_{org} in Late Quarternary sediments of different types varies in the range between 0.0002 – 0.0005 to 0.0360 – 0.0523 g cm^{-2} per 1000 years. The minimal mean rates of C_{org} accumulation are typical of the red abyssal clays, while high rates are characteristic of terrigenous sediments. The eupelagic clays in the Pacific Ocean that were deposited at a distance from the land in the central parts of the ocean accumulated, characteristically, far more slowly than any of the sediments in the Atlantic Ocean. (It should again be stressed that the mean and not the individual values for different types of sediments are considered). The Late Quarternary sediments in the Pacific and Atlantic Oceans reveal a similar consecutive order of decrease in the rates of accumulation of the absolute masses of C_{org}: in the Pacific Ocean: terrigenous > terrigenous-carbonate + carbonate > carbonate diatomaceous (Antarctic region) + diatomaceous-radiolarian + radiolarian-foraminiferial > miopelagic (red) clays > eupelagic (red) clays; in the Atlantic Ocean: terrigenous > carbonate-terrigenous + carbonate > siliceous (Antarctic region) + red clays (miopelagic). The rate of C_{org} accumulation in the volcanogenic deposits in the area of Iceland is similar to that for the terrigenous deposits. In the Late Quarternary the average rate of accumulation of absolute masses of C_{org} was about 2.5 times greater in the Atlantic Ocean as compared to the Pacific Ocean (0.0265 and 0.0106 g cm^{-2} per 1000 years, respectively). In the Pacific Ocean the rates of accumulation of absolute masses of C_{org} in sediments of the continental margin, marginal and central parts of the ocean floor were 0.0472, 0.0047, 0.0006 g cm^{-2} per 1000 years, respectively. Similar figures were obtained for the continental margin and ocean floor of the Atlantic – 0.0750 and 0.0056 g cm^{-2} of C_{org} per 1000 years; i.e. these values were found to be greater than those for the continental margin and marginal parts of the ocean floor in the Pacific.

The total amount of C_{org} buried in the sediments of the Pacific and Atlantic Oceans (excluding the seas) was estimated for the Late Quarternary period at $(17.1-17.9)\times10^9$ tons of C_{org} and $(22.5-26.2)\times10^9$ tons C_{org} per 1000 years, respectively, averaging 17.5×10^9 and 24.4×10^9 tons of C_{org}, respectively[4]. In both oceans the major bulk of this material was deposited within the submarine continental margins: 15.5×10^9 tons of C_{org} per 1000 years in the Pacific Ocean, and 22.8×10^9 tons in the Atlantic. Assuming that the average rate of accumulation of the absolute masses of C_{org} in the Indian Ocean is the same as in the Atlantic (0.0265 g cm^{-2} per 1000 years), and in the Arctic Ocean (without seas) the same as in the Pacific (0.0106 g cm^{-2} per 1000 years), the total amount of C_{org} deposited in these oceans amounts to 21×10^9 tons and 0.5×10^9 tons per 1000 years, respectively. Summation of the above-cited figures shows that the amount of C_{org} that undergoes burial in the world ocean (without seas) estimates 63×10^9 tons per 1000 years.

No reliable estimates are as yet available on the average rates of accumulation of the absolute masses of sedimentary materials deposited in the seas. The database is insufficient for such calculations, due to the fact that the rates of sedimentation and physical properties of the sediments in the seas are subjected to greater variations than in the oceans (e.g., Lisitzin 1974). On the slopes and within the basinal parts of the seas, where the major portion of sedimentary matter is deposited, the rates of sedimentation per 1000 years range in the Okhotsk Sea from 2 to 55 cm and more, in the Bering Sea from 2 to 50 cm (mainly 3 – 20 cm), in the Mediterranean from several cm to 54 cm (mainly 3 – 20 cm); in the Red Sea the rates of sedimentation are usually 0.1 – 0.3 cm, in the Andaman mainly

4 The seas and their areas are listed in the work by Stepanov (1961). In the calculations presented above the Pacific Ocean proper includes the Coral, Fiji, Tasman, Rossa, Amundsen, Bellinshausen Seas and the Gulf of Alaska. The Atlantic Ocean proper includes the Carribean, North, Norwegian Seas, the Mexican Gulf, Bay of Biscay, Gulf of Guinea. The Indian Ocean proper includes the Arabian Sea, and Bay of Bengal (78.3×10^6 km^2 in area). Estimations for the Arctic Ocean (without seas) were made for the central part (5.03×10^6 km^2). The total area occupied by seas, estimates $\sim20\times10^6$ km

10 – 25 cm, but frequently up to 200 cm, in the Black Sea 1 – 40 cm, in the Azov about 240 cm, in the Caspian mainly 30 – 100 cm.

From analysis of the available data, the author derived the following mean values for the Late Quarternary sediments of seas: C_{org} content 1% (based chiefly on the data from Table 26); rate of sedimentation 10 – 20 cm per 1000 years; volumetric weight 1.50 g cm^{-3}; moisture content 50%. Calculations based on these data, show that the rate of accumulation of absolute masses of C_{org} is 0.188 g cm^{-2} per 1000 years, and the total C_{org} buried in the seas, amounts to 37×10^9 tons per 1000 years. In particular basins the rate of accumulation of absolute masses of C_{org} is far greater than these figures, as, for example in the Caspian Sea, where it reaches 1.1 g of C_{org} cm^{-2} per 1000 years (Bordovsky 1969). The areas of this sea is, however, small, whereby the contribution to the total C_{org} mass is not so great. Thus on an average $\sim 100 \times 10^9$ tons of C_{org} per 1000 years is buried in the world ocean sediments. About one third of this amount settles at the bottom of the seas.

4.4.3 The Amount of Organic Carbon and Its Fossilization Coefficients in Various Zones of the Ocean

The amount of OM that accumulated in the Late Quarternary sediments is estimated as follows (in 10^9 tons of C_{org} per 1000 years): in the area of the continental margin of the Pacific Ocean – 15.53, in the marginal and central parts of the ocean floor – 1.77 and 0.57, respectively; in the Atlantic Ocean – 22.79 (submarine continental margin), 3.39 (ocean floor) (Table 35). From this it is apparent, that in the Pacific Ocean 86.9%, 9.9% and 3.2% of the total C_{org} which is buried in the sediments is deposited within the areas of the continental margin, marginal and central parts of the ocean floor, respectively. Similar relative values were obtained for the Atlantic Ocean: 87% (continental margin), 13% (ocean floor).

According to other data (Table 34), 97.5% of the total C_{org} in the sediments is deposited within the continental margin of the world ocean; of this, 10% was deposited on the shelf, 87.5% on the continental slope and rise, whereas only 2.5% of the total C_{org} is deposited in the sediments of the ocean floor.

It is conceivable that over 85 – 90% of the total OM in the world ocean is deposited within the region of the submarine continental margin. The bulk of OM is buried in the sediments of the continental slopes. Hence, the circumcontinental localization of the absolute masses of OM in the Ocean is consistent with the distribution patterns of the masses of sedimentary matter and C_{org} concentrations (in %).

The C_{org} percentage concentrations and the rates of accumulation of its absolute masses are extremely irregular within the continental margin. In particular troughs on the shelf, continental slopes and at the bottom of the nearcontinental deep trenches, where natural settling basins occur and the sinking rate of POC maintains a high level, thick mud layers accumulate, which are characterized by high percentage concentrations and high absolute masses of C_{org}.

Very prominent in this respect in the ocean are the highly productive coastal upwellings, for instance, off the coast of South West Africa, where thick layers

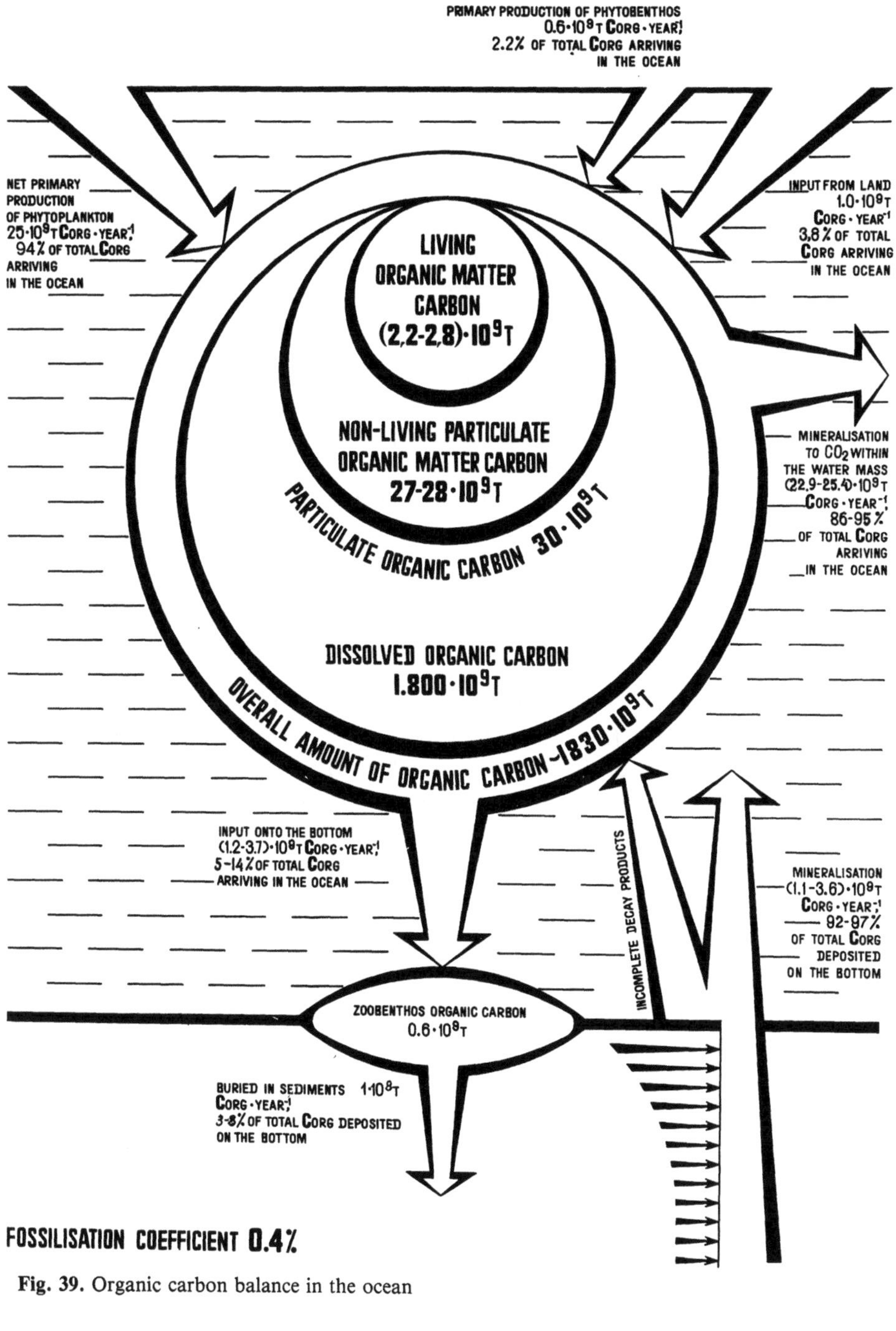

Fig. 39. Organic carbon balance in the ocean

of mud with a high planktonic OM content accumulate, characterized initially by high degrees of bituminization and simultaneous humification. Such facies are exceptionally favorable for the accumulation of large masses of OM which serve as the source of petroleum and natural gas, and are evidently of considerable petroleum-generating potential (Romankevich and Baturin 1974).

The annual photosynthesis productivity of the ocean and the OM input from the land total some 26.6×10^9 tons of C_{org} (see Chap. 2). Of this $1.2-3.7 \times 10^9$ tons of C_{org} per year settles at the bottom in particulate form, and 100 million tons of C_{org} per year are buried in the sediments (Romankevich and Liutsarev 1983). Thus, at the surface of the ocean floor and in the topmost layer of sediments about $1.1-3.6 \times 10^9$ tons of C_{org} per year undergo combustion in biochemical and chemical reactions, partial dissolution, and are contributed to the deep-sea OM reserves; this approximates 92–97% of the particulate OM that sinks to the ocean floor (Fig. 39). The OM fossilization coefficients, or the amount of buried OM, relative to the sum of primary productivity plus the input from the land, calculated for the ocean, as a whole, equals 0.40%. Thus approximately 0.4% of the energy created by photosynthesis is transfered each year to the bottom sediments.

The OM fossilization coefficients for the sediments of the inland seas, marginal seas and submarine continental margin are on average higher than the above-cited mean values; whereas in the sediments of the ocean floor they are lower. Considering that about 50% of the OM originates within the continental margin of the ocean, and 97% of the total undergoes burial in the sediments, the calculated mean values of fossilization coefficients are 0.74% (continental margin) and 0.02% (ocean floor), respectively. The OM fossilization coefficients, calculated on the basis of the above-cited values of photosynthetic productivity (see Table 6), and the data on the absolute masses of OM (see Table 34) were found to be 1.66% for the continental margin and 0.04% for the ocean floor (Gershanovich et al. 1974a).

No matter which figures are nearer to the actual fossilization coefficients, it is clear that the degree of equilibrium between productivity and decomposition of OM in the ocean water column and upper sedimentary layer is much higher in the region of the ocean floor than at the continental margin. Variations in this equilibrium are a result of the effects of biological, physico-geographical, lithological-geochemical processes, and the cause of the circumcontinental localization of the percentage concentrations and absolute masses of OM in the ocean sediments.

5. Nitrogen and Phosphorus in the Process of Sedimentogenesis

Nitrogen and phosphorus exist in the ocean in the form of both inorganic (mineral) and organic compounds, which may occur in dissolved and suspended state in the ocean waters, and in the sediments. Inorganic and organic forms of these elements are the most essential biogenic constituents of ocean waters, and being functionally involved in the processes of living organisms, are subjected to a rapid turnover.

These elements, like carbon, display rather intricate biogeochemical behavior in the ocean, owing to their ability to form a great variety of organic compounds, each differing in their reactivity.

The changes in the N_{org}/C_{org} and P/C_{org} ratios in the sequence plankton – surface water suspension – deep ocean suspension – bottom sediments are important characteristics that serve as implicit indicators of the trend of OM transformation in the ocean.

5.1 Nitrogen and the Nitrogen-Carbon Ratio

5.1.1 Nitrogen in Marine Organisms. Regeneration of Nitrogen

Nitrogen is an integral element, indispensable to all living organisms. It is incorporated in the structure of all proteins and nucleic acids and plays an important role in the metabolic processes of living organisms and their exchange with the surrounding environment; 70–85% and more of the N_{org} contained in phyto-, zooplankton and bacteria is associated with proteins.

The nitrogen content in contemporary marine organisms varies considerably (Table 36), and depends upon its content in the waters (this applies particularly to the phototrophic types and bacteria), nutritional conditions, the physiological state of living organisms, and also upon the relation between OM and the residual ash. The latter depends largely upon the phylogenetic level of development of the mineral skeleton of the test organism.

The N_{org} content in the phytoplankton amounts to 8.0–22.6 parts (on the average, ca. 16, Table 22) per 100 parts of C_{org}. The phytobenthos OM is poorer in nitrogenous substances (C:N 100:5.3), whereas the crustacean plankton is richer (C:N 100:23, see Table 22). The average weight ratio C:N:P for the marine plankton may be assumed to be 100:20:3, and in atomic terms 100:17:1. The latter relationship is close to that given by Redfield, Ketchum and Richards (1963), namely, C:N:P 106:16:1.

Table 36. Content of nitrogen and phosphorus in contemporary marine plants and animals (%, on a dry weight basis)

Species studied	N	P	Source
Contemporary marine plants			
Diatoms	(2.5)	(0.6)	Vinogradov (1953)
	2.3	0.3	this volume
	2.7 – 5.9(4.6)	0.4 – 2.0(1.4)	Parsons et al. (1961)
	4.5	0.7	Harris and Riley (1956, quoted by Strickland 1965)
	5.4	0.8	Ketchum and Redfield (1949, quoted by Strickland 1965)
	2.4 – 4.7(3.0)	–	Krey (1958)
	1.6 – 4.8(3.1)	–	Vinogradova (1967b)
Peridineans	2.5 – 7.2(4.5)	(0.6)	Vinogradov (1953)
	4.5 – 5.0	1.1 – 1.3	Parsons et al. (1961)
	3.1 – 5.3(4.2)	–	Krey (1958)
	7.8	–	Vinogradova (1967b)
Golden algae	7.8 – 9.0	1.2 – 3.0	Parsons et al. (1961)
Blue-green algae	(8.0)	(1.0)	Vinogradov (1953)
	5.8	1.4	Parsons et al. (1961)
	4.3	–	Fay and Fogg (1962, quoted by Strickland 1965)
	5.1 – 5.8(5.3)	–	Vinogradova (1967b)
Green algae	up to 8.0	0.5	Vinogradov (1953)
	8.3 – 9.1	3.3	Parsons et al. (1961)
	2.4 – 7.1(5.4)	0.9 – 2.7(2.0)	Ketchum and Redfield (1949, quoted by Strickland 1965)
	0.9 – 5.0	–	Kizevetter et al. (1981)
Red algae	1 – 6(2.5)	(0.3)	Vinogradov (1953), Kizevetter et al. (1981)
Brown algae	(1.8)	(0.3)	Vinogradov (1953), Kizevetter et al. (1981)
Bacteria	(10.0)	(2.0)	Vinogradov (1953), Kizevetter et al. (1981)
Contemporary marine animals			
Copepods	7.0 – 10.4(8.4)	–	Krey (1958)
	8.2 – 11.2(9.6)	0.7 – 0.9(0.8)	Beers (1966)
	8.4 – 10.4(9.7)	–	Vinogradova (1967b)
	5.1 – 13.1	–	Omori (1969)
Euphausiids-mysiids	9.4 – 10.5(10.0)	1.4 – 1.6(1.5)	Beers (1966)
	10.0 – 10.7	–	Omori (1969)
Chaetognathae	6.3 – 9.4(7.8)	0.5 – 0.7(0.6)	Beers (1966)
	10.7 – 11.1	–	Omori (1969)
Fishes and their larvae	8.3 – 10.6(9.7)	0.9 – 1.8(1.4)	Beers (1966)
Polychaets	4.4 – 11.2(8.9)	0.4 – 1.8(1.0)	Beers (1966)
Pteropods	2.7 – 4.2(3.2)	0.2 – 0.4(0.3)	Beers (1966)
	1.5 – 6.0	–	Omori (1969)
Fish meat	14.3 – 15.0	0.7 – 3.2	Kizevetter (1973)

The figures in parenthesis show mean values

Nitrogen is involved in biogeochemical cycles. The biological cycle of nitrogen involves assimilation of its organic and inorganic compounds, including gaseous N_2, its transfer along food-chains, its liberation resulting from excretion by living organisms and the breakdown of the cellular substance. In recent years, the biogeochemical cycle of nitrogen and phosphorus is subject to intense anthropogenic influence. It is believed that by the year 2000 the production of nitrogenous fertilizers will exceed 120×10^6 t per annum. This will have a considerable impact on the quantity of different forms of nitrogen reaching the ocean.

The nitrogen uptake by phytoplankton in the world-ocean approximates 4×10^9 tons per year (calculated on the basis of a primary productivity of 25×10^9 tons of C_{org} and N/C = 16%). Most of the N-containing OM is readily hydrolyzable under alkaline conditions in the sea waters and bottom sediments, both with and without hydrolytic enzymes of the protease type. Besides planktonic and benthonic animals, aerobic and anaerobic bacteria have a paramount role in the mineralization of proteins (e.g., Kuznetsov 1970, Rheinheimer 1973, Sorokin 1982). Among the organisms that cause the breakdown of proteins are fungi and actinomycets containing proteolytic enzymes. Enzymic hydrolysis of proteins results in the formation of peptides, amino acids, purine and pyrimidinal bases and urea. During complete OM mineralization nitrogen is liberated in the form of ammonium salts, which undergo further oxidation to free nitrogen, nitrite- and nitrate ions. The conversion ammonium into nitrites and nitrates is due chiefly to bacterial activity, but part of the nitrites is released by phytoplankton as extracellular products during the assimilation of nitrates (Kuznetsov 1970, Parsons and Takahashi 1977). In the process of feeding of the zooplankton, a certain amount of nitrogen is liberated in the form of ammonium salts, also in forms of various soluble organic compounds. For instance, the actively feeding calanuses utilize in their growth only 26.8% of N_{org}, whereas the major portion of nitrogen is liberated with the feces (37.5%) and in the form of various soluble compounds (35.7%), that are contributed to the reserve products of the particulate and dissolved OM in the ocean waters (Butler et al. 1970, quoted by Parsons and Takahashi 1977).

5.1.2 Organic Nitrogen in Sea Waters and the Particulate Fraction

Like the C_{org}, most of the N_{org} in sea waters, occurs in dissolved forms. In the euphotic zone the amount of dissolved nitrogen (N-DOM) varies from tens to hundreds, less frequently, several thousand micrograms per litre (Table 37). The typical range of fluctuation for the photic layer, embracing the major part of values, is 30 – 300 $\mu g\ l^{-1}$. The N-DOM/C-DOM ratios in seas and oceans are in the order of 4 – 57%. In some areas the values vary in a more narrow range; in the Santa Catalina Basin the N-DOM/C-DOM ratio within the interval of 0 – 1300 m is 6 – 16.5%; in the Peruvian area – 5.7 – 18.1% (Holm-Hansen et al. 1966, Williams 1967). Most of the mean values of the N-DOM/C-DOM ratios are 8 – 14%. According to the scarce available data, the ratio shows a tendency to decrease in the vertical profile, although no general systematic decrease with depth has been observed.

Table 37. Organic nitrogen and phosphorus and their ratios to organic carbon in the waters and particulate fraction of seas and oceans

Sampling area	Depth, m	N_{org}	P_{org}	N_{org}/C_{org}	P_{org}/C_{org}	Source
		μg l^{-1}		weight ratios (%)		
Sea waters						
Western Tropical Pacific	0 – 1000	–	8 – 23 (15)	–	0.4 – 1.3 (0.8)	Romankevich (1979)
Sea of Japan	0 – 2000	–	0 – 42 (14)	–	0 – 3.1 (0.9)	Liutsarev and Romankevich (1975)
Eastern Equatorial Pacific	0 – 3000	–	(4.5 – 6)	–	(0.4 – 0.7)	Liutsarev et al. (1975)
North-eastern Pacific	0 – >1000	(60 – 80)	(6 – 8)	–	–	Armstrong et al. (1966, quoted from Williams 1975)
Area off California	Film	–	–	5.9 – 10.0 (7.8)	0.7 – 1.1 (0.8)	Williams (1967)
	10	–	–	8.5 – 10.4 (9.2)	0.6 – 3.4 (1.6)	Williams (1967)
Mediterranean Sea	0 – >2000	(45 – 70)	(1.3 – 2.2)	(11.5 – 25)	(0.3 – 0.7)	Banoub and Williams (1972, quoted from Williams 1975)
Area off Peru	Film	–	–	9.7 – 18.1 (13.0)	0.5 – 1.0 (0.8)	Williams (1967)
	20 – 35	–	–	5.7 – 11.6 (9.2)	0.8 – 2.5 (1.4)	Williams (1967)
	>200	–	–	(6.7)	–	Williams (1967)
	0 – 50	>100 – 300	–	–	–	Sviridova (1980)
Santa Catalina Basin	0 – 1300	~25 – 110	~2 – 12	(6 – 16.5)	(0.8 – 1.6)	Holm-Hansen et al. (1966)
San Diego Basin	0 – 50	(60)	(7)	(7.6)	(0.9)	Williams (1969)
	100 – 500	(16)	(7)	(4.3)	(1.9)	
High-productive Eastern Pacific	0 – 100	154 – 308 (210)	9.6 – 19.2 (16)	–	–	Ivanenkov (1979)
	100 – 200	140 – 308 (168)	6.4 – 19.2 (12.8)	–	–	Ivanenkov (1979)
	200 – 500	112 – 210 (140)	3.2 – 16.0 (9.6)	–	–	Ivanenkov (1979)
	500 – 1000	84 – 168 (126)	3.2 – 12.8 (6.4)	–	–	Ivanenkov (1979)
	1000 – 2000	56 – 140 (98)	1.6 – 4.8 (3.2)	–	–	Ivanenkov (1979)
	2000 – 4000	56 – 84 (70)	traces-3.2 (1.6)	–	–	Ivanenkov (1979)

South of Greenland	>200	–	–	33 – 50(37)	–	Duursma (1965)
High-productive Eastern Tropical Atlantic	0 – 100	84 – 364(164)	6.4 – 28.8(16)	–	–	Ivanenkov (1979)
	100 – 200	84 – 280(126)	6.4 – 22.4(9.6)	–	–	Ivanenkov (1979)
	200 – 500	84 – 168(112)	3.2 – 9.6(6.4)	–	–	Ivanenkov (1979)
	500 – 1000	42 – 98(84)	1.6 – 4.8(3.2)	–	–	Ivanenkov (1979)
	1000 – 2000	42 – 98(70)	0.1 – 3.2(1.6)	–	–	Ivanenkov (1979)
	2000 – 4000	56 – 84(70)	traces-1.6(1.6)	–	–	Ivanenkov (1979)
White Sea	35 – 100	(310 ± 20)	7 – 19	(8.3)	–	Datsko (1939, quoted from Datsko 1959)
Greenland Sea	0 – 750	(220 ± 20)	–	(10.5)	–	Datsko (1939, quoted from Datsko 1959)
Azov Sea	0 – 10	(620)	28 – 52	18.0 – 20.3(18.3)	–	Datsko (1959)
Black Sea	0 – 300	(230)	6 – 41(16)	8.8 – 11.7(9.9)	–	Datsko (1959)
	10 – 2000	(180 – 248)	(2.5 – 15)	–	–	Skopintsev (1975)
	10 – 2000	–	(4.34)	–	–	Skopintsev (1975)
Caspian Sea	25 – 940	(490)	up to 30	8.3 – 9.2(8.8)	–	Datsko (1959)
Suspended Matter						
North-eastern Pacific	15 – 65	(35.0)	–	15.3 – 18.5(16.9)	–	Parsons and Strickland (1962)
	1500 – 3000	(11.8)	–	7.4 – 49.0(29.6)	–	Parsons and Strickland (1962)
Tasmanian Sea	0 – 4210	37.4	–	14.1 – 71.4(30.3)	–	Dal Pont and Newell (1963)
Santa Catalina Basin	0 – 75	~4 – >24	~0.8 – 3.9	(18)	(3)	Holm-Hansen and Strickland (1966) Williams (1966)
	75 – 450	3 – 5.5	~0.6 – 0.9	(9 – 11)	(1.7 – 2)	Holm-Hansen and Strickland (1966) Williams (1966)
	~450 – 1300	2 – 7	~0.6 – 1.2	(7 – 10)	(1.5 – 4)	Holm-Hansen and Strickland (1966) Williams (1966)

Table 37 (continued)

Sampling area	Depth, m	N_{org}	P_{org}	N_{org}/C_{org}	P_{org}/C_{org}	Source
		$\mu g\ l^{-1}$		weight ratios (%)		
World-Oceans	0–7	–	0.1–8.7	–	2–125	Bogdanov et al. (1971)
Area of California	Film	–	–	10.5–17.0(13.4)	1.2–4.1(2.5)	Williams (1967)
	>200	–	–	(7.7)	–	Williams (1967)
Area of Peru	Film	–	–	14.9–33.6(27.0)	<0.1–3.4	Williams (1967)
	>200	–	–	(12.5)	–	Williams (1967)
Area of California	0	(4.1)	–	11.8	–	Holm-Hansen (1969)
	200– ~1000	(0.6)	–	(~8)	–	Holm-Hansen (1969)
Gulf of California	Euphotic layer	(23.4)	–	(12.2)	–	Zeitschel (1970)
North-western Pacific	0–150	–	–	10–18	–	Handa (1970)
	200–1000	–	–	4.8–9	–	Handa (1970)
	0–4000	0–65.0	–	2–50	–	Nakajima (1973)
Western sector of the Sea of Japan	0–200	0.5–6.5	–	12–20	–	Ichikawa (1982)
	>200	~1	–	5–20	–	Ichikawa (1982)
Eastern sector of the Sea of Japan	0–200	2.0–6.5	–	7–14[a]	–	Ichikawa (1982)
	>200	0.5–2.5	–	3.6–6.7[b]	–	Ichikawa (1982)
South China Sea (typical station)	0–300	<1–9	–	6–12[c]	–	Ichikawa (1982)
	>300	0–2.3	–	5–10[d]	–	Ichikawa (1982)
Sulu Sea	0–200	2–18	–	–	–	Ichikawa (1982)
	>200	0–9	–	–	–	Ichikawa (1982)
Celebes Sea	0–200	3–5	–	7–14	–	Ichikawa (1982)
	>200	1–3	–	6–14	–	Ichikawa (1982)
Western Equatorial-Tropical Pacific	0–100	–	0.58	–	0.3–2.7(1.4)	Finenko and Ostapenia (1971)
	1000–3000	–	0.21	–	0.7–1.2(0.8)	Finenko and Ostapenia (1971)
North of the Hawaiian Islands	0–250	(2.04)	–	(13.3)	–	Gordon (1971)
	1000–2000	(0.21)	–	(8.1)	–	Gordon (1971)
	3000–4000	(0.19)	–	(6.6)	–	Gordon (1971)

Sea of Japan	0 – 1000	(2.7)	(0.23)	11 – 16 (14)	0.9 – 1.8 (1.1)	Liutsarev and Romankevich (1975)
Bering Sea	Coastal waters	6 – 79	–	6.6 – 33.3 (14.3)	–	Tanoue and Handa (1979)
	Deep waters	1 – 26	–	5 – 20 (9)	–	Tanoue and Handa (1979)
	Kuril current	1 – 85	–	6.6 – 33.3 (14.3)	–	Tanoue and Handa (1979)
South-western Atlantic	0 – 200	–	–	(~14.1)	–	Hobson and Menzel (1969)
	200 – 5200	–	–	(~9.1)	–	Hobson and Menzel (1969)
Baltic Sea	1	–	–	(12 ± 0.9)	–	Ehrhardt (1969)
	230	–	–	(10.8 ± 1.3)	–	Ehrhardt (1969)
	0 – 230	–	0.4 – 10.6	–	–	Jurkovski and Fominykh (1980)
Indian Ocean	0 – 7	–	0.1 – 8.7	–	(3)	Lisitzin (1964)
Northern Atlantic	–	–	–	6.5 – 12.8	–	Gordon (1970b)
Area of South-western Africa	0	–	–	14.5 – 20.0 (17.3)	–	Hobson (1971)
Java-Freemantle profile	0 – 100	(1.64)	–	(10.2)	–	Newell and Kerr (1968)
	1100 – 5000	–	–	(8.3)	–	Newell and Kerr (1968)
Trophical and subtropical part of the Indian Ocean	Surface waters	4 – 8	0.6 – 1.0	(~15)	(~2)	Copin-Montegut and Copin-Montegut (1978)
Indian Ocean, Antarctic region	Surface waters	11 – 14	1.8 – 2.8	(~20)	(~3.5)	Copin-Montegut and Copin-Montegut (1978)
	Deep waters	0.2 – 0.3	0.03	(~19)	(~2)	Copin-Montegut and Copin-Montegut (1978)

Figures in parenthesis show mean values
[a] For depth interval 0 – 450 m. [b] For depth interval ~450 – 2700 m. [c] For depth interval 0 – 150 m. [d] >150 m

The particulate organic nitrogen N-POM, like the N-DOM, is subjected to large fluctuations; it varies from analytical zero to several hundred micrograms per litre. In the euphotic layer the N-POM content is in the range from a few micrograms to tens, or less frequently hundreds of micrograms per litre, and in deeper waters from analytical zero to several micrograms per litre (see Table 36). In the Pacific Ocean, according to the data obtained by the author, the N-POM content varies from 0.1 to 83 $\mu g\, l^{-1}$. The maximum concentrations are typical of the shallow waters, and of surface waters in areas distant from the land. The N-POM decreases in the vertical profile, but below 300 – 500 m only slight changes are apparent (Fig. 40; Holm-Hansen et al. 1966, Nakajima 1973, Gordon et al. 1979, Ichikawa 1982). However, at certain stations large fluctuations were noted

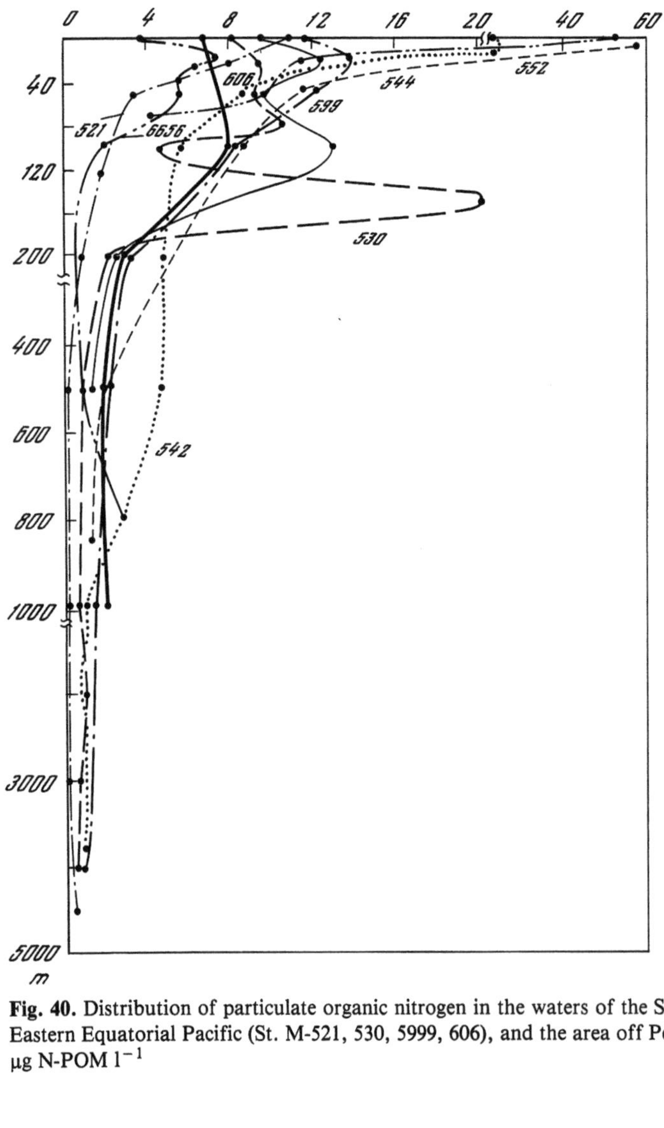

Fig. 40. Distribution of particulate organic nitrogen in the waters of the Sea of Japan (St. V-6656). Eastern Equatorial Pacific (St. M-521, 530, 5999, 606), and the area off Peru (St. M-542, 544, 552), μg N-POM l^{-1}

in the N-POM content at depths below 500 – 1000 m, for instance, in the waters of the Oyashio current in the northern part of the Pacific Ocean (Nakajima 1973, Ichikawa 1982). For some regions, the distribution curves of POC, ATP and particle volume have been found to have a symbatic shape (e.g., Gordon et al. 1979).

In the suspension where diatoms, peridineans and copepods predominante, the N-POM mean values, calculated on a dry sediment basis, were found to be 2.41 – 3.14% (Table 38; amended from Romankevich 1974b). A positive and commonly high degree of correlation is apparent between the C-POM and N-POM contents in the surface and deep water suspension. This was established through the seasonal observations carried out in the North Atlantic (Menzel and Ryther 1964), as well as in the Pacific (Nakajima 1973, Romankevich 1977, 1979, Ichikawa 1982). Investigations carried out in the Pacific Ocean showed that the particulate matter in the surface waters inherits the N-POM/C-POM ratios that reflect the make-up of many marine organisms contained in the suspension. According to mean values, the maximum N-POM/C-POM ratios occur characteristically in the suspension where the planktonic Crustacean remains are predominant (21.3%); the diatom suspension reveals smaller ratios (19.0%) and the peridinean the lowest (16.3%) (see Table 38). This can be ascribed mainly to

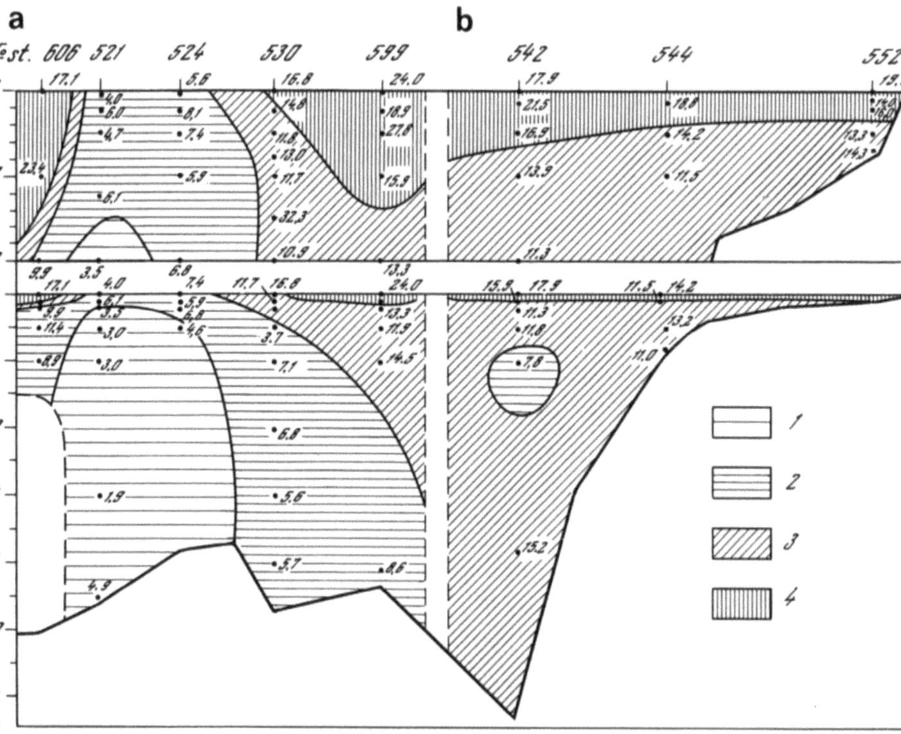

Fig. 41 a, b. Content of particulate organic nitrogen in the Eastern Equatorial-Tropical Region of the Pacific Ocean (**a**) and in the area off Peru (**b**). N-POM/C-POM (weight ratios), in %; *1* = <5; *2* = 5 – 10; *3* = 10 – 15; *4* = >15

Table 38. Biogeochemical composition of surface water suspended material in the north-western Pacific[a]

Station number	C_{org}	N_{org}	P	SiO_2 (amorph)	N/C	P/C	Si/C	Protein equivalent of nitrogen	Protein-like compounds	CHC	Lipids[b]	HA	ASS	WSP	WISP
	%, on a dry solid basis				weight ratios, %			% of POC[c]					% of total CHC		
Diatom suspension (areas off the Kuril Islands and Peru)															
39-1	16.8	3.71	0.53	–	22.1	3.2	–	69.0	–	14.3	–	5.3	13.0	8.4	78.6
39-2	21.51	4.33	0.50	–	21.3	2.3	–	62.9	–	7.2	14.6	5.0	–	–	–
39-3	12.37	2.21	0.24	16.20	17.8	1.9	131	55.8	–	12.6	16.5	4.3	–	–	–
K-298[d]	15.65	2.33	0.34	25.80	14.9	2.2	165	46.5	–	14.5	–	–	–	–	–
Average	16.58	3.14	0.40	21.00	19.0	2.4	148	58.6	–	12.2	15.6	4.8	–	–	–
Peridinean suspension (north-western part of Pacific Ocean)															
46-1, 2	17.61	3.38	0.38	2.70	19.1	2.1	15	59.8	45.4	11.9	22.6	4.3	14.8	13.6	66.7
46-3	16.67	2.64	0.38	1.83	15.8	2.3	11	49.6	34.2	15.7	14.7	5.1	7.3	12.4	77.9
46-4	13.05	1.67	0.20	0.87	12.8	1.5	6.7	40.2	14.3	17.1		–	5.3	11.3	80.9
46-5	15.11	2.49	0.28	1.59	16.5	1.9	10.5	51.5	24.3	14.8	17.9	–	5.4	13.1	79.1
46-6	15.79	2.30	0.29	1.55	14.6	1.9	10	45.6	23.8	15.5	17.3	–	7.2	11.1	77.4
46-7	18.73	3.55	0.41	2.22	19.0	2.2	12	59.3	32.1	13.1		–	7.8	13.7	72.6
Average	16.16	2.67	0.32	1.79	16.3	2.0	11	51.0	29.0	14.7	18.1	4.7	8.0	12.5	75.8
Sestone with a predominance of crustacean plankton (north-western part of Ocean)															
V-6170	10.81	2.41	0.23	–	21.3	2.1	–	69.7	69.4	4.6	–	–	9.0	8.1	76.7

[a] All samples were taken by separators, with the exception of the zooplankton (St. V-6170) and suspended matter from off the coast of Peru (St. K-298), which were screen samples

[b] Mixed chloroform-methanol (2:1) extraction at 25 °C

[c] Carbon content taken equal to 50% in proteins, 40% in carbohydrates, 75% in lipids, and 55% in humic acids

[d] Sestone, consisting practically entirely of diatoms of *Chaetoceros* sp

the protein content, which is greater in the zooplankton (~60–70% of the OM) than in the phytoplankton (45–60% of the OM).

In the sea water suspensions the N-POM/C-POM ratios vary mostly in the range between 2 to 50%. In the euphotic layer, judging by the mean values for various regions, the ratios vary in a narrower range of 10–20%. The N-POM/C-POM ratios, taking their mean values, decrease in the vertical profile of the water column, which suggests that the degradation of N-organic compounds is relatively enhanced. This was found to be the case in many regions of the Pacific, Atlantic and Indian Oceans (Table 37, Fig. 41), and is a characteristic feature of the degradation of particulate OM in natural waters. At certain stations the N-POM/C-POM ratios were variable, and the weighted average values within some layers differed considerably (Fig. 41). These discrepancies in the P-POM/C-POM ratios can only partly be ascribed to errors in the method of determination of N-POM and C-POM content. An alternative explanation is that the variations in the N-POM/C-POM ratios reflect differences in the composition of intermediate products that are formed during the complex OM degradation.

According to the data obtained through our investigations in the southeastern Pacific, the POM in the area of the continental slope (below 200 m) is richer in nitrogenous compounds (the N-POM/C-POM ratios in the 200–1000 m average 12.3% while in layers below 1000 m, they average 10.2%) than at the same depths on the ocean floor (7.5 and 7.1%). In the surface layers of the ocean (0–100 m) these differences in the N-POM/C-POM ratios become narrower: shelf – 15.3%, continental slope – 16.1%, ocean floor – 13.5%.

5.1.3 Nitrogen in Bottom Sediments and Interstitial Waters

The N_{org} content of bottom sediments in seas and oceans varies from less than 0.01 to 1.22% (Table 39). A direct correlation is obvious between the N_{org} and C_{org} contents in the sediments, which is most pronounced in the areas of the submarine margin of continents and marginal parts of the ocean floor. According to the data available from the Puget Sound and Gulf of Maine, the C_{org} and N_{org} in the sediments reveal a linear logarithmical correlation, approaching a functional relationship (Bader 1955). Beyond the limits of the continental margin the dependency between these values decreases considerably. In the halistatic areas of the deep ocean basins this relationship is not reflected (Romankevich 1974a, 1977).

On the basis of the data obtained in the Institute of Oceanology of the USSR Academy of Sciences, and other data available from earlier publications (e.g., Trask 1932, Lisitzin 1966), a map was developed, showing the distribution of N_{org} in the sediments of the Pacific Ocean (Fig. 42). Comparison of this map with that of the C_{org} (see Fig. 30), reveals closely similar features, as well as certain differences in their distribution patterns. As in the case of the C_{org}, the N_{org} content increases towards the coastal areas, particularly in the zones of strong upwellings of deep waters. The differences in the N_{org} distribution are reflected in the relative enhancement of the N_{org} concentrations in the red abyssal clays of the tropical areas in the Pacific Ocean (predominantly, in the eupelagic types), that were deposited under conditions of extremely low rate of sedimentation.

Table 39. Organic nitrogen content and nitrogen-carbon ratios in sediments of seas and oceans[a]

Sampling area, object studied	N_{org} (%) on a dry solid basis	N_{org}/C_{org}, %	Source
Sediments in seas and oceans			
Pacific Ocean, (pelitic muds)			
submarine continental margin	0.05 – 0.94 (0.34)	6.3 – 23.7 (11.9)	this volume
marginal part of ocean floor	0.04 – 0.28 (0.13)	13.0 – 23.0 (17.2)	this volume
central part of ocean floor	0.006 – 0.12 (0.06)	2.2 – 46.2 (13.9)	this volume
Pacific Ocean (aleuritic muds)			
submarine continental margin	0.05 – 0.48 (0.15)	8.4 – 20.7 (12.2)	this volume
marginal part of ocean floor	0.06 – 0.10 (0.07)	13.9 – 17.8 (16.0)	this volume
central part of ocean floor	0.01 – 0.05 (0.03)	3.0 – 8.5 (5.7)[b]	this volume
Okhotsk Sea	–	11.4 – 14.9 (13.0)	Weber et al. (1956)
Bering Sea	0.028 – 0.34	4.5 – 125 (11.4)	Lisitzin (1966)
Sea of Japan	0.10 – 0.39 (0.21)	13.7 – 32.2 (19.2)	Ohta (1953)
Area of Hokkaido	0.05 – 0.81 (0.21)	5.2 – 52.6 (10.6)	Kato (1956)
	0.03 – 0.23 (0.13)	4.2 – 20.8 (10.0)	Kato (1956)
Gulf of Ishikari	0.04 – 0.40 (0.13)	4.0 – 83.3 (12.2)	Kato (1956)
Pacific Ocean, area of Hokaido	0.03 – 0.22 (0.11)	4.0 – 47.6 (10.9)	Kato (1956)
Japan Trench	–	5.9 – 19.2	Kawahara et al. (1966)
Kuril-Kamchatka Trench	0.04 – 0.14 (0.06)	9.9 – 16.1 (12.5)	Romankevich and Artemiev (1969)
Bering Sea	0.01 – 0.2	11 – 16	Tanoue and Handa (1980)
Area of California			
0 – 10 m	–	(12.5)	Emery (1960)
0 – 35 m	–	(8.9)	Emery and Rittenberg (1952)
275 – 325 m	–	(7.7)	Emery and Rittenberg (1952)
Gulf of California	–	5.5 – 16.4 (11.4)	Van Andel (1964)
Tropical Pacific	–	(7.4)	Arrhenius (1952), Olausson (1960)
Western Africa (shelf, slope)			
– humid zone	0.04 – 0.26 (0.15)	6.2 – 10.6 (9.1)	this volume
– arid zone			
– sand	0.11 – 0.84 (0.32)	7.1 – 12.3 (10.0)	this volume
– aleurites	0.07 – 0.42 (0.21)	8.4 – 12.3 (10.6)	this volume
– pelites	0.40 – 1.22 (0.73)	9.4 – 12.2 (10.6)	this volume
Falkland Islands	0.01 – 0.09 (0.04)	5.5 – 15.0 (12.2)	this volume

Table 39 (continued)

Sampling area, object studied	N_{org} (%) on a dry solid basis	N_{org}/C_{org}, %	Source
Maine and Puget Sounds	0.19	3.9 – 40.0	Bader (1955)
Black Sea	–	5.9 – 12.5 (8.3)	Starikova (1961)
Azov Sea	0.11 – 0.28 –	11.8 – 13.2 (12.5) 9.1 – 11.1	Starikova (1961) Gorshkova (1970)
Caspian Sea	–	7.9 – 33.3 (14.7)	Weber et al. (1956)
Northern Caspian	0.02 – 0.46 (0.14)	7.0 – 25.6 (14.9)	Pakhomova (1961)
Central Caspian	0.03 – 0.48 (0.25)	11.4 – 30.3 (15.6)	Pakhomova (1961)
Southern Caspian	0.09 – 0.52 (0.26)	12.5 – 25.0 (14.9)	Pakhomova (1961)
Norwegian Sea	0.04 – 0.22 (0.12)	4.1 – 16.4 (11.1)	Jocteur-Monrozier (1977)
Region of Cap	0.02 – 0.39 (0.12)	5 – 17 (12.7)	Jocteur-Monrozier and Jeanson (1979)
Gulf of Aden, Oman Sea	0.19 – 0.97 (0.43)	11.5 – 17.8 (14.5)	Jocteur-Monrozier and Jeanson (1981)
Barents Sea	–	11.55 – 20.0 (14.3)	Gorshkova (1975)
Norwegian Sea	–	12.5 – 20.0 (15.4)	Gorshkova (1962)
East of Long Island	0.08 – 0.13 (0.11)	13.3 – 16.4 (14.7)	Emery and Ross (1968)
Mexican Gulf			
– shelf	(0.063)	5.0 – 25.0 (11.8)	Trask (1953)
– slope	(0.119)	5.9 – 25.0 (11.6)	Trask (1953)
– Sigsby Basin	(0.068)	5.0 – 16.7 (10.5)	Trask (1953)
North of Venezuela	up to 0.42	6.8 – 19.2	Okuda and Gomez (1964)
Paria Bay	–	14.7; 7.5	Van Andel and Postma (1954)
Lagoons of Cananey (Brazil)	0.26 – 0.81 (0.36)	6.1 – 24.4 (12.6)	Kutner (1964)
Between Recife and Buenos Aires	–	10.0 – 20.8 (15.4)	Hobson and Menzel (1969)
Gulf of Thailand	0 – 0.16	(7.7)	Emery and Niino (1963)
Red Sea	0.016 – 0.091 (0.041)	12.2 – 50.0 (19.2) (25.0)	Mohamed (1949), Emery (1964)
Arabian Sea	–	2.9 – 20.4 (6.9)	Wiseman and Bennet (1940)
Area of Karachi-Cochin	0.02 – 0.83 (0.23)	7.0 – 15.9 (10.0)	Dietrich et al. (1966)

Table 39 (continued)

Sampling area, object studied	N_{org} (%) on a dry solid basis	N_{org}/C_{org}, %	Source
Sedimentary rocks			
Pliocene and Miocene of the Los Angeles Basin			
0–610 m	–	(7.6)	Emery (1960)
610–1220 m	–	(7.2)	Emery (1960)
1220–1830 m	–	(7.1)	Emery (1960)
1830–2440 m	–	(7.0)	Emery (1960)
Ancient sediments	–	(6.8)	Emery (1960)
Bituminous substances, oil shales, crude oils			
Bitumoids from sediments	–	1.0–1.4	Bordovsky (1964)
Tertiary oil shales	–	1.1–9.4(2.8)	Weber et al. (1956)
Crude oils	–	(0.5)	Zobell (1945)
Dictyonemite shales	–	(2)	Manskaya and Drozdova (1964)
Kuckersite	–	(0.3–0.7)	Manskaya and Drozdova (1964)

Figures in parenthesis show mean values
[a] Values for upper sedimentary layer (0-5-10 cm); the areas where data is given for deeper layers are specified
[b] Only carbonate sediments

The N_{org}/C_{org} ratios for the sediments in the seas and oceans vary considerably from 2.2 to 12.5% and, according to our data, even up to 46.2% in the Pacific and Atlantic Oceans. In most types of sediments the range of the N_{org}/C_{org} ratios is much narrower. The minimum values are characteristic of carbonate sediments, averaging 3.0–8.5% for different types in the Pacific Ocean. This implies that the OM entrapped in the sediments is poor in nitrogenous compounds. The maximum values of the N_{org}/C_{org} ratios are typical of the red abyssal clays: miopelagic (11–31%, averaging 22.4%) and eupelagic (14.7–46.2%, averaging 25.4%), where nitrogenous compounds are preferentially preserved. The major portion of them is tightly sorbed by the mineral base of the sediment and incorporated in heteropolycondensed structures of the type of humins, whereas the minor component is represented by stable biological structures, such as cell-walls (Jocteur-Monrozier 1980, Suess and Müller 1980).

The role of benthos in the relative enrichment of OM of deep red clays with nitrogenous compounds, as well as the impact of differences in the composition of clay minerals of unlike sorption capacity (illite, montmorillonite), seem to be small.

The N_{org}/C_{org} ratios in the pelitic muds of the Pacific Ocean, within the continental margin (6.3–23.7%), marginal parts of the ocean floor (13–23%) and

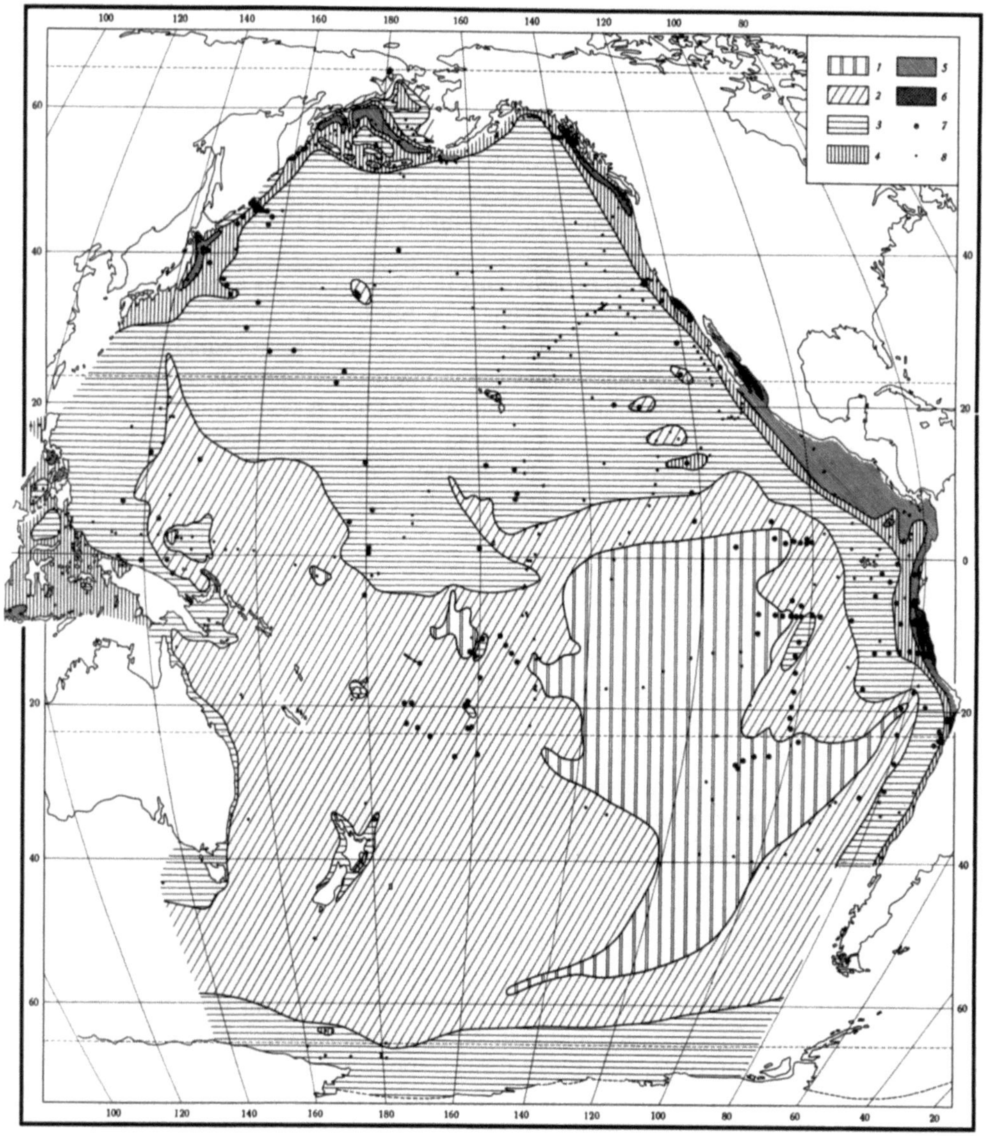

Fig. 42. N_{org} distribution pattern in upper layer of bottom sediments in the Pacific Ocean, in % on a dry sediment basis. *1* = <0.03; *2* = 0.03 – 0.06; *3* = 0.06 – 0.10; *4* = 0.10 – 0.20; *5* = 0.20 – 0.40; *6* = >0.40; *7* probe data obtained by the Institute of Oceanology of the USSR Academy of Sciences; *8* data from foreign expeditions

in its central part (2.2 – 46.2%) average 11.9, 17.2 and 13.9%, respectively. The N_{org}/C_{org} mean value for all types of sediments in the Pacific Ocean approximates 11.8% (C_{org}/N_{org} = 8.5; Romankevich 1974a).

The level of the N_{org}/C_{org} ratio may increase or decrease with the increase of C_{org}, dependent upon the rate of supply of fresh nitrogen-containing OM, conditions and rate of OM decayal during its burial. The tendency towards a reversed relationship between the N_{org}/C_{org} and C_{org} values, observed in the sediments of the Pacific Ocean, reflects the Uspensky-Vassoyevich law of the sum of nitrogenous compounds. The fact that the Kjeldahl combustion procedure allows for the complete or nearly complete extraction of N_{org} from the sediments, again serves to indicate the existence of structures resistant to decay among the nitrogenous compounds, which remain relatively intact during the degradation of OM.

As a rule, the N_{org}/C_{org} ratios of sediments in seas and oceans are lower than in phyto-, zooplankton, zoobenthos and bacteria, but higher than in kelp, rockweeds and angiospermae. Assuming that the N_{org}/C_{org} mean values equal 20% in the plankton, and 12% in the sediments of the submarine continental margin, the estimated amount of OM that sinks to the bottom sediments was found to be on the average 40% poorer in N_{org}, as compared to the initial substance; which implies that the breakdown of the nitrogenous compounds is 40% more enhanced, as compared to the destruction of the OM as a whole. A gradual decrease in the level of N_{org}/C_{org} values has been observed in the depth profile of sediments of the submarine margin, the Pliocene and Miocene deposits in the Los Angeles Basin, and in certain other types of sediments, combustible shales and crude oils, which serve to indicate the relatively accelerated rate of degradation of nitrogenous compounds in the process of diagenesis.

Evidence supporting this is found in the interstitial waters, which have been extensively investigated by Starikova (1961, 1970) in different seas and oceanic areas. Here the N_{org}/C_{org} ratios are higher in the upper layers, and decrease with increasing depth of the sediments (Table 40). Not only the C_{org} and N_{org} contents, but also the ammonium content, increases concurrently with the destruction of OM. The order of magnitude of these processes is much higher in the sediments of the submarine margin of continents than in the ocean floor deposits, which are poor in OM. The degradation of OM, which is primarily a biochemical process and results in the formation and release of free H_2S and NH_3, gives rise to reducing environments for certain elements, thereby enhancing the alkalinity in the sediments. As a consequence, the rate of OM degradation is retarded, owing to the depletion first of free oxygen and then of combined oxygen which is necessary for the respiration of the microorganisms responsible for decay.

The N_{org}/C_{org} ratios in the eupelagic clays and certain types of pelagic diatomaceous-radiolarian ooze in the Pacific Ocean, increase with depth, suggesting that the nitrogenous compounds are exceptionally resistant to destruction.

The specific nature of the OM composition in eupelagic clays as reflected in the high N_{org}/C_{org} level, applies not only to the sediments, but also the interstitial waters (up to 59%). Such high N_{org}/C_{org} values are, however, not typical of terrigenous, and even less of carbonate sediments, while they are extremely rare in the particulate fraction.

Very high N_{org}/C_{org} ratios occur characteristically in such metabolic products, as urea (200%), creatine and guanidine (100%), uracil and glycine (50%), that are present in sea waters. Although, the N_{org}/C_{org} ratios generally seem to be lower in sea waters than in the solid phase and interstitial waters of red abyssal clays, high values have also been reported here (see Table 37). This can

Table 40. Organic nitrogen and nitrogen-carbon ratios in interstitial waters and host sediments

Layer (in m)	Interstitial waters		Host sediments		Source
	N_{org}, mg l^{-1}	N_{org}/C_{org} (%)	N_{org} (%)	N_{org}/C_{org} (%)	
Black Sea					
0–0.3	2.3–5.0	14.3–47.6(15.0)	–	5.9–12.5(8.3)	Starikova (1961)
5.4–8.3	–	3.6–9.3(6.1)	–	7.1–12.0(9.7)	Starikova (1961)
Azov Sea					
0–0.3	5–6.9	20.8–23.8(22.7)	0.11–0.28	11.8–13.2(12.5)	Starikova (1961)
2–6.5	–	10.3–23.8(16.5)	–	11.2–14.1(12.6)	Starikova (1961)
Peruvian shelf; terrigenous and phosphate-terrigenous muds (St.M-552, depth – 70 m)					
0–0.1	3.26	27.0	0.58	11.8	this volume
1.0–1.5	3.07	24.4	0.38	11.6	this volume
2.5–3.0	3.80	18.2	0.44	10.4	this volume
Peruvian continental slope; terrigenous muds (St.M-545, depth – 550 m)					
0–1.0	1.69	15.7	0.38	10.8	this volume
2.2–2.5	1.90	12.5	0.21	9.7	this volume
2.9–3.4	2.12	10.8	0.25	11.1	this volume
Peruvian Oceanic Basin; metalliferous clays (St.M-599, depth – 4460 m)					
0–0.2	0.92	52.6	0.072	23.2	this volume
0.6–1.8	0.72	40.0	0.049	35.2	this volume
3.6–4.2	0.88	11.8	0.024	20.0	this volume
Continental slope of Honshu Island; terrigenous muds (St.V-6163, depth – 7360 m)					
0–0.1	2.51	22.0	0.180	12.0	Starikova et al. (1974)
2.6–3.0	9.81	11.0	0.144	10.4	Starikova et al. (1974)
6.1–6.4	12.0	10.1	0.164	12.9	Starikova et al. (1974)
North-western Oceanic Basin of the Pacific; eupelagic clays (St.V-6174, depth – 5915 m)					
0–0.1	0.88	19.2	0.067	23.9	Starikova et al. (1974)
3.3–3.5	1.11	16.6	0.043	47.8	Starikova et al. (1974)
6.0–6.3	0.85	14.2	0.047	58.8	Starikova et al. (1974)

Figures in parenthesis show mean values

probably be attributed to the fact that in the red abyssal clays the OM is in most part adsorbed on to the mineral particles. Part of the nitrogen present in the red abyssal clays exists in associated ammonium forms.

In the pelagic deposits the N_{org}/C_{org} ratios are inherited from the sediments of the surface layer, whereas the enclosed ferrous-manganese nodules do not inherit these values (Table 33). The N_{org}/C_{org} ratios in the external crust and orebody of the concretions are characteristically very low, which indicates, firstly, the enhanced carbonization of the OM, and, secondly, the different genesis of the OM in the nodules and enclosing red clays. The N_{org}/C_{org} ratios of the cores (which are also rich in these oxides, and are not younger than the ore-body proper), are close to those of the red clays. From this the inference can be drawn that the formation of the oceanic concretions is little related to the process of diagenesis, during which the host sediments maintain high N_{org}/C_{org} values.

5.2 Phosphorus and the Phosphorus-Carbon Ratio

5.2.1 Phosphorus in Marine Organisms. Regeneration of Phosphorus

Phosphorus, like nitrogen, is an indispensable element, invariably present in all living organisms. It is essential in the structure of nucleic acids, phospholipids, ATP, certain proteins, and actively participates in a wide spectrum of biochemical reactions, involving energy transfer to the living tissues, and the exchange between the organisms and their surrounding environments.

The content of phosphorus in contemporary marine plants and animals was found to be 0.3 – 3.3 and 0.2 – 3.2%, respectively (Table 36). In marine animals the highest phosphorus concentrations occur in the tough body tissues, bones and chitinous exoskeletons containing up to 6 – 16% (Kizevetter 1973); the soft tissues have a lower phosphorus content (0.3 – 3.3%). The phosphorus content in marine organisms depends upon its concentration in sea waters, the physiological state of living organisms, and the nature of their organo-mineral skeleton. A rise or fall in the level of phosphorus in sea waters may cause manifold increases or decreases in its concentration in living organisms, compared with their normal phosphorus content. Phytoplankton contains between 1.6 – 5.4 parts of phosphorus per 100 units of C_{org}. The C:P mean ratios are 100:3.6 in the diatom algae, 100:2.5 in peridineans, 100:2.9 in blue-green algae, and 100:3 in planktonic Crustacea. These values are closely similar to the mean atomic ratios of C:P for marine plankton (100:1).

In the phytoplankton phosphorus occurs mostly in phosphorus-organic compounds; a smaller portion is in the form of inorganic polyphosphates. The latter are most predominant in bones of animals, chitinous exoskeletons and scales. According to its degree of biochemical activity in the organisms, labile and tightly linked phosphorus can be distinguished. Most of the phosphorus contained in the phytoplankton can be readily hydrolyzed, whereas in the calanuses it is tightly bound and requires more rigid hydrolysis for its release (Goldberg et al. 1951, Conover 1961, 1978, Parsons et al. 1977).

Phosphorus participates in the active biogeochemical cycling of matter. The annual uptake of phosphorus by phytoplankton approximates 0.75×10^9 tons (the estimates are based on phytoplankton P/C ratios = 3.0%). About 99% of this undergoes recycling during OM mineralization in the water column. In the metabolic processes of zooplankton phosphorus is liberated as salts of orthophosphoric acid and in the form of organic compounds. The calanuses, which thrive actively on phosphorus, utilize in their growth only 17.2% of P_{org}, whereas a considerable proportion is released with the feces (23.0%) and in the form of various soluble compounds (59.8%) (Butler et al. 1970, quoted from Parsons et al. 1977). Thus, with the active feeding, calanus releases into the environment more than 80% of phosphorus contained in the consumed phytoplankton.

The factors controlling the regeneration of phosphorus are manifold. Its regeneration is promoted by the increase in the density of the phytoplankton and bacteria populations, increase in temperature, and decrease in the phosphorus content of the waters. The turnover rate of phosphorus in the Azov Sea is 1.5 months, in the productive coastal waters it is 1.5 days, and in certain eutrophic lakes only a few minutes (Pomeroy 1960, Redfield et al. 1963, Maksimova 1977). The turnover of phosphorus, as compared to that of organic nitrogen, more nearly approaches a closed cycle. Bacteria and other microorganisms have a paramount role in the degradation of organically bound phosphorus by means of phosphatase. The rate of microbial mineralization of organically combined phosphorus depends upon the temperature, pH of the medium, and aeration conditions.

Experiments carried out on algal cultures established the fact that after that the death of living organisms, autolysis and microbial degradation attain such intensive development, that in the course of a few days 80% of the phosphorus is regenerated, most of which is released into solution in the form of phosphates (e.g., Kuznetsov 1970, Golterman 1972). Experimental investigation of the microbial degradation of diatom plankton OM showed that over a period of a year, 64% of the phosphorus was converted to inorganic phosphates, 32% to the suspension, and 4% to organically bound phosphorus (Grill and Richards 1964). In the experiments described by Skopintsev (1950, 1981), the rate of removal of phosphorus during the decomposition of the particulate fraction was three times higher than that of nitrogen, and after a half year their content was reduced to 51% and 18% of the initial value (respectively). Among the organophosphorus compounds, most easily subjected to dephosphorization are nucleic acids (more readily of bacterial DNA, then RNA) followed by lecitines, and, finally, phytine. The rate of microbial mineralization of organic phosphates increases in the following order: OM released into solution by weak acid treatment, nucleic acids, proteins and phospholipids.

5.2.2 Organic Phosphorus in Sea Waters and the Suspended Material

The content of dissolved organic phosphorus (P-DOM) in oceans varies from analytical zero to 52 $\mu g\,l^{-1}$ (Table 37). In the euphotic layer the concentrations are commonly from $2-30\,\mu g\,l^{-1}$, whereas in the deep waters they are $0.5-10\,\mu g\,l^{-1}$. Although in waters at depths below 500 – 1000 m, the P-DOM content is lower than in the upper layers, evidence has been found of concentrations up to $30-60\,\mu g\,l^{-1}$ in the deep waters. Their occurrence at depth cannot as yet be feasibly explained. In the upper water layer of inland seas the P-DOM content is higher on average, than in the open ocean areas. As in the case of nitrogen, the distribution of various forms of phosphorus in seas and oceans, characteristically exhibits the following pattern: in the upper few tens of meters P-DOM predominates, whereas in deeper waters reactive phosphorus (mainly

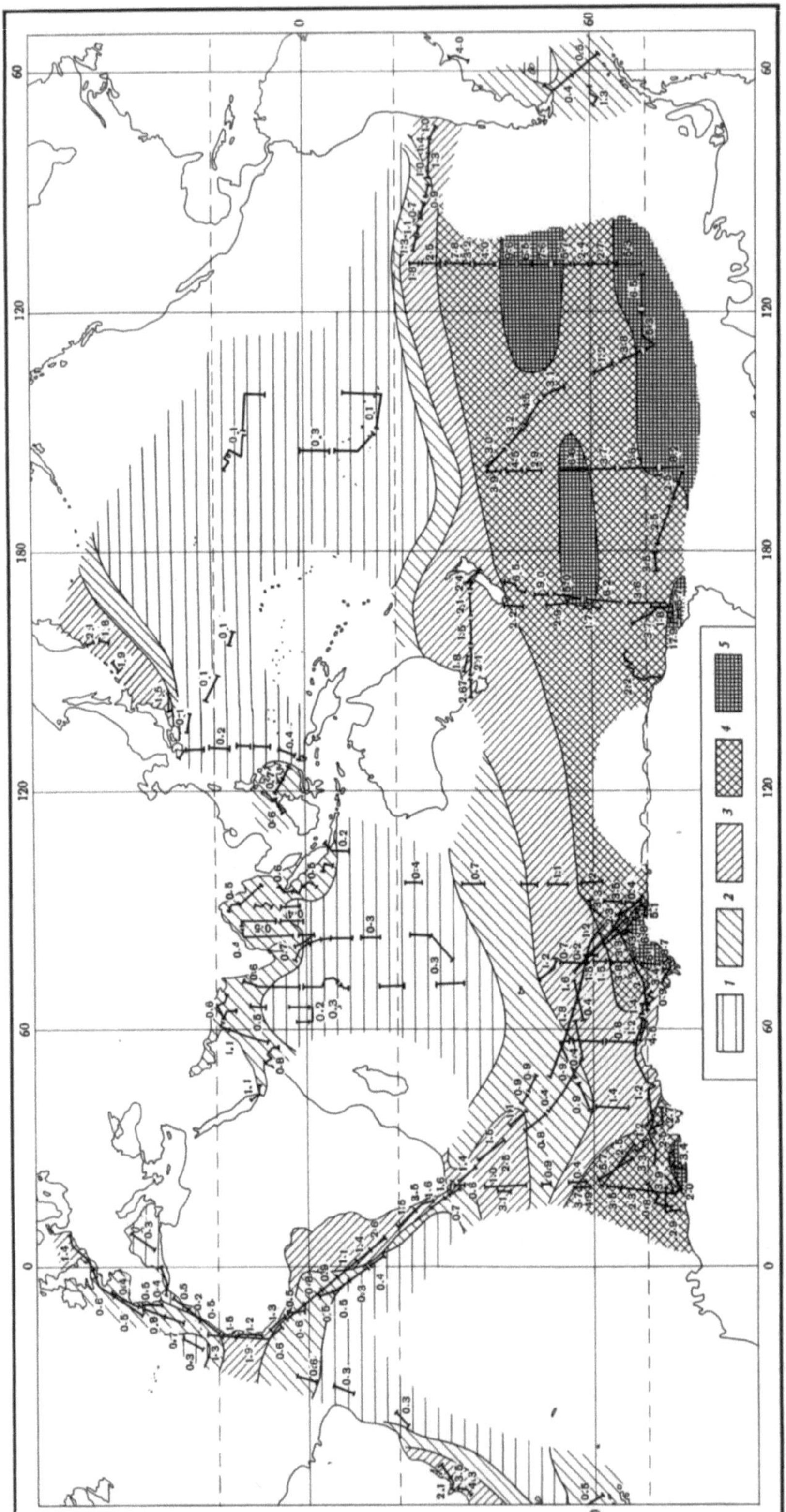

Fig. 43. Distribution pattern of phosphorus in suspended matter and surface waters of the world ocean ($\sim 0-7$ m layer) $\mu g\,l^{-1}$: *1* = <0.5; *2* = 0.5 – 1.0; *3* = 1.0 – 2.5; *4* = 2.5 – 5.0; *5* = >5.0. (After Bogdanov et al. 1971, complemented by the author)

P-PO_4) does so. In general, the P-DOM averages only a few percent of the bulk of phosphorus in the ocean.

The P-DOM/C-DOM ratios in the oceans are between 0.5 to 3.4% (see Table 37). Most of the mean values of the layers are between 0.8 – 1.4. The phosphorus content of the dissolved OM decreases slightly in the vertical profile (1.1 – 3-fold decrease, judging from the weighted average values for the layers between 0 – 100 m and below 1000 m), or even increases. The latter has been observed in the western tropical Pacific Ocean, in the areas off California and Peru. This might show that in the ocean deeps some dissolved phosphorus-organic compounds are resistant to decomposition, inert to bio-utilization and chemical breakdown. In this respect these compounds differ essentially from the P-POM. In general, there is not a high correlation between DOC, N-DOM, P-DOM (Williams 1975).

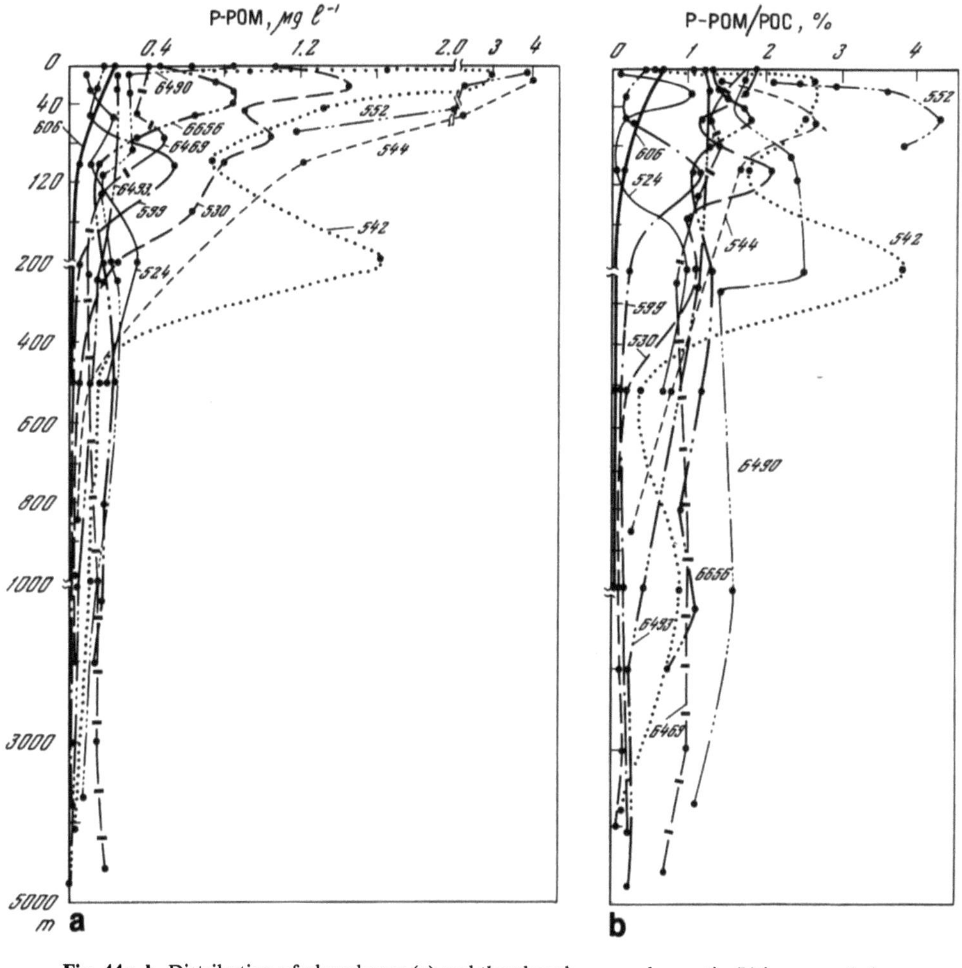

Fig. 44 a, b. Distribution of phosphorus (**a**) and the phosphorus-carbon ratio (**b**) in suspended matter along vertical profile in the Pacific. *Figures near curves* show Station Nos

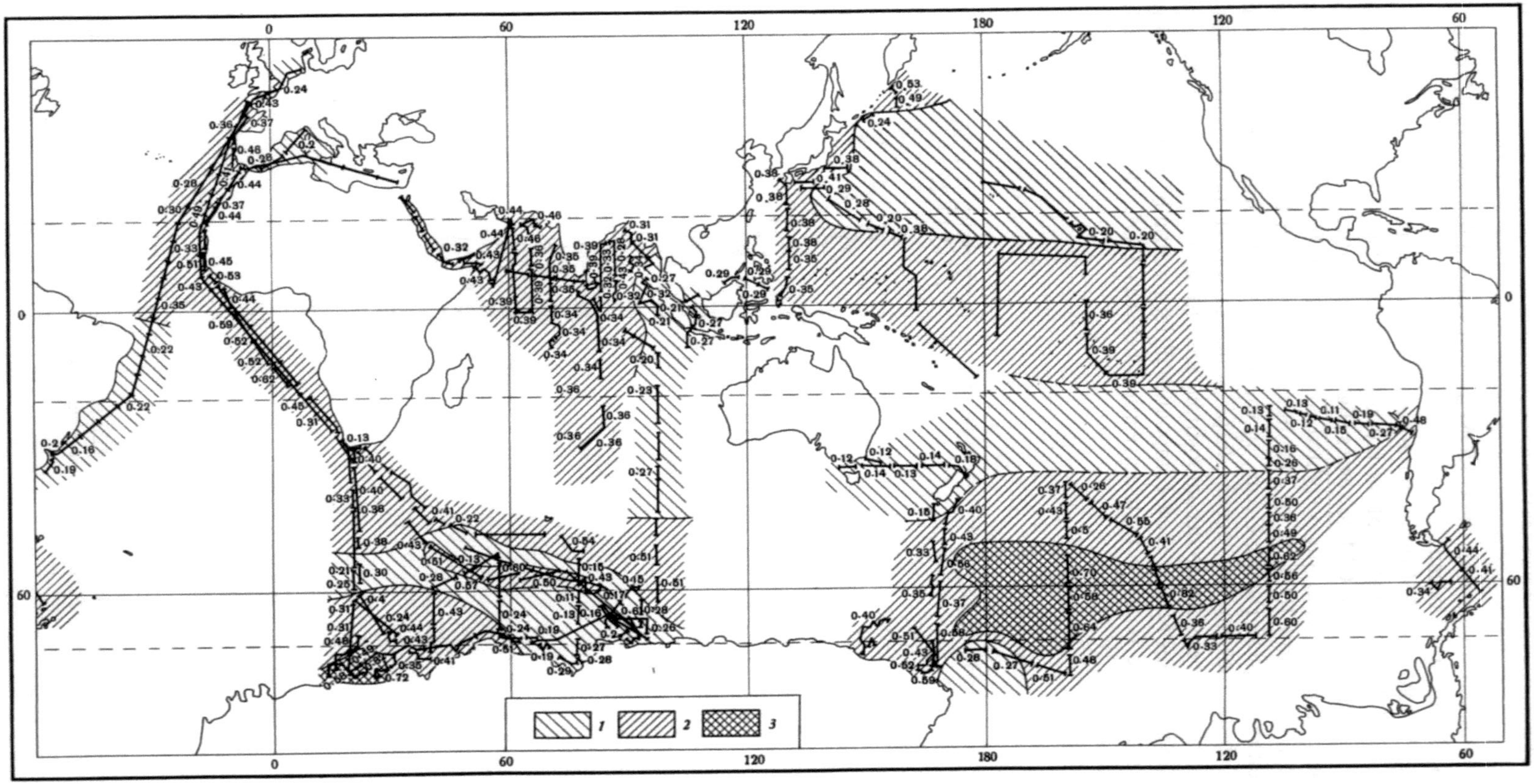

Fig. 45. Phosphorus content of suspended matter and surface waters in the world ocean (~0–7 m layer), in % of dry weight. *1* = 0.1–0.3; *2* = 0.3–0.6; *3* = 0.6–1.0. (After Bogdanov et al. 1971, complemented by the author)

The organic phosphorus content in the surface layer particulate fraction in seas and oceans varies from 0.1 – 8.7 μg l^{-1} (Fig. 43). The maximum P-POM concentrations, like those of C-POM and N-POM, are restricted to the near-continental areas and upper layers of the water column (Fig. 44). Comparison of the areal distribution patterns of P-POM and C-POM, also the correlation between the latter, strongly suggest that the major portion of phosphorus in the suspended matter is associated with the OM. In the particulate fraction of the euphotic layer, during the predominant growth of phytoplankton, a direct linear relationship is apparent between the P-POM, C-POM, N-POM and chlorophyll contents, which are specific elements of living OM (e.g., Menzel and Ryther 1964). At the same time, the strict correlation between the total content of suspended matter and P-POM is, as a rule, lacking (e.g., Jurkovskis and Fominykh 1980).

In the world-ocean the concentration of P-POM in the dry solid particulate matter, was found to be 0.11 – 0.82% (Fig. 45). In the Indian Ocean the phosphorus content of the particulate fraction averages 0.30% (Lisitzin 1964). The P-POM content in the dry solid particulate matter of the Northwestern Pacific Ocean is 0.20 – 0.53%. The P-POM shows a 3 – 100-fold decrease in the vertical profile (comparison was made of weighted average values for layers between 0 – 100 m and below 1000 m, as depicted on Fig. 44). At depths below 100 m, and at certain stations below 200 – 500 m, the changes in the P-POM content are only slightly reflected.

The P-POM/C-POM ratios in the diatom (2.4%) and peridinium (2.0%) suspensions of the surface layer in the Pacific Ocean are only slightly smaller, than in the diatom (3.6%) and peridinium (2.5%) algae. These differences increase in the vertical profile. For instance, the weighted average values for layers from 0 – 100 m, 100 – 200 m, 200 – 1000 m and below 1000 m in the Pacific Ocean are 1.6%, 1.5%, 0.9% and 0.6%, respectively (see Fig. 44). The sharp drop in the P-POM content, as well as the P-POM/C-POM ratios in the vertical profile, evidence the lability of the phosphorus-containing compounds in the suspension. The particulate fraction at depths below 1000 m is 60 – 70% poorer in phosphorus, as compared to that of the surface waters. The breakdown of phosphorus-containing compounds is 1.5 – 2 times more rapid than that of nitrogenous compounds in the particulate fraction. The accelerated rate of decomposition of the P-containing compounds is a typical feature of the world ocean in all climatic zones, although in cold waters this process is slow (Maksimova 1977).

According to the P-POM/C-POM ratios, the particulate fraction at the respective depths in the areas of the ocean floor is on average poorer in organically-combined phosphorus, as compared to that in the areas of the continental slope and shelf. In the euphotic layer the POM is richer in P-organic compounds than the DOM, whereas in deep waters it is often poorer.

5.2.3 Phosphorus in Bottom Sediments

The phosphorus content of bottom sediments in seas ranges from 0.01 to 0.90%, whereas in the open ocean sediments it varies from traces to 1 – 2%, and averages 0.07% in terrigenous sediments, 0.05% in calcareous, 0.04% in siliceous sediments, 0.11% in red abyssal clays (Baturin 1982). The mean phosphorus content of ocean sediments is 0.07% (El-Wakeel and Riley 1965). In shallow water areas of the ocean, beyond the zones of contemporary phosphorite-formation, the fine-grained sediments are rich in phosphorus, whereas the abyssal muds are poor in it. However, most types of sediments contain less phosphorus than the particulate fraction in the surface and deep water layers. The distribution pattern of phosphorus in the particulate fraction shows that the distance from the land and the depth are the controlling factors that determine the amount of P-POM that settles at the bottom. No large concentrations of organically combined phosphorus can be expected in recent sediments at depths below 200 m, even within zones of high productivity.

Inorganic phosphorus is the major form in which phosphorus occurs in the sediments (in contrast to suspended material). This is evident from the high P/C_{org} ratios that have been reported for the bottom sediments and their individual size-fractions. The P/C_{org} values here were found to be most commonly 10–50% (Lisitzin 1966), which is not typical of the POM. Organically-combined phosphorus is either entirely absent from the sediments, or present in very small quantities in shallow waters (e.g., Ostroumov and Volkov 1957, Victor and Anantha 1973).

At depths of less than 200 m, during the mineralization of the phosphorus-containing OM, phosphorus accumulates in the residue, and also forms conspicuous concentrations as a result of diagenetic migration processes and precipitation from interstitial waters which is concurrent with the removal of certain organic and inorganic components from the phosphates. The process of phosphatization of authigenic aggregates and their conversion into consolidated authigenous phosphorite concretions is concomitant with the breakdown of a significant portion of OM (Romankevich and Baturin 1972, Baturin 1982).

The amount of phosphorus that is deposited within the shelf, hemipelagic and eupelagic sediments, were calculated by Horn and Adams (quoted by Baturin 1975), to be 4.85×10^{14}, 10.1×10^{14} and 5.99×10^{14} tons, respectively; i.e. the bulk of absolute masses of phosphorus (about 71%), like that of C_{org}, accumulates within the submarine margin of continents.

5.3 Certain Features of the Cycling of Nitrogen and Phosphorus During Sedimentogenesis

Analytical treatment of the results serves to suggest that the loss of nitrogen- and phosphorus-containing radicals reflects the major trend in the alteration of the OM element composition in the process of sedimentogenesis (Figs. 46 and 47). As a consequence of the consumption of the dominant diatom plankton by zooplankton and fishes, as well as of the microbial degradation of OM, and the feeding regime of the benthos, the living organisms become enriched in organic nitrogen and phosphorus, and, conversely, the non-living matter becomes poorer in these compounds. This implies that in the process of sedimentogenesis nitrogen and phosphorus undergo large-scale differentiation, which occurs mainly in the upper layers of the water column.

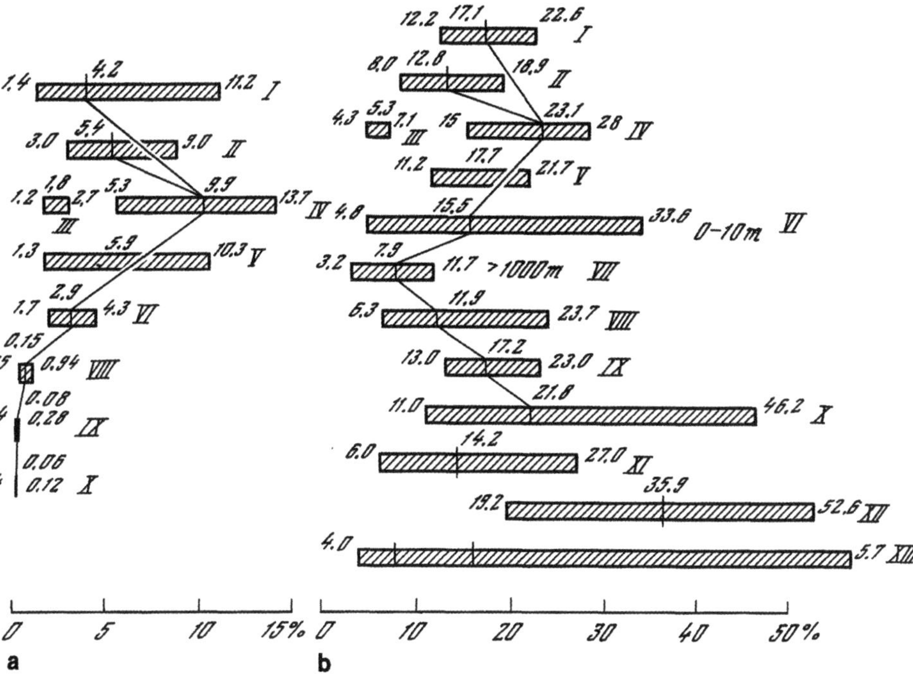

Fig. 46a, b. Changes in the N_{org} content in the process of sedimentogenesis. **a** N_{org}, in %, on a dry solid basis; **b** N_{org}/C_{org}, in %. *I* diatom algae; *II* peridinium algae; *III* phytobenthos; *IV* zooplankton (chiefly copepods); *V* eolian suspension; *VI* suspended matter of upper water layer; *VII* water suspension at depths below 500 – 1000 m; *VIII* terrigenous sediments of submarine continental margins; *IX* terrigenous hemipelagic sediments; *X* red abyssal clays; *XI* interstitial waters of terrigenous sediments; *XII* interstitial waters of red abyssal clays; *XIII* surface layer of sea water. Values for *V* – *XII* apply mainly to the Pacific Ocean

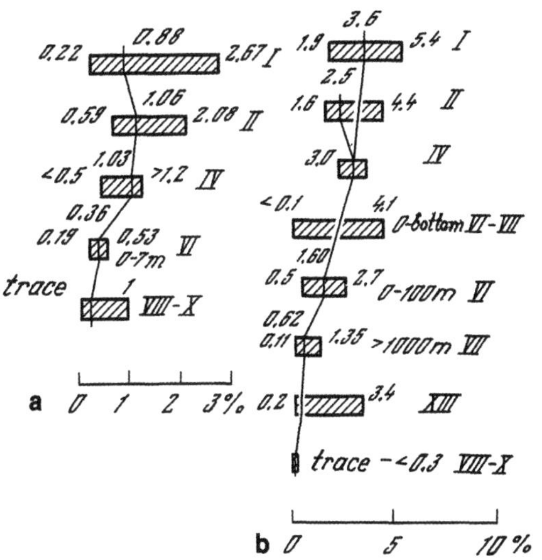

Fig. 47a, b. Variations of the phosphorus content in the process of sedimentogenesis. **a** P, in %, on a dry solid basis; **b** P/C_{org} ratio in %; other symbols are as in Fig. 46

The most essential result of the cycling of nitrogen and phosphorus in the ocean is their turnover in not closed cyclic systems under conditions of steady-state dynamic equilibrium between the constituents: between the inorganic and organic nitrogen- and phosphorus-compounds, existing in living organisms, in the particulate fraction, solution and bottom sediments.

In some cases the interchange of nitrogen and phosphorus between the sediments and bottom waters plays an important role in the gross balance of recycling of these elements (Lerman 1979, Blackburn 1980, Jurkovskis and Fominykh 1980, Wollast 1980, Rosenfeld 1981). The regeneration of phosphorus in the bottom sediments causes a diffusion flux in the system bottom-water at the rate of 20 – 2000 mg P $m^{-2}y^{-1}$ in different ocean areas (Baturin 1983). In some cases, the flux of NH_4 (the first mineral decomposition product of the nitrogen-containing OM) out of the bottom deposits into the near-bottom water may provide 20 – 30% of diurnal nitrogen requirements of plankton (Aller and Benninger 1981). From calculations based on the total amount of POM that sinks to the bottom, but does not undergo burial in the sediments ($1.1 - 3.6 \times 10^9$ tons of C_{org}), and the mean values of N_{org}/C_{org} and P_{org}/C_{org} in the deep water particulate fraction, which equal 10 and 1%, respectively, it was found that approximately 110 – 360 million tons per year of organic nitrogen and 11 – 36 million tons per year of phosphorus undergo mineralization in the uppermost layer of sediments.

6. Proteinaceous Compounds and Amino Acids

Most of the data on the total content of proteins in the plankton and benthos was obtained by means of the Biuret and Lowry methods, or was calculated from the nitrogen content (N×6.25, e.g., Parsons et al. 1961, Vinogradova 1967b, Raymont et al. 1971, Kizevetter 1973). Bogorov (1967) reviewed the data available from various publications on the content of the major classes of OM in marine organisms, and derived their mean values (Table 41). Using these N/C ratios, the estimated total content of nitrogenous compounds in OM of marine organisms is as follows: for diatom plankton – 38–71% (averaging 53%), peridinium – 25–59% (averaging 40%), plankton with a predominance of blue-green algae – 33–40% (averaging 36%), crustacean zooplankton – 46–88% (averaging 72%), zoobenthos – 72–84%, bacteria – about 75%, phytobenthos – 7–36% (averaging 17%). Proteins constitute only 70–85% of the nitrogenous substances in marine organisms. These seem to be overestimates, although on the whole they correctly depict the total content of amino acids in marine organisms (Fig. 48).

Proteins are present in organisms in forms of colloidal solutions, as elastogels they are incorporated in the tough tissues (bones, scales, etc.). As a result of the metabolism of living organisms and posthumous OM decay, amino acids and peptides of various degrees of polymerization are released into the waters; amino acids account for 30–70% of the OM that is liberated by living phyto- and zooplankton in aqueous environments (Zlobin et al. 1975). The primary agencies in

Table 41. Biochemical composition of marine organisms

Organisms	Content (%), in substance										Energy, $kcal\ g^{-1}$, of wet weight
	wet			dry[a]				organic matter[a]			
	water	OM	ash	ash	proteins	lipids	carbohydrates	proteins	lipids	carbohydrates	
Phytoplankton	80	11	9	45	30	5	20	55	9	36	0.35
Phytobenthos	80	15	5	25	15	0.5	60	20	0.7	80	0.40
Zooplankton	80	18	2	10	60	15	15	67	17	16	0.50
Zoobenthos	63	14	23	62	27	3	8	71	8	21	0.45
Nekton	73	24	3	11	70	15	4	78	17	5	1.1

[a] Author's estimations

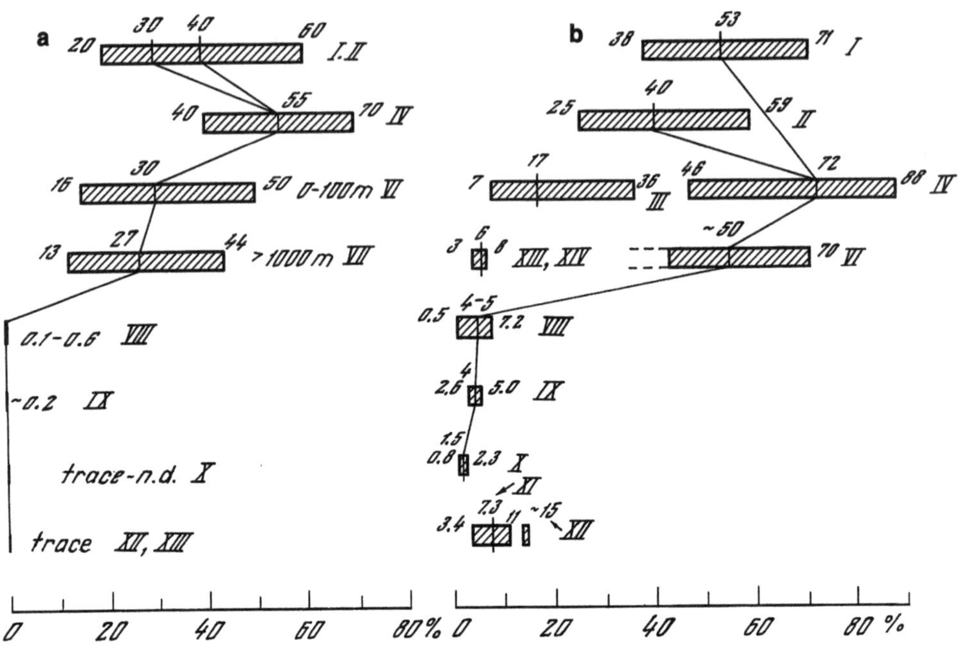

Fig. 48 a, b. Changes in the content of proteinaceous compounds (**a**) and amino acids (**b**) in the process of sedimentogenesis, in % of OM. *XIII:* sea waters at depth between 9 and 3000 m. Other symbols as in Fig. 46

the deamination of proteins are enzymes present in the living tissues and putrifactive bacteria.

Polymers with molecular weights up to 200,000 were recovered from coastal waters in the Barents and Black Seas by means of chloroform extraction and gel-chromatographic separation. Among these (mainly in the >50,000 molecular weight fraction) peptides and polysaccharides were identified (Khailov 1971). At two stations in the zone of the Peruvian upwelling (see M-542, M-544), the >1000 molecular weight fraction was isolated by means of gel-chromatographic separation. Measurements of the DOC showed that in the 0 – 50 m layer this fraction accounted for 38 – 69% (on average 50%) and, at a depth of 3800 m, 25% of the total DOC. By means of Lowry's method a small amount of proteinaceous compounds was identified in the OM concentrates from the upper layers of the water column; the tryptophan content in the >1000 fraction did not exceed 10^{-7} mol l^{-1}.

In the analytical procedures of solvent extraction and gel-chromatographic separation, the proteins are subjected to denaturation, and cannot be completely extracted from the waters. Because of these deficiencies, ultrafiltration methods have found increasingly wide use in recent years in the fractionation, purification and concentation of biological liquids, involving filtration through a series of inert membrane filtres (e.g., Baturina 1980). The fractionation of dissolved OM by means of successive filtration through ultrafiltres carried out on samples from the equatorial Pacific and off Peru, served to establish that the relative content of the >300,000 and 300 – 10 thousand molecular weight fractions decreases from the surface waters to the deep waters (the decrease of each averaging from 6 to 1%), while at the same time the content of OM with molecular weights <10 thousand increases (see Table 14). The major portion of dissolved OM (over 50%) in the surface and deep waters was contained in the <1000 molecular weight fraction.

Peptides constitute the dissolved OM and under hydrolysis treatment produce urea and a uniform spectrum of amino acids. However, in different areas various individual monomers may predominate among the dissolved free, combined and total amino acids. For example, in the Black Sea among the prevailing dissolved amino acids are glycine, serine, cysteine; in the area of the Japan deep trench – valine, isoleucine and leucine, aspartic acid; in the Buzzards Bay – glycine, alanine, serine, at times, ornithine; in the Eastern Equatorial Pacific – serine, alanine, at times aspartic acid (Degens 1970, Starikova and Korzhikova 1973, Lee and Bada 1975) (Table 42).

The total dissolved amino acid concentration in sea waters is commonly 50 – 300 $\mu g\, l^{-1}$ (Table 43). In some areas the concentration of amino acids differs by a factor of 5 to 10, as, for instance, in the Norwegian Sea, where it is 1510 – 4490 $\mu g\, l^{-1}$, and in the Northeastern Atlantic up to 7990 $\mu g\, l^{-1}$ (Zlobin et al. 1975). Low concentrations of the total dissolved amino acids (tens of $\mu g\, l^{-1}$) have been found in the waters of the eastern equatorial Pacific (Lee and Bada 1975). Such large fluctuations (many hundreds of $\mu g\, l^{-1}$) in the content of amino acids can be attributed, to a large degree, to the different analytical techniques employed for their determination. However, environmental factors are also largely responsible for the variations of concentrations of the total dissolved amino acids involving tens, or even some hundreds of $\mu g\, l^{-1}$. Of special importance among these factors is the productivity level that most strongly affects the concentration of the combined dissolved amino acids. In winter, the ratio of combined dissolved to free dissolved amino acids in different parts of the Atlantic is less than one. During the phytoplankton development period it increases to 2.5 – 4, and in the Bellingshausen Sea even to 5 – 8.5 (Naletova 1979b, 1982). The dissolved combined amino acids usually prevail over the dissolved free ones. The latter are promptly utilized by the hydrobiotic community; they are distinguished by a high cycling rate, owing to which their concentrations vary less over a year than the concentrations of all the dissolved amino acids.

The concentration of dissolved free and dissolved combined amino acids undergoes the strongest fluctuations in the 0 – 100 m layer. Below 200 m the content of amino acids falls and the distribution becomes more uniform. No great differences have been established for Atlantic and Pacific areas within the 100 – 1200 m layer. Thus, the average concentration of total amino acids in this layer in the Sea of Scotia is 209 $\mu g\, l^{-1}$, off the atlantic coast of South America 216 $\mu g\, l^{-1}$, in the Bellingshausen Sea 178 $\mu g\, l^{-1}$, in the Chilean area of the Southeastern Pacific 200 $\mu g\, l^{-1}$. Some regions differ, however, in having markedly enhanced amino acids concentrations. Sometimes they can be used as an indicator of water masses, the position of the pycnocline, the boundary between the

Table 42. Amino acid composition of plankton, sea waters, suspended matter and bottom sediments (% of total amino acids)

Amino acid	Plankton[a]	Sea waters		Suspended matter			Bottom sediments		Interstitial waters
		I	II	I	II	III	IV	V	IV
Glycine	*11.3–16.4* 14.0	*1.8–5.1* 3.5	*9.4–24.3* 18.9	*abs.–6.0* 1.7	*11.2–13.9* 12.1	*7.8–12.0* 9.8	*3.4–4.9* 4.1	*6.6–8.9* 7.4	*abs.–6.3* 3.4
L-Alanine	*8.6–10.8* 9.9	*3.9–6.4* 5.1	*6.5–12.3* 10.0	*9.0–13.2* 11.3	*7.1–9.5* 8.8	*3.3–5.8* 4.5	*8.2–11.4* 9.6	*9.6–21.0* 11.9	*5.3–9.1* 6.3
Valine	*4.4–6.1* 5.6	*15.6–25.7* 19.6	*1.7–5.0* 2.7	*6.4–12.1* 8.7	*1.7–6.0* 4.5	*3.5–5.3* 4.3	*35.9–52.9* 44.5	*7.2–11.4* 9.4	*13.0–27.4* 19.2
Izoleucine	*2.9–4.4* 3.8	*9.0–11.7* 10.7	*1.1–6.5* 4.7	*10.1–18.5* 13.1	*2.7–4.4* 3.4	*0.7–4.8* 3.3	*5.4–10.4* 8.1	*13.0–19.6* 15.6	*10.2–17.0* 12.5
Leucine	*6.5–7.6* 7.1		*5.8–8.7* 7.6		*5.8–7.7* 6.7	*5.2–7.5* 6.6			
Serine	*3.6–6.9* 5.3	*4.2–8.5* 6.4	*6.4–12.3* 8.9	*tr–9.3* 4.7	*2.5–10.8* 7.6	*6.0–11.0* 8.3	*1.3–4.0* 2.1	*3.5–5.6* 4.9	*4.3–7.3* 5.6
Threonine	*3.5–10.1* 5.8	*3.4–4.9* 4.1	*2.2–4.4* 3.9	*3.1–6.3* 5.4	*2.5–6.7* 4.8	*3.9–8.1* 5.3	*1.5–4.1* 2.4	*3.1–5.6* 4.5	*4.1–9.1* 5.1
Proline	*2.7–6.0* 4.3	*2.3–3.7* 4.4	*3.0–10.2* 6.6	*tr–13.2* 9.1	*1.5–9.5* 4.0	*1.3–4.2* 3.0	*1.7–10.0* 4.1	*4.6–14.1* 7.2	*1.5–10.7* 8.0
Phenyl-alanine	*1.5–3.4* 2.6	*3.5–10.9* 6.4	*1.2–5.1* 3.1	*10.0–20.2* 14.4	*1.5–5.3* 3.4	*0.9–6.6* 3.5	*tr–6.7* 4.5	*2.5–13.8* 7.0	*6.5–10.1* 8.5
Tyrozine	*2.4–3.8* 3.1	*2.3–7.4* 5.4	*0.3–2.8* 1.4	*tr–13.2* 9.0	*0.7–4.2* 2.7	*0.8–3.0* 2.1	*1.9–10.8* 4.5	*1.6–4.8* 3.0	*1.5–10.8* 7.9
Arginine	*1.3–8.2* 4.5	*tr–11.6* 4.4	*1.4–10.0* 6.0	*abs.–15.2* 3.1	*4.5–9.7* 7.8	*4.6–13.0* 10.4	*2.0–9.0* 5.7	*2.0–20.3* 9.7	*tr–6.4* 3.6
Lysine	*2.5–7.9* 6.3	*3.0–13.7* 5.8	*4.3–11.0* 8.1	*n.d.–3.8* 0.6	*6.4–12.5* 8.9	*8.8–14.0* 10.6	*3.2–7.5* 4.9	*9.8–13.0* 11.5	*tr–8.6* 4.2
Histidine	*0.5–3.0* 1.6		*1.1–13.7* 5.4		*1.2–4.1* 2.7	*0.1–5.2* 1.4			

Ornithine	–	*tr – 6.0* 4.3	*2.9 – 11.1* 6.2	*n.d. – tr* tr	*0.2 – 3.3* 0.8	*tr – 3.8* 1.5	*0.5 – 3.8* 1.4	–	*n.d. – 4.8* 2.5
Aspartic acid	*9.0 – 12.0* 10.3	*5.2 – 22.3* 14.8	*1.5 – 2.4* 1.8	*5.4 – 26.2* 13.6	*3.5 – 9.2* 7.8	*1.3 – 9.5* 7.4	*1.5 – 3.3* 2.1	*n.d. – 12.4* 5.4	*3.9 – 9.9* 6.9
Glutamic acid	*11.3 – 13.8* 12.1	*3.9 – 5.7* 4.6	*1.3 – 2.1* 1.6	*3.1 – 6.3* 5.3	*9.7 – 15.9* 12.6	*13.0 – 15.0* 13.9	*n.d. – 7.1* 1.8	*n.d. – 10.0* 2.9	*4.5 – 12.1* 6.3
Cysteine	*0.2 – 3.5* 1.4	*abs. – tr.* tr.	*0.5 – 3.4* 1.6	*traces*	*0.8 – 1.4* 1.1	*1.0 – 5.2* 3.3	*n.d. – 1.3* 0.6	*n.d. – 1.7* 0.5	*n.d. – tr.* tr.
Methionine	*1.9 – 3.1* 2.3	*n.d.*	*0.1 – 3.3* 1.3	*n.d.*	*0.1 – 2.2* 1.2	*0.8 – 1.4* 1.1	*n.d. – 1.8* 0.5	*n.d. – tr.* tr	*n.d. – tr.* tr
Total[b]	*14 – 48* 35	*104 – 202* 155	*129 – 401* 218	*9.9 – 48.0* 28.1	*144 – 221* 186	*10.3 – 83.0* 25.1	*50 – 1782* 928	*232 – 9949* 3180	*964 – 2367* 1678
C_{am}/C_{org} (%)	–	*3.8 – 7.5* 5.8	–	–	–	–	*0.8 – 50* 3.3	*0.7 – 7.2* 3.5	*3.4 – 14.7* 8.6

I = Japan Deep Trench (St. V-6163, 6 samples, 0 – 5000 m; Starikova and Korzhikova (1973), the 7000 m layer is excluded, because the water bottle was muddy). II = Buzzards Bay; surface layer (Degens 1970). III = Sargasso Sea (St. 649; 7 samples; 0 – 2500 m; Degens 1970). IV = Japan Trench and northwestern Basin (7 samples from bottom sediment surface layer; Starikova and Korzhikova 1973); V = Shelf and continental slope off Peru and Peruvian Basin (12 – 13° S; 7 samples from sediment surface layer; Korzhikova and Romankevich, unpublished data)

Italics = extreme values; Roman = mean values of total amino acids

[a] Areas off Peru and Ecuador (18 samples). In all samples hexoseamine was reported (0.54 – 3.02% of dry weight; $\delta^{13}C$ – between 16.5 to 21.2% (Degens 1970)

[b] For plankton – % of dry weight; for sea and interstitial waters and particulate matter – in $\mu g\ l^{-1}$; for sediments – in μg per g of dry solid

Table 43. Dissolved amino acids in sea water

Area	Occurrence form of amino acids[a]	Concentrations[b], $\mu g\ l^{-1}$	Source
Carribbean Sea and Mexican Gulf	FDAA + CDAA	180	Tatsumoto et al. (1961)
Irish Sea	FDAA	100	Chou and Riley (1966)
	FDAA	4.8 – 31.2	Riley and Segar (1970)
	CDAA	2.1 – 111.9	Riley and Segar (1970)
Northern Atlantic	FDAA	6 – 47	Pocklington (1971)
English Channel	FDAA	9.8 – 80	Andrews and Williams (1971, cited by Williams 1975)
Saragasso Sea	FDAA	38 – 147	Liebezeit et al. (1980)
Northern Atlantic	FDAA + CDAA	1510 – 7990	Zlobin et al. (1975)
Heligoland Sound	FDAA	8 – 80[c]	Bohling (1970)
Black Sea	CDAA	11 – 643 (256)	Starikova and Korzhikova (1972)
Eastern Pacific	FDAA	16 – 124	Degens et al. (1964)
Kuril-Kamchatka trench	FDAA	2.5 – 21.1	Starikova and Korzhikova (1970)
	CDAA	8.5 – 141.4	Starikova and Korzhikova (1970)
Northwestern Pacific	FDAA + CDAA	104 – 202 (155)	Starikova and Korzhikova (1973)
Japan trench	FDAA + CDAA	150 – 177	Starikova and Korzhikova (1973)
Buzzards Bay	FDAA + CDAA	129 – 401 (218)	Degens (1970)
Auke Bay, Alaska	FDAA + CDAA	~ 50 – 200[c]	Coughenower and Herbert (1975)
Hood Canal, Washington	FDAA + CDAA	< 10 – > 100[c]	Coughenower and Herbert (1975)
Panama basin	FDAA	~ 4[c]	Lee and Bada (1975)
	CDAA	12 – 36[c]	Lee and Bada (1975)
Guatemala basin	FDAA	2 – 11[c]	Lee and Bada (1975)
	CDAA	14 – 50[c]	Lee and Bada (1975)
Gulf of Marseille	FDAA + CDAA	394 – 2826 (900 – 1200)	Daumas (1976)
Sea of Scotia	FDAA	32 – 276	Naletova (1979)
	CDAA	22 – 585	Naletova (1979)
SE of Montevideo	FDAA	21 – 225	Naletova (1979)
	CDAA	12 – 405	Naletova (1979)
Bellingshausen Sea	FDAA	35 – 212	Naletova (1982)
	CDAA	72 – 364	Naletova (1982)
Mediterranean Sea	FDAA + CDAA	20 – 46	Naletova (1982)
Chilean area of the Southeastern Pacific	FDAA	59 – 160	Naletova (1982)
	CDAA	102 – 282	Naletova (1982)

[a] FDAA – free dissolved amino acid, CDAA – combined dissolved amino acid
[b] Figures in parenthesis are average concentrations
[c] Calculated assuming a mean molecular weight of 100

oxic and anoxic layers (Lee and Bada 1975, Liebezeit et al. 1980, Michaelis et al. 1980, Mopper and Lindroth 1982). Amino acids account for 5 – 10% of the DOC (rarely more) in the DOM. The greatest proportion of combined amino acids in sea waters is present as phenol-chinoidal complexes and humic acid type of compounds. The combined amino acids possess molecular weights of 400 – 1000.

Most of the data available on the content of proteinaceous compounds in the particulate fraction was obtained by means of the Biuret and Lowry methods, or calculated on the basis of the nitrogen content (e.g., Krey 1961b, 1967a, Parsons and Strickland 1962a, Derenbach 1969, Finenko and Ostapenia 1971, Agatova and Bogdanov 1972, Handa et al. 1972, Daumas 1976, Barlow and Swart 1981).

In the particulate matter from the Northwestern Pacific the content of proteinaceous substances, extractable by 0.2 N NaCl, was determined by the difference between the 215 and 225 nm absorption bands in the > 10 thousand molecular weight fractions, that were isolated in the gel-chromatographic separation (see Table 38). In the peridinean suspension the content of the proteinaceous compounds with molecular weights > 10 thousand was 14.3 – 45.4% (on the average 29.0%) of the POC. The nitrogen content of these proteinaceous compounds was found to be in the range of 36 – 76% (averaging about 56%) of the total N_{org} in the particulate fraction.

In the particulate fraction of the 0 – 100 m layer in the tropical Pacific, the content of proteinaceous compounds possessing molecular weights > 10 thousand was from 9 – 87 $\mu g\, l^{-1}$ (16 – 50% of the OM). In the euphotic layer their maximum concentrations showed a correlation with the layers of zoo- and bacterial-plankton accumulation, enhanced POC content and maximum light scattering (Romankevich 1974b, 1979). The observed direct correlation with living matter feasibly explains why the content of the proteinaceous compounds in the suspended matter is higher at the respective depths in the shelf and slope areas than in the area of the ocean floor.

The content of proteinaceous compounds with molecular weights > 10,000 decreases in the vertical profile. These proteinaceous substances have a slower rate of decomposition in the water mass, and their destruction, re-synthesis and incorporation in non-extractable complexes occur in such proportions, that they maintain a relatively high level (13 – 44%) in the OM. A portion of these proteinaceous compounds is apparently incorporated in the non-living OM. High molecular weight proteinaceous compounds are not the equivalents of living OM, but indicate the presence of labile compounds in the particulate fraction, which were probably generated recently by living organisms. This holds for the bottom sediments, where determinations of the proteinaceous compounds were made by the same analytical techniques that were applied to the suspended matter; their content was found to be between 0 and 233 ± 44 μg per g of dry sediment, or 0 – 0.6% of the OM (Fig. 49).

In the terrigenous sediments of the submarine margin east of Japan and west of Peru, proteinaceous compounds of > 10 thousand molecular weights and ATP were encountered only in small quantities within the depth of the sediments penetrated (3.5 m for ATP and 8 m for proteinaceous substances). This suggests that the lower boundary of biogenic degradation of OM occurs here at a deeper level. In the red abyssal clays of the Pacific Oceans proteinaceous compounds

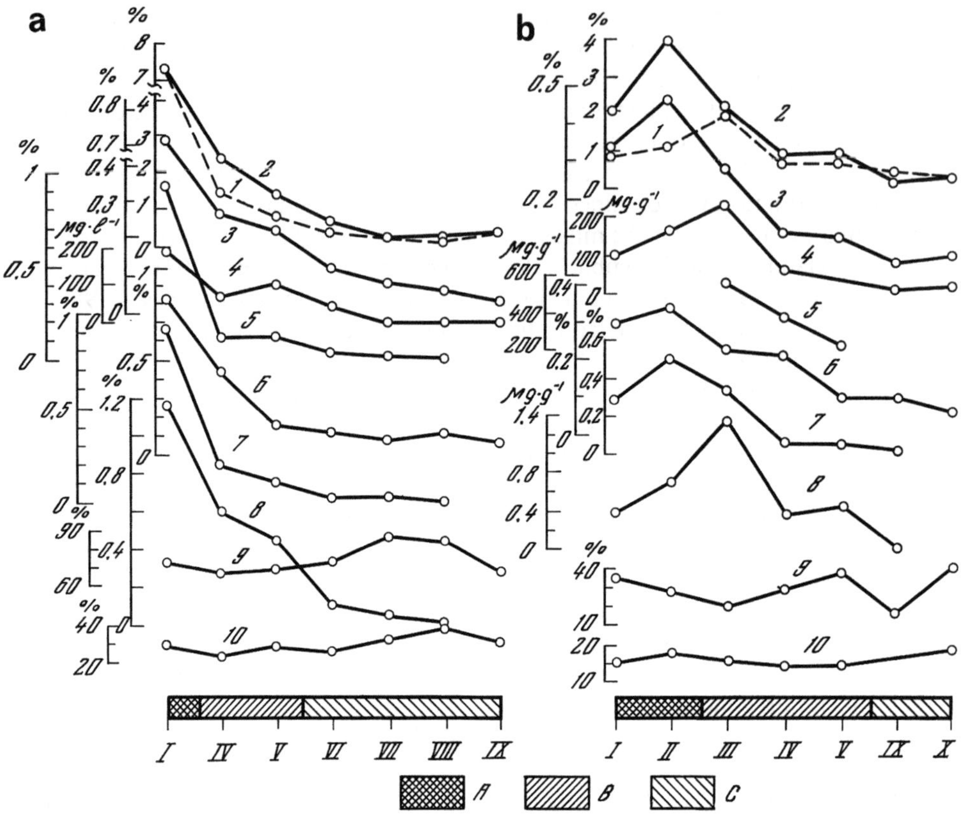

Fig. 49 a, b. Biogeochemical composition of various lithofacies in the Pacific Ocean. **a** pelitic muds; **b** sandy-aleuritic sediments.
I terrigenous sediments in shelf areas; *II* carbonate benthogenous (mostly foraminiferal) sediments of the shelf; *III* glauconite sediments in the upper parts of continental slopes; *IV* terrigenous sediments in the upper parts of continental slopes; *V* terrigenous sediments in the lower parts of slopes and at the bottom of abyssal trenches; *VI* terrigenous hemipelagic muds; *VII* miopelagic red clays and diatomaceous-radiolarian silt; *VIII* eupelagic red clays; *IX* pelagic carbonate foraminiferal sediments; *X* volcanogenic sediments.
Content in dry sediment: *1* C_{org} (%) derived from all sample data; *2* C_{org} (%) for samples in which the OM composition was determined; *3* N_{org} (%); *4* proteins (μg per g); *5* amino acids (μg per g); *6* carbohydrates (%); *7* free lipid-bitumoids (%); *8* humic acids (%); *9* <0.01 mm fraction; *10* <0.001 mm fraction. *A* reducing environment (negative Eh values, excluding the topmost layer, 1–2 cm thick); *B* moderate oxidation environment (predominantly positive, not high Eh values in 0–10 cm layer; below: negative values); *C* oxidation and strong oxidation environments (positive Eh values throughout the penetrated sedimentary column) (Romankevich 1974a)

with molecular weights >10,000 have not been encountered in the upper layer, whereas ATP is present in traces only within the 0.3 cm layer. This holds true for the proteinaceous compounds in most of the investigated samples of pelagic sediments (types VII–IX on Fig. 49). The destruction of the proteins and their incorporation into non-extractable complexes (these processes cannot as yet be separately evaluated) terminate mainly at the water-sediments interface and within several centimeters of the upper sediments. The OM content of proteinaceous compounds with molecular weights >10,000 tends to be 150 – 200 times less in the sediment surface layer than in the plankton. At the water-sediment interface, the proteinaceous compounds are an important source of energy for the benthos. In the sedimentary columns they are markers of the microbial OM degradation processes, giving only an indirect indication of the order of magnitude of these processes (Romankevich 1974a, b, 1976).

At present, information is available on the amino acid composition of the suspended matter and bottom sediments in different areas of seas and oceans (Erdman et al. 1956, Rittenberg et al. 1963, Stevenson and Tilo 1966, 1970, Bada et al. 1970, Degens 1970, Wehmiller and Hare 1971, Starikova and Korzhikova 1972, 1973, Litchfield et al. 1974, Drozdova 1977, Andreyeva and Agatova 1981, 1982, Setchell 1981, Zaslavsky 1981, Romankevich et al. 1982). The amino acid content in the suspended matter of the surface layer varies from tens to a few hundred micrograms per litre (Table 42) and decreases in the vertical profile; for example, in the Japan Trench (0 – 5000 m) – from 43.8 to 9.9 – 21.5 $\mu g\,l^{-1}$; in the Black Sea, at two deep water stations – from 91.0 – 43.8 to 26.8 – 31.0 $\mu g\,l^{-1}$, at a near-shore station (0 – 700 m) – from 108.2 to 19.8 $\mu g\,l^{-1}$; at two stations in the Sargasso Sea – from 27 – 83.0 to 5 – 10 $\mu g\,l^{-1}$; in the area of the experimental "Mohole" drillings (0 – 3000 m) – from 20.6 to 6.4 $\mu g\,l^{-1}$ (Rittenberg et al. 1963, Degens 1970, Starikova and Korzhikova 1972, 1973). Most of the particulate amino acids are present throughout the water column in combined forms, for instance, 80 – 98% in the Black Sea, and cannot be extracted prior to acid hydrolysis. The total content of nitrogenous compounds in the suspended matter, which consist predominantly of amino acids, averages 50% of the OM (see Table 38). The amino acid nitrogen in the suspended matter of the 0 – 5000 m water column over the Japan trench, amounts to 42% of the N-POM (Starikova and Korzhikova 1973), whereas in the suspended matter of the northwestern Pacific (0 – 7 m layer) its share is 35 – 76% (an average of 55.7%) (Romankevich 1974b).

The uppermost (0 – 0.5 cm) layer of various types of shallow-water sediments has a characteristically high content of protein-like compounds, determined with different modifications of the Lowry method. Thus, in the muds and sands of the bays of Posiet and Peter the Great, the content of protein-like compounds in this layer (with the bottom depth being 4 – 69 m) is 21 – 67%, averaging 39% (Andreyeva and Agatova 1982).

The concentration of protein-like compounds (as well as that of CHC, lipids, nucleic acids) in the 0.05 cm layer is found to be in direct proportion to the OM content. The underlying sedimentary layers inherit high concentrations of protein-like compounds if there has been no change in sediment accumulation rates, the granulometric and material composition of deposits, the supply of OM to the bottom or the intensity of the stirring of sediments by the benthos (bioturbation).

The investigation of sediments from the Sea of Japan, Bering and Chukchee seas, the Canari and Peruvian regions has made it possible to distinguish the detrital fraction of OM (pellets, other aggregates, fragments of organisms, isolated by means of flotation in a heavy liquid, 1.5 $g\,cm^{-3}$). The content of protein-like compounds in the detrital OM (making up 0.6 – 4.1% of OM) determined from the ninhydrine reaction, amounts to 16–54% and, apparently, decreases with the change-over from sands to silts and muds, as well as with the growing depth (Zaslavsky 1981).

The amino acid content in the upper layer of deepwater sediments of the Pacific is 14–9949 $\mu g\,g^{-1}$, or 0.5–7.2%, an average of 3.0% of the OM (Starikova and Korzhikova 1970, 1973 and unpublished data from 21 samples from the southeastern part of the Pacific). The maximum amino acid concentrations (both in terms of µg per g and % of OM) are restricted to the shallow water muds, that are rich in OM, whereas the lowest values are found in the eupelagic red clays (0.8% of the OM; north-western Basin) and the abyssal sands (0.5% of the OM; 2020 m in depth; Japan Trench). This refers also to the amino acid nitrogen content, which in the investigated samples was found to be 1.2–20.9% (averaging 7.3% of the total N_{org}).

In the terrigenous sediments the amino acid content, estimated in the dry sediment and in the OM, decreases with depth, whereas in the red abyssal clays no significant variations have been observed. The amino acid nitrogen content, calculated from the total N_{org}, reveals a similar distribution pattern. It is conceivable that the rate of destruction of amino acids is higher in the terrigenous sediments than in red abyssal clays, owing to their more labile OM composition and overload in respect to oxygen of individual groups of compounds, which makes them unstable to enzymic hydrolysis and spontaneous destruction (Andreyev 1956).

The level of amino acid concentrations in sediments of the seas is, in general, higher than in the oceanic sediments; for example, in the Black Sea it is in the range between 1551–7373 $\mu g\,g^{-1}$, averaging 3650 $\mu g\,g^{-1}$. However, in the Black Sea muds, same as in the ocean sediments, amino acids are only minor constituents, accounting for 0.7–5.4% and 1.8–6.3% of the total OM, respectively. Although many chemical reactions are known by which proteins can be altered to paraffinic, naphtenic and aromatic hydrocarbons, no great importance should be attached to proteins, or to amino acids, as precursor materials for petroleum.

During the decomposition of proteins, amino acids are released into the interstitial waters, where in the presence of ammonification and nitrification bacteria, they undergo deamination resulting in the formation of ammonium and nitrates. The level of amino acid concentrations in the interstitial waters is 10 times higher than in the sea waters, averaging 1.7 $\mu g\,l^{-1}$ in the sediments of the northwestern Pacific (Starikova and Korzhikova 1973). In the interstitial waters of the deep pelagic muds the amino acid content, in contrast to the C_{org}, does not show such a significant decrease. As a consequence, the amino acid concentration in the OM of the interstitial waters has a tendency to increase from coastal sediments to pelagic muds (this is characteristic also of the total nitrogenous compounds (Fig. 46). The specific amino acid content (the number of amino acid units in the interstitial waters relative to 100 amino acid units in the solid phase of the sediments) simultaneously increases. In general, the OM of the interstitial waters contains relatively more amino acids than that of the enclosing sediments (Table 41). This applies also to the amino acid nitrogen calculated from the total N_{org} (average 16%).

It is conceivable that amino acids, released into the interstitial waters, are resistant to decomposition and can be preferentially preserved in pelagic sediments under conditions of extreme oxidation. Several factors are responsible for this phenomenon. Besides the inherent nature of the complexes, in which the amino acids are incorporated, important factors are the low level of bacterial activity in pelagic sediments, and the apparently low rate of removal of amino acids from the liquid phase of the sediments.

During the process of sedimentogenesis and early diagenesis, the level of amino acid concentrations undergoes a 10–20-fold decrease in the sequence plankton – particulate fraction – sediments (see Fig. 48). A similar sharp drop, but higher by a factor, is obvious in the content of the high molecular proteinaceous compounds. From this it is inferred that the OM composition undergoes essential reconstruction in the water column, especially in the upper layers and at the water-sediment interface, as a result of which only a small amount of the proteins and amino acids is buried in the sediments. Only 2–7% of the OM that sinks to the bottom undergoes burial, which to a great extent, is due to the mineralization of the proteïnaceous compounds. In aqueous environments (chiefly at the sediment surface) strong interreactions occur between the degradation products of proteins and other forms of OM, and a portion of these are preserved in forms of complex compounds (Degens 1970, Stevenson 1973, 1974). Evidence supporting this is found in the decrease in the proportion of free amino acids in the sediments and interstitial waters, which shows a general trend towards the pelagic sediments, and also in the presence of amino acids in the humic acids of the sediments (see Chap. 9).

The suspended matter, sediments, sea- and interstitial waters contain the same array of amino acids, which closely resembles the amino acid spectrum of the plankton (Table 42). However, the proportions of various groups and individual amino acids do not remain constant throughout the water column and underlying sediments. The composition of amino acids may vary in different areas. For example, in the particulate fraction of the waters over the Japan Trench, there is no indication of a systematic increase with depth in basic amino acids and glycine, and a simultaneous decrease in the alanine and serine contents, as was observed in the Sargasso Sea (Degens 1970). By contrast, the serine content is enhanced in the deep water particulate fraction over the Japan Trench. No feasible explanation has as yet been offered for the high valine content in the ocean waters, sediments and interstitial waters of the Japan Trench. No such concentrations have been reported for the plankton, suspended matter, sea waters and sediments of other oceanic areas.

The mean values for amino acid composition in the suspended matter and in solution are closely similar. The particulate matter, in contrast to the dissolved, has a low ornithine content. The latter is present in conspicuous quantities in nearly all samples of sea waters. Ornithine and urea are easily produced from arginine, and, like glycine and serine, are metabolically related excretory products. From the suspended matter to the sediments a general increase is observed in the content of neutral amino acids (which results from the decrease of monoaminomonocarbonic acids) and basic amino acids, whereas there is a decrease in the content of acidic, sulphur-containing, and, apparently, also aromatic amino acids. The underlying column of sediments shows a tendency to maintain this trend in the change of the amino acid composition. Diagenetically the most stable amino acids, and of most wide occurrence in ancient sediments and meteorites, are glycine, α-alanine, izoleucine, leucine, serine (all neutral), ornithine and acidic amino acids. Thus the increase of neutral amino acids from the suspended matter to the bottom sediments can be attributed to their inherent properties and preferential preservation. The increase in the content of labile basic amino acids in the surface sediment layer, as compared to the suspended matter, which has been observed in discrete areas, is probably a result of OM modification by the burrowing organisms.

7. Carbohydrates

Carbohydrates and their derivatives are among the most essential organic compounds of the biosphere. They occur as early products of photosynthesis, and are the major source of energy for living organisms, accounting for over 60% of their energy requirements. Carbohydrates exist in living organisms in combined forms, being associated with proteins, lipids and nucleic acids, and constitute intricate high molecular complexes which form the basis of living matter. They are incorporated in the cells in forms of various derivatives, and are functionally involved in the structure, energy- and growth-regulating processes of all living organisms.

Carbohydrates include compounds of highly variable molecular weights; from low molecular substances, containing only several carbon atoms (mono- and oligosaccharides) to polysaccharides, with molecular weights up to several millions. Among the various groups of carbohydrates, monosaccharides (polyhydroxycarbonyl compounds – polyhydroxy aldehydes, polyhydroxy ketones – open chain carbon-carbon linkages) are of the highest reactivity, owing to which free sugars are of rare occurrence in nature. Oligosaccharides are intermediate products between mono- and polysaccharides, and their building blocks are the remains of monosaccharides, connected by o-glycosodic linkages. Arbitrarily they are defined as compounds containing from 2 to 10 monosaccharide units. In living organisms oligosaccharides are usually found as reserve substances. Polysaccharides are high molecular polycondensation products of monosaccharides which are linked with each other by oxygen atoms. Some of the polysaccharides contained in living organisms are functionally involved in the skeleton (structural polysaccharides, including, for instance, cellulose and chitin), whereas others constitute sources of energy (reserve polysaccharides, for example, starch-like polysaccharides, laminarine, mannane).

Most carbohydrates are unstable outside of living organisms and undergo degradation to end-products. This causes a drop in the pH, Eh of the surrounding environments and affects the metabolic processes of the aqueous organisms. Interreactions between the intermediate decomposition products of high molecular complexes, consisting of carbohydrates, and with block structures of other biopolymers, lead to the formation in natural environments of geopolymers, containing carbohydrate carbonyl units. In chemical-analytical studies, such products are traditionally defined as carbohydrate-like compounds (CHC). Among them are highly stable compounds, which, as will be shown below (Chap. 9), are incorporated in the structure of specific geopolymers, such as humic and fulvic acids and humins.

Numerous publications are devoted to studies of carbohydrates in various natural sources. The first comprehensive study, based on extensive information obtained on carbohydrates (in living organisms, waters, bottom sediments and ancient rock deposits) was published by Vallentyne (1963). During two last decades a wealth of information accumulated on the behavior of carbohydrates in the process of sedimentogenesis, and this has been the subject of several reviews (for example, Uspensky 1970, Artemiev 1976, Degens and Mopper 1976, Parsons et al. 1977, Romankevich 1979, Mopper 1980).

7.1 Carbohydrates in the Major Producers and Consumers of Organic Matter in the Ocean

The content of carbohydrates in marine organisms varies between a fraction of a percent and 60–70%, on a dry weight basis. Brown and red algae contain the maximum amount of carbohydrates (up to 74% of the OM). Planktonic algae commonly contain less carbohydrates. The lowest content of carbohydrates is found in the zooplankton and zoobenthos: between a fraction of a percent to several percent (Table 44). However, in certain individual representatives of the zooplankton the carbohydrate content coupled with the chitin content can reach 10% (Raymont et al. 1969, Barashkov 1972, Bordovsky et al. 1974, Romankevich 1979). During certain periods, when synthesis of proteins and lipids is retarded, the concentration of carbohydrates in zooplankton may reach 20% and more, of the dry weight of the organism.

Among the unicellular algae, the group- and monosaccharide composition of the carbohydrates have been best investigated with the diatoms (Table 44; Serenkov and Pakhomova 1959, Parsons et al. 1961, 1977, Handa 1969, Handa and Yanagi 1969, Barashkov 1972, Haug and Myklestad 1976). In these studies it was established that the content and composition of carbohydrates depend strongly on the phase of growth (Haug and Myklestad 1976; Table 45). Polysaccharides account for about 90% of the total carbohydrates in the diatom algae, the WSP and WISP being of equal occurrence (Table 46; Handa and Yanagi 1969). The content of alcohol-soluble carbohydrates (oligosaccharides) is not high, and free simple sugars have been encountered in very small quantities, only in singular diatom species and in the golden and blue-green algae.

Among the carbohydrates found in diatom algae are cane sugar, maltose, dextrane, starch, hemicellulose and cellulose. Hemicellulose was found to contain glucose, galactose, mannose, rhamnose, thyvelese, also cellulose and other polysaccharides, which consist of glucose, arabinose and xylose (Serenkov and Pakhomova 1959). The most labile carbohydrate fraction soluble in ethanol-acetone mixtures, that are present in the diatom algae (*Sceletonema costatum*), are glucose and related oligosaccharides. Glucose is also the major sugar of water- and acid-soluble carbohydrates, which consist predominantly of β-1,3-glucane, which serves as reserve material for diatom algae (Handa 1969, Myklestad 1974, Haug and Myklestad 1976). The fact that other sugars, besides glucose, are present in the WSP fraction, demonstrates its more complex composition compared to the fraction which is soluble in alcohol-acetone mixtures. The structure of the WISP fraction of the diatoms is unknown, but mannose may

Table 44. Content of carbohydrates in phytoplankton, phytobenthos, zooplankton and bacteria

Sample studied	Carbohydrate content (%)		Source
	of dry substance	of OM	
Phytoplankton			
Diatoms			
Chaetoceras sp. (screen sample from off Peru)	5.69	14.5	this volume
Chaetoceros sp. (culture)	6.6	7.3(9.0)[a]	Parsons et al. (1961)
Sceletonema costatum (culture)	20.8	26.6(29.1)[a]	Parsons et al. (1961)
Coscinodiscus sp. (culture)	4.1	8.7(11.2)[a]	Parsons et al. (1961)
Phaeodactylum tricornutum (culture)	24.0	19.4(19.9)[a]	Parsons et al. (1961)
Sceletonema costatum (culture)	–	16.5 – 25.8	Handa (1969)
Rhabdonema adriaticum (outwash from Cystosyra, Black sea)	–	13.63	Serenkov and Pakhomava (1959)
Chaetoceros decipiens (screen sample, Danish Strait)	–	13.20	Serenkov and Pakhomava (1959)
Chaetoceros affinis	–	22.5; 42.6[e]	Haug and Myklestad (1976)
C. curvisetus	–	29.9; 34.0[e]	Haug and Myklestad (1976)
C. decipiens	–	– 72.5[e]	Haug and Myklestad (1976)
C. debilis	–	14.3; 68.5[e]	Haug and Myklestad (1976)
C. socialis	–	30.9; 47.3[e]	Haug and Myklestad (1976)
Thalassiosira gravida	–	18.7; 40.2[e]	Haug and Myklestad (1976)
Corethron hystrix	–	12.8; 63.3[e]	Haug and Myklestad (1976)
Average	–	~15	
Peridinium			
Amphidinium carteri (culture)	30.5	25.9(26.8)[a]	Persons et al. (1961), Strickland (1965)
Exuviella sp. (culture)	37.0	31.3(42.6)[a]	Persons et al. (1961), Strickland (1965)
Cachonia niei	–	14.0 – 29.4	Strickland et al. (1969)
Gonyaulax polyedra	–	12.8	Strickland et al. (1969)
Average	–	~23	
Golden			
Monochrysis lutheri (culture)	31.4	28.2(29.1)[a]	Parsons et al. (1961)
Syracosphaera carterae (culture)	17.8	20.2(20.6)[a]	Parsons et al. (1961)
Average	–	~24	

Table 44 (continued)

Sample studied	Carbohydrate content (%)		Source
	of dry substance	of OM	
Blue-green			
Agmenellum quadruplicatum	31.5	28.0(33.0)[a]	Parsons et al. (1961)
Aphanizomenon flosaquae, Anabaena fiosaque	–	7.4–11.4	Serenkov and Pakhomova (1965)
Marine and freshwater	30–55	–	Holm-Hansen (1968)
Phytobenthos			
Brown algae	–	~60–70	Mori (1953), Hirst (1958), both quoted by Barashkov (1972)
Red algae	–	~60–70	Mori (1953), Hirst (1958), both quoted by Barashkov (1972)
Zooplankton			
Undinula darwini	–	1.9[b]	this volume
Rhincalanus cornutus	–	2.8[b]	this volume
Euchaeta marina	–	1.4[b]	this volume
Candacia aethiopica	–	1.9[b]	this volume
Calanus sp.	<1	~1[b]	this volume
Decapoda	1–3[c]	2.2(7.0)[d]	Raymont et al. (1969)
Mysidea	2–3[c]	2.7(7.0)[d]	Raymont et al. (1969)
Euphausidea	2–3[c]	2.8(4.0)[d]	Raymont et al. (1969)
Deep water decapods and mysiids	3.4–3.8	–	Vinogradov et al. (1970)
Average	2–3	2	
Bacteria	10–30	–	Kuznetsov (1970)

[a] Figures in parenthesis represent total carbohydrates, soluble in acids and alkali
[b] Carbohydrate content in OM of individual organisms, in %. The analyses were made during the "Vityaz" cruise (1971). The samples of the organisms were procured by T. S. Petipa
[c] Mean values
[d] Figures in parenthesis represent the chitin content
[e] First figure equals total carbohydrate content during the logarithmic growth phase; second figure equals same during the stationary stage

be the dominant monosaccharide (Table 47; Parsons 1963, Handa and Yanagi 1969).

The alkali-soluble polysaccharides of the diatom algae, which were isolated after the removal of acid-soluble carbohydrates, contained a broad spectrum of monosaccharides, the relative amount of which varied with the species of diatom algae and their growth stages (Table 45). This implies a complex heteropolysaccharide composition of the alkali-soluble carbohydrates occurring mostly in the algal cellular walls.

Table 45. The composition of carbohydrates of diatom algae (Haug and Myklestad 1976)

Species	Phase[a]	Acid-soluble (0.1 N H_2SO_4) fraction (% of OM)	Alkali-soluble (0.1 N NaOH) fraction (% of OM)	Composition of alkali-soluble fraction, % of the total fraction						
				Rha	Fuc	Rib	Xyl	Man	Gal	Glu
Chaetoceros affinis	S	29.4	14.2	52	11	n.f.	7	6	18	2
	L	9.7	12.5	25	7	15	7	11	22	3
C. curvisetus	S	24.8	10.3	16	32	10	4	6	31	1
	L	20.4	8.5	13	14	18	8	12	30	5
C. decipiens	S	60.6	9.9	25	4	4	10	32	7	6
	L	63.6	8.4	27	5	10	2	24	10	n.f.
C. debilis	S	60.8	4.7	22	11	23	15	n.f.	23	6
	L	9.3	8.1	22	8	31	10	2	23	4
C. socialis	S	43.8	5.8	23	18	11	8	9	16	3
	L	28.1	7.1	10	10	20	12	15	19	7
Thalassiosira	S	34.4	6.6	n.f.	43	27	7	7	12	4
	L	12.4	8.8	n.f.	25	27	12	12	17	7
Corethron hystrix	S	49.6	8.2	1	60	8	2	9	9	11
	L	4.5	8.7	3	31	17	12	12	13	12

[a] S and L: stationary and logarithmic phases, respectively

Table 46. Group composition of carbohydrates in diatom algae (% of total carbohydrates) (Handa and Yanagi 1969)

Diatom algae	ASS	WSP	WISP
Phaeodactylum tricornutum	3.5	40.2	46.4
Chaetoceros sp.	6.7	55.9	37.4
Skeletonema costatum	3.5	42.6	55.8
Average	4.6	46.2	46.5

The nature of the carbohydrates in peridinium algae has been little studied. It is commonly considered that they consist of cellulose-like and pectine-like polysaccharides. Among the reserve carbohydrates a polysaccharide was found which is closely similar to starch and is a structural polymer containing β-1,3 and β-1,4 linkages, rendering it quite resistant to hydrolysis.

The major carbohydrates present in blue-green algae are gel-like polysaccharides, which have more affinity to the carbohydrates in red and brown algae than those in green algae, in that they are not readily hydrolyzed, incapable of acetylization, are hydrophobic, and contain esters of polyoles and residual sulphuric acid. In blue-green algae tregalose, saccharose and fructoses were found (Goriunova et al. 1969, Barashkov 1972).

Table 47. Monosaccharide composition of carbohydrates in diatom algae (% of total carbohydrates)

Diatom algae	Carbo-hydrate fraction	Hexoses				Pentoses				Uniden-tified
		glucose	galactose	mannose	rhamnose	arabinose	xylose	ribose	fucose	sugars
Phaeodactylum tricornutum	Water-extract	61.4	3.1	8.1	6.7	4.1	13.8	1.8	0.3	0.7
	WISP	10.8	7.2	54.2	6.9	2.5	11.9	2.7	3.8	0.0
Chaetoceros sp.	Water-extract	86.1	2.2	7.5	1.1	0.2	2.7	0.1	0.1	0.0
	WISP	0.0	17.7	66.6	5.1	0.2	0.7	0.5	2.4	0.0
Sceletonema costatum	Water-extract	90.0	0.0	1.5	1.0	6.0	0.5	0.6	0.5	0.0
	WISP	6.5	2.8	50.5	13.0	0.8	15.7	0.8	10.4	0.0
Chaetoceros sp.	–	34.7	15.8	8.3	29.5	–	4.2	7.5	–	–
Sceletonema costatum	–	74.1	8.1	3.9	4.5	–	–	5.4	4.0	–
Coscinodiscus sp.	–	51.1	9.7	10.0	17.0	–	–	–	12.0	–

The composition of carbohydrates in brown and red algae has been investigated in greater detail. The red algae were found to contain tregalose, the sugar alcohols dulcite and sorbite, a polysaccharide, agar, which may reach to 60% of the dry weight of the algae, and also karaginan in the cell-wall. In the brown algae mannitol, alginic acid, laminarine are most abundant (Barashkov 1972, Combaut et al. 1981).

Net plankton carbohydrates (0 – 200 m layer, net No. 10, Mopper 1980) are characterised by the high variability of their content and the lack of a dominant sugar. In this respect the plankton studied sharply differs from the land plants, the hydrolyzate of which shows the prevalence of glucose among the monosaccharides. The common polysaccharides in marine organisms are rhamnans, xylans, mannans, glucuromannans, chitans (Mackie and Preston 1974, quoted by Mopper 1980).

A large amount of carbohydrates is liberated by algae into the water, as a result of autolytic division of normally functioning cells, extracellular excretion of metabolites of low molecular weight, and other specific activities. The quantity of organic products liberated depends upon many factors: physiology, age of organisms, their food supply and light intensity, and is highly variable (Hellebust 1965, Handa 1969, Khaylov 1971, Myklestad 1974, Haug and Myklestad 1976, Burney et al. 1981, Ittekkot et al. 1981, Tambiyev and Kirikova 1981). Khailov (1971) estimated that extracellular excretion products account for 15% of the total OM contributed by primitive, and 37% of that from more complex algae. The bulk of carbohydrates is liberated at the termination of phytoplankton bloom (Ittekkot et al. 1981). Investigations have shown, that the dissolved carbohydrates found in ocean waters form only a portion of the total liberated by living organisms, which is due to the rapid utilization of the carbohydrates by marine organisms. The quantity of dissolved carbohydrates synthesized by phytoplankton in bacterial-sterile cultures may be considerably higher than that produced in natural environments (Wangersky 1959).

In the excretory products of algae, carbohydrates occur primarily as polysaccharides. For instance, the diatom algae *Thalassiosira fluvialis* releases into the water fibres, consisting of microfibrilla poly-N-acetyl-α-glucosamine. However, in short-duration experiments only low molecular compounds have been detected – the earliest products of photosynthesis, such as glycolic acid (Lewin 1956, Hellebust 1965). The products of excretion of some marine algae may also include sugar spirits, such as glycerin and mannitol. Extracellular carbohydrates have a rather complex composition differing, apparently, from that of cellular walls and plasm. In some diatom species studied, among the extracellular polysaccharides there were found six monosaccharides with the prevailing ramnose, fucose, galactose (Haug and Myklestad 1976).

Extracellular carbohydrates are liberated by living organisms during all stages of their growth, but the major mass is released into the surrounding environment during the stationary stage of growth of the organisms and after their death. In the dark, the excretion of carbohydrates probably ceases (according to the investigations on diatom algae); this refers primarily to low-molecular oligo- and water-soluble polysaccharides, which are utilized as the substrate for photorespiration (Handa 1969). Studies of the algal metabolites of the carbohydrate type using ^{14}C tracer techniques revealed that they are taken up by marine invertebrates, and become functionally involved in the respiratory and energy exchange of these organisms. In these experiments the labelled carbon-14 was traced among all the major biochemical constituents of the marine invertebrates – in the proteins, lipids, calcite skeleton and the respired CO_2 (Yerokhin 1972). This implies that during the metabolic processes new compounds can be formed using carbohydrates in different types of OM, and serves to indicate the ecological significance of carbohydrates.

Research involving the enzymic breakdown of green and blue-green algae established that the first products that undergo decomposition are the alcohol and water-soluble carbohydrates, and of much slower rate of decomposition, hemicellulose and cellulose (Oparin and Bezinger 1950, Shabarova 1950). As a consequence of the rapid utilization of the WSP, particularly of glucane, by the algae, and their decomposition outside the living organisms, the content of the WSP in the particulate fraction, as will be shown below, is much less than in the plankton.

The cell tissue is the most stable structural element of the algae, bacteria, protozoa, fungi, leaven. In certain types of bacteria and fungi, 90% of the cell tissue is comprised of carbohydrates (glucane, mannane). The carbohydrate content in the cell tissue of gram-positive bacteria is 35 – 60%. The total carbohydrate content in the cell walls of gram-negative bacteria, which are of dominant occurrence in the ocean, is 15 – 20% (Rose 1968, Kuznetsov 1970), i.e. practically the same as in the particulate OM, sea water and in the OM of the pelagic sediments (see below).

The cell wall in microorganisms consists of the matrix and microfibrilla which are immersed in the latter and toughen it. The wall tissues account for 10 – 50% of the total dry mass of the microorganism, 50 – 80% of which is comprised of microfibrilla. Cytoplasm, other adventives of the cell tissues (cilia, tendons, hairs) and mucous membranes also contain carbohydrates (about 10%). But they undergo rapid decomposition, partly pass into solution, and their possible long-time preservation in sea waters, in the particulate fraction is doubtful. The structural polysaccharides are highly stable only in relative terms: for instance, enzymes of the lysozyme type are able to split the glycopeptide matrix of the cell walls of some microorganisms.

Several types of polymers were detected in the cell matrix of algae; most of these are heteropolysaccharides. In the *Chlorella pyronoidosa*, which has been studied in most detail, the microfibrilla consists of cellulose, whereas the matrix consists of galactoramnan with β-linkages, and also of a more complex heteropolysaccharide. The microfibrilla in other types of algae consists of xylenes and mannanes. The cell wall of the brown algae contains alginic acid, fuccidine (sulphated polymer, comprised of the residue of L-fucoses). The cell walls of the blue-green algae contain uronic acid (pectine-like). Glycopeptides occur as structural elements of most types of bacteria, and on acid hydrolysis yield glucosamine and amino acids (D- and L-alanine, D-glutamic acid, diamine acids). Structural units of the glycopeptides are N-acetyl glucosamine (chitinous structures) and N-acetyl muramic acid (derivatives of N-acetyl glucosamine), which are connected by β-1,4 linkages. The cell walls of gram-negative bacteria contain, besides glycopeptides, lipopolysaccharides, and the walls are of more complex composition compared with those of the gram-positive bacteria.

The most stable forms in which carbohydrates are preserved are diatoms with cell walls containing silica in the form of silica gel ($SiO_2 \cdot nH_2O$) or opal ($SiO_2 \cdot 2H_2O$), and also carbonate-inoculated cell walls, or the cell walls, of golden algae which have silicified parts. Carbohydrates of cellular walls of diatom algae are seemingly represented by heteropolysaccharides resistant to bacterial and chemical decomposition. Most of the structural polysaccharides form the exterior shell which, owing to its hydrophilic nature, facilitates the mutual exchange between the cell and the outer environment, and protects the deeper lying layers of glycolipids and proteins, as well as the silica frustule, from dissolution and decomposition. A wide array of sugars (rhamnose, furose, ribose, arabinose, xylose, mannose, galactose, glucose) have been found among the monomers of diatom cellular walls; their total content averages 0.6 – 3.7% on the dry weight basis, depending upon the algal species and its development stage (McLachland et al. 1965, Hecky et al. 1973, Haug and Myklestad 1976).

Among the stable compounds which are most likely to be preserved in the environment of the organisms are starch grains of algal origin, such as those that were recovered from the Arctic waters off Canada, the Baltic, Mediterranean Seas, and the Atlantic and Indian Oceans. They are synthesized by pyrrophyte algae, macrophytes and flower plants, more seldom by green planktonic algae. These starch grains retain their forms after the breakdown of the protoplast. Their content in surface waters (0 – 50 m) may reach to 100 – 3155 $\mu g\,l^{-1}$. The fact that they occur in conspicuous amounts in fecal pellets serves to suggest that they are not readily assimilated by phytophages (Bursa 1968), with the result that they may be selectively preserved in the ocean water column and settle at the bottom.

It is conceivable that living organisms, including the major producers and consumers of OM (planktonic algae, crustacean plankton), contain carbohydrates in structures of variable degrees of resistance to decomposition. The carbohydrates which have accumulated during the growth of the organisms, are released as a result of decomposition of the cells and tissues as the organisms become senescent or after their death. They pass into the sea water where they form new polymeric associates, including the organo-mineral geopolymers which are resistant to biodegradation.

7.2 Carbohydrates in Sea Waters

The first data on the content of carbohydrates in sea waters was obtained through investigations carried out by Collier and his collaborators (Collier et al. 1950) in the coastal waters of the Gulf of Mexico. In later years extensive studies were made on the carbohydrates in the oceans, as a result of which considerable information is available now on the distribution of (mainly) dissolved carbohydrates, their group- and monosaccharide composition (Lewis and Rakestraw 1955, Wangersky 1959, Koyama 1962, Parsons and Strickland 1962b, Kawahara et al. 1967, Handa 1970, Khailov 1971, Starikova and Yablokova 1972, Yerokhin 1972, 1975, Artemiev and Melnikov 1974, Artemiev and Scherbinin 1975, Williams 1975, Artemiev 1976, Michaelis et al. 1980, Mopper 1980, Mopper

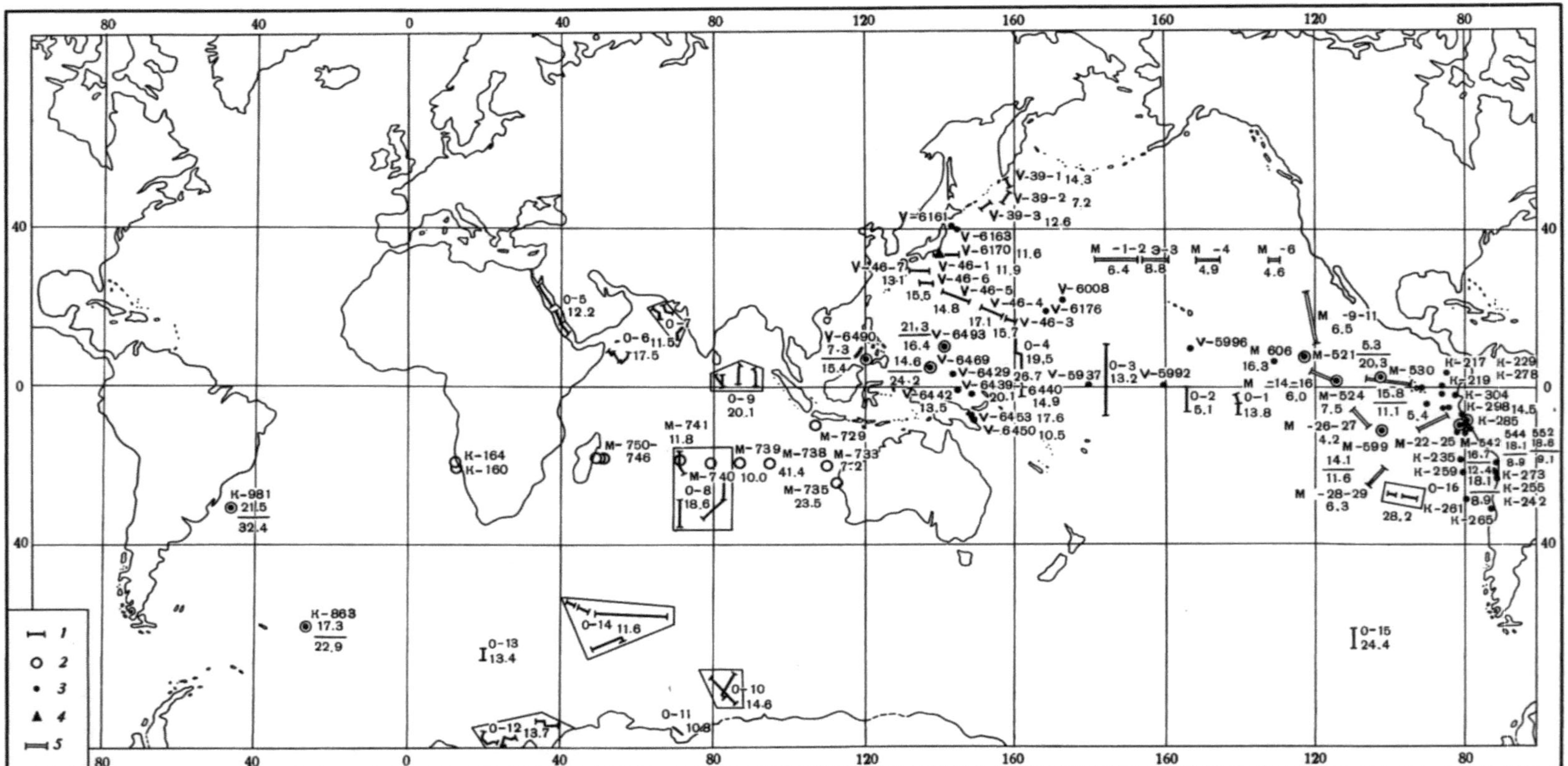

Fig. 50. Locations of sea water and particulate matter probes, which were analysed for their CHC content during the cruises of the r/vs of the Institute of Oceanology (*V* = Vityaz; *K* = Akademik Kurchtov; *M* = Dmitrii Mendeleyev; *O* = Ob). *1* separation suspensate samples; *2* filtrated sea water; *3* filtration suspensate samples; *4* sestone; *5* eolian suspension. *Figures near station numbers* show CHC content in OM (C_{CHC}/C_{org} 100) of surface water layer; *figures above line:* in the POM; *below line:* in DOM

et al. 1980, Burney et al. 1981, Ittekkot et al. 1981, Ittekkot and Degens 1982), as well as on the particulate CHC (Parsons and Strickland 1962a, Newell and Kerr 1968, Handa 1970, Romankevich 1971, 1974a, 1975, Romankevich and Urbanovich 1971, Handa et al. 1972, Artemiev 1973, 1976, Artemiev and Melnikov 1974, Hitchcock 1977, Ittekkot and Degens 1982).

These investigations have shown that 90 – 95% of the carbohydrates in the ocean waters are in solution. Extensive analytical studies of the carbohydrates extracted by means of the phenol-sulphuric acid method, were made during the 1966 – 1978 cruises and provided a wealth of information on the concentration and distribution of the dissolved and particulate CHC, as well as the CHC in bottom sediments in various areas of the Pacific, Atlantic and Indian Oceans (Fig. 50).

According to these data, the content of dissolved CHC in seas and oceans varies between traces to several milligram per litre (Table 48). The concentrations of carbohydrates in the near-shore areas of seas and oceans were found to be both very low ($<0.1 - 0.4$ mg l^{-1}) and very high ($2 - 7.9$ mg l^{-1}). Exceptionally high concentrations of CHC were encountered in the shallow waters, in those areas where large quantities of OM, liberated by living organisms and after the death of algae, are contributed to the waters. Enhanced CHC concentrations occur also in the waters near the estuaries of certain rivers, as, for instance, the Gang, Limpopo, Zambezi, which results from the additional supply of carbohydrates from land with the river run-off (our unpublished data). In the near-coastal upwelling areas (Walvis bay, the Peruvian region) the content of CHC is higher than in the open ocean but lower than in some estuaries. The monomer composition of CHC and the markedly enhanced glucose-to-rybose ratio in some estuaries bear evidence of the terrigenous source of a considerable proportion of CHC.

In the surface layer of the deep pelagic areas of the Pacific, Indian and Atlantic Oceans, the concentration of dissolved CHC is $0.2 - 1.4$ mg l^{-1}, that is essentially less than in the near-shore zones of seas and oceans. Beyond the limits of the coastal shallow waters no recognizable variations have been detected in the CHC concentrations, i.e., – in the shelf areas, continental slope and deep basins of the ocean floor (Artemiev 1976).

In the vertical profile of the studied columns, no general trend is apparent in the variations of the content of dissolved carbohydrates, whether measured in mg l^{-1}, or in terms of the OM composition (Fig. 51). Besides, the variations detected in the vertical profile at several stations were small. At the same time it was noted that the fluctuations in the content of dissolved CHC in the water column, observed at two stations in the Southern Atlantic, could be linked, first of all, to different water masses (Artemiev 1976). No such relationship has been established in other areas of the ocean. The mean values of the content of dissolved CHC in the Pacific Ocean, in the 0 – 100 m, 100 – 200 m, 200 – 1000 m layers and below 1000 m, are practically the same – 0.64, 0.60, 0.57 and 0.59%, respectively.

The distribution pattern of the dissolved CHC in the pelagial oceanic areas is, in general, similar to that of DOC. Notwithstanding the wide spread in the values, a direct relationship is obvious between the DOC and dissolved CHC contents (Fig. 52). Like the DOC, the content of dissolved CHC in the western

Table 48. Concentrations of total dissolved carbohydrates in sea water

Sampling site	Method[a]	Concentration mg l^{-1}	Source	Standard
Coastal zone of seas and oceans				
Florida waters	II	2–25	Collier et al. (1950)	Arabinose
Gulf of Mexico	II	0–20	Collier et al. (1953)	Arabinose
South Atlantic coast, USA	II	0–12	Anderson and Gehringer (1958)	Arabinose
Gulf of Mexico, shelf	II	0–3	Collier (1958a)	Arabinose
Gulf of Mexico, western shore	II	0–2.6	Collier (1958b)	Arabinose
Near Scripps Institution pier, USA	I–II	0.14–0.38	Lewis and Rakestraw (1955)	Sucrose
– during phytoplankton bloom	I–II	0.15–0.36	Lewis and Rakestraw (1955)	Sucrose
La Jolla, USA (luxuriant growth of brown algae)	I–II	0.11–0.45	Lewis and Rakestraw (1955)	Sucrose
San Diego coastal lagoons	I–II	0.3–7.9	Lewis and Rakestraw (1955)	Sucrose
Pier at Naples, Florida	II	0.2–4.1	Dragovich (quoted by Yerokhin 1972)	Arabinose
Long Island Sound	II	0–2.5	Wangersky (1959)	Glucose
Cape Cod waters	I	0.40–3.17	Walsh (1965)	Sucrose
– estuaries	I	1.16–3.17	Walsh (1965)	Sucrose
– other waters	I	0.40–1.00	Walsh (1965)	Sucrose
Dalne-Zelenetskaya Inlet	III	0.5–3.0	Yerokhin (1972)	Glucose
– luxuriant growth of brown algae	III	2.5–7.0	Yerokhin (1972)	Glucose
Shelf of Southwest	III	2.20	Artemiev (1976)	Glucose
Africa	III	1.7–1.8	Artemiev (1976)	Glucose
	III	1.3–1.6	Artemiev (1976)	Glucose
Shelf off Peru	III	0.32	this volume	Glucose
	III	0.23–0.40	this volume	Glucose
Kojima and Osaka, Japan	IV	0.48–1.13	Hirayama (1974)	Xylose and glucose
Gulf of Cariaco, Venezuela	III	0.35–3.25	Cedeño and Fermin (1972, quoted by Burney and Sieburth 1977)	Glucose
Narrangansett Bay	V	0.45–2.02	Burney and Sieburth (1977)	Various CHC
Salt Marsh of Narrangansett Bay	V	~0.3–1.2	Burney et al. (1981)	Various CHC
Abyssal and pelagic areas of seas and ocean				
Continental slope	III	0.32–0.37	this volume	Glucose
off Peru	III	0.20–0.40	this volume	Glucose

Table 48 (continued)

Sampling site	Method[a]	Concentration mg l^{-1}	Source	Standard
Eastern Equatorial-Tropical Pacific	III	0.32 – 0.71	this volume	Glucose
	III	0.10 – 0.91	this volume	Glucose
Sulu Sea	III	1.14	Liutsarev et al. (1972)	Glucose
	III	0.50 – 1.22	Liutsarev et al. (1972)	Glucose
Western Equatorial-Tropical Pacific	III	0.99 – 1.15	this volume	Glucose
	III	0.55 – 1.40	this volume	Glucose
South-western Atlantic	III	1.21 – 1.34	Artemiev and Melnikov (1974)	Glucose
	III	0.48 – 2.14	Artemiev and Melnikov (1974)	Glucose
Southern Tropical part of the Indian Ocean	III	0.21 – 1.21	Artemiev (1976)	Glucose
	III	0.10 – 1.31	Artemiev (1976)	Glucose
Black Sea	VI	0.93 – 2.99	Starikova and Yablokova (1972)	Glucose
NW Pacific	II	0.24 – 1.68	Koyama (1962)	Sucrose
NE Pacific	I	0 – 0.4	Antia and Lee (1963)	Glucose
Sargasso Sea	I	0.18 – 0.75	Walsh and Douglass (1966)	Sucrose
NW Pacific	III	0.09 – 0.46	Handa (1967)	Glucose
Indian Ocean	III	0.19 – 0.66	Handa (1966)	Glucose
Off Southern Hokkaido, Japan	I	0 – 1.3	Kawahara et al. (1967)	Sucrose
Off Honshu, Japan	III	0.24 – 0.37	Handa (1970)	Glucose
Northern North Sea		0.04 – 0.40	Ittekkot et al. (1981)	Glucose

[a] I = Anthrone, II = N-ethylcarbazole, III = Phenol-sulphuric acid, IV = Fluorimetric, V = 3 methyl-2-benzothiazolinone hydrazone hydrochloride (MBTH) method, after hydrolysis, VI = Paper chromatography after hydrolysis

equatorial-tropical Pacific is higher than in the eastern areas. Similar regional differences are characteristic of their content in the OM. As a consequence, the content of the dissolved CHC in the OM generally varies over a wide range; in the 0 – 100 m layer 4.7 – 38.8%, and below 1000 m from 8.0 – 38.9%, averaging 16.9% within the studied water column ~0 – 4000 m. The mean values of the CHC content in the OM of different layers are practically the same (16.0 – 19.5%). The small fluctuations in the mean values of the dissolved CHC content (both in mg l^{-1} and in the OM), serve to suggest that the carbohydrates are one of the most conservative constituents of sea waters.

Only limited information is available on the composition of the total dissolved carbohydrates in ocean waters (Degens et al. 1964, Khailov 1971, Starikova and Yablokova 1972, Yerokhin 1972, Williams 1975, Michaelis et al. 1980,

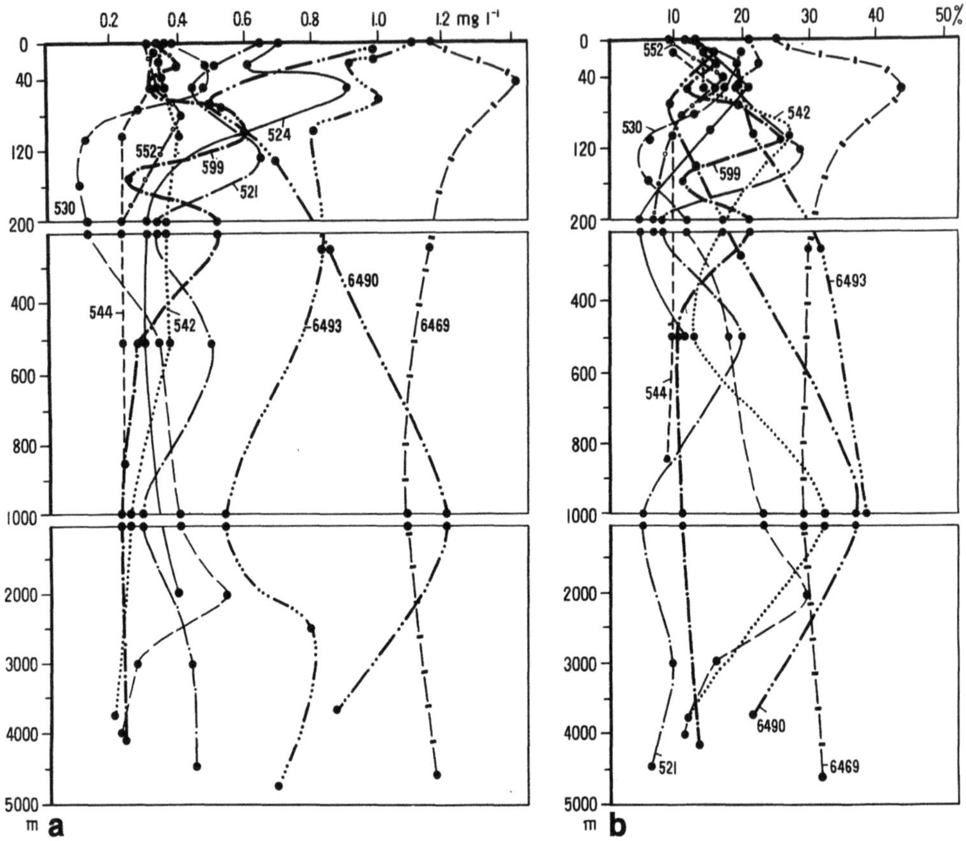

Fig. 51 a, b. Distribution of dissolved carbohydrates in the Pacific Ocean water column. **a** mg l^{-1}; **b** in OM (C_{CHC}/C-DOM 100). *Figures near curves* show Station Nos.

Mopper 1980, Ittekkot et al. 1981). It is generally assumed that in natural waters the carbohydrates are incorporated into high-molecular compounds, occurring in the form of polysaccharides with a mixture of inorganic polymers. These were found to be present in heteropolycondensation humic acids, and in the "yellow stuff" ("gelbstoff"). Studies of the monosaccharide composition of the total dissolved CHC give evidence of the complexity of their structures. Samples taken in the region of California and the Black Sea showed that they consist predominantly of glucose, galactose and mannose, to a less extent of rhamnose, arabinose and, in traces, of xylose and ribose (Table 49). In the dissolved combined CHC (total dissolved minus dissolved free CHC) of the Norwegian Sea during the phytoplankton bloom the glucose made up more than 60% of the fraction concerned; it was followed by mannose, galactose and xylose (Ittekkot et al. 1981). The monosaccharide spectrum of the total dissolved CHC is narrower than that of the plankton or the particulate fraction and is practically invariable throughout the entire water column; the same monosugars predominate to a depth of 2000 m in the Black Sea and 3890 m in the Pacific Ocean, as in the

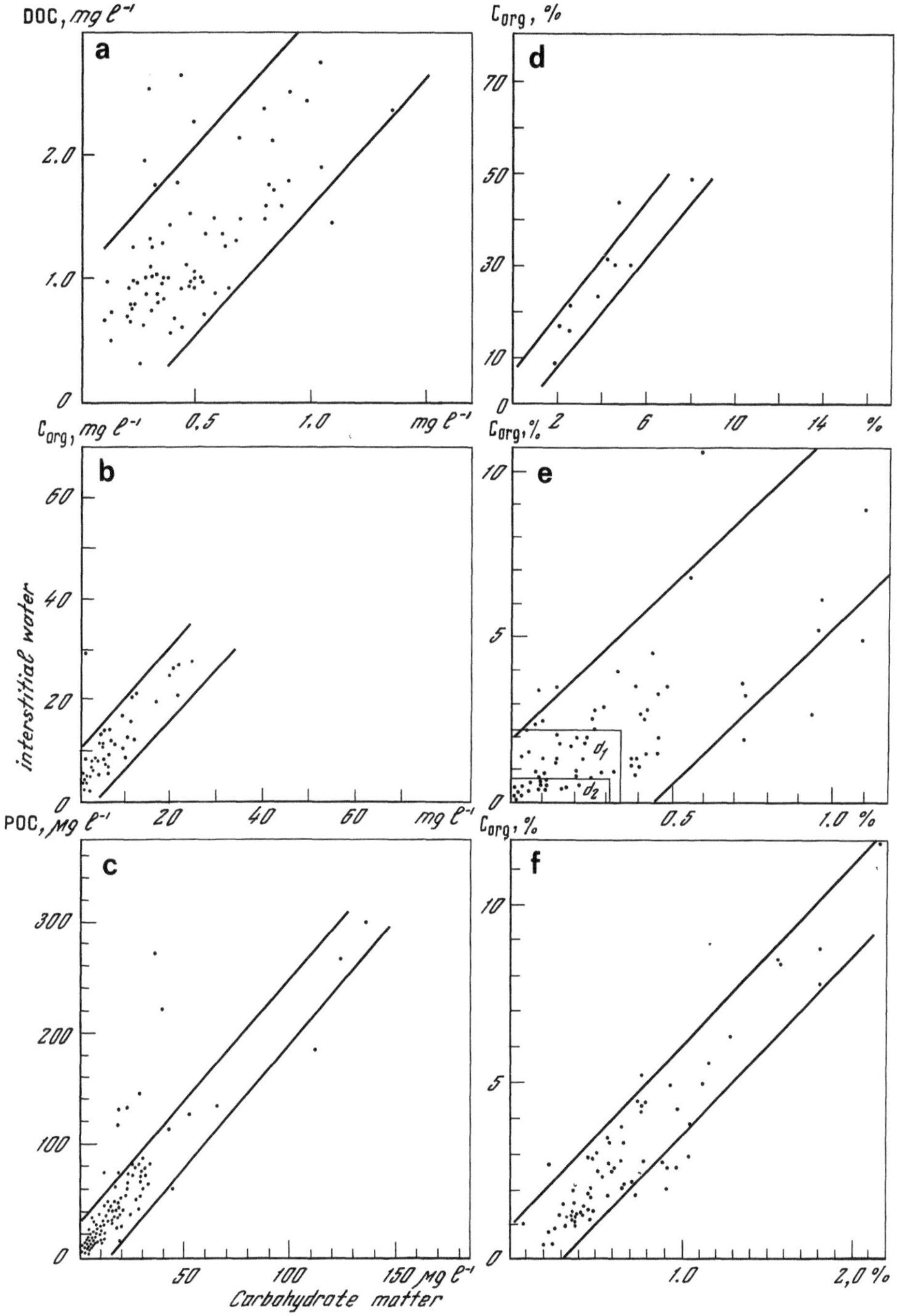

Fig. 52 a – f. Relationship between CHC and C_{org} contents. **a** filtrated sea water from the Pacific; **b** interstitial waters from the Pacific; **c** water suspension of the Pacific Ocean; **d** eolian suspension over the Pacific; **e** sediments of the submarine margin of the Pacific; **d_1** hemipelagic; **d_2** pelagic; **f** sediments of the submarine margin off southwestern Africa

Table 49. Monosaccharide composition of sea waters, suspended matter and bottom sediments (after hydrolysis, % of total sugars)

Sampling area	CHC fraction[a]	Hexoses				Pentoses			
		glucose	galactose	mannose	rhamnose	arabinose	xylose	ribose	fucose
Sea water (solution)									
Area off Caifornia									
0 m	–	40.0	40.0	20.0	–	–	–	–	–
1990 m	–	43.4	9.6	38.2	5.1	3.7	–	–	–
3890 m	–	42.7	8.3	49.0	–	–	–	–	–
Black Sea (0 – 200 m)[b]	–	54.3	traces	45.6	traces	traces	traces	traces	–
Suspended matter									
East of Honshu Island, 0 – 7 m (St. V-46-1)	ASS	79.0	11.8	–	–	–	–	–	9.2
	WSP	46.0	13.5	–	5.4	13.5	13.5	8.1	–
	WISP	35.3	18.5	8.0	7.6	8.7	15.2	6.3	–
Average		40.2	16.2	5.5	5.8	7.6	12.0	5.4	1.4
East of Hokkaido 0 – 20 m	Water-extract	72.7	2.8	7.7	1.1	3.2	12.3	0.5	0.0
	WISP	30.4	10.7	24.0	5.2	10.5	15.6	1.6	2.5
10 – 20 m	Water-extract	84.2	2.8	3.9	1.7	2.6	2.2	1.3	1.2
	WISP	40.3	14.8	14.0	7.0	5.7	8.3	2.3	7.5
South of the Aleutian Islands (0 – 20 m)	Water-extract	67.0	10.3	4.8	0.2	4.6	12.8	0.0	0.6
	WISP	40.9	10.7	11.0	2.1	4.1	23.0	2.7	4.1
Black Sea (0 – 2000 m)[b]	–	31.9	50.9	traces	7.1	–	8.5	traces	–
Bottom sediments									
Shelf off south-western Africa (St. K-209)[c]	ASS	traces	traces	n.f.	traces	n.f.	n.f.	n.f.	n.f.
	WSP	12	18	23	18	traces	23	n.f.	n.f.
	WISP	15	15	19	14	traces	21	5	n.f.
Continental slope off Chile (St. K-252)[d]	WISP	n.f.	25	28	11	n.f.	37	n.f.	n.f.
Peruvian Basin (St. K-231)[e]		n.f.	20	25	26	n.f.	29	n.f.	n.f.

Kuril-Kamchatka Basin									
(St. V-5603)[c]		15	20	18	12	n.f.	16	2	traces
(St. V-5622)[f]		n.f.	55	29	17	n.f.	n.f.	n.f.	n.f.
San Diego Basin									
0 m	Free	–	–	100.0	–	–	–	–	–
	Combined	11.2	13.5	29.5	8.5	14.9	10.1	12.3	–
0.5 m	Free	43.4	–	56.6	–	traces	traces	traces	–
	Combined	9.0	19.6	24.9	7.8	13.8	14.4	10.5	–
Black Sea									
~1500 years	–	13.3	20.0	23.7	4.8	4.6	16.3	9.6	7.6
~5000 years	–	18.8	18.3	10.0	15.1	8.9	12.6	5.5	10.9
New York Bright Hundson Canyon (recent)	–	21.2	22.7	16.7	9.5	7.3	10.3	4.1	8.3
Argentine basin									
– recent	–	19.9	23.7	13.7	8.8	10.2	9.5	6.2	8.1
– 5000 years	–	17.1	24.3	13.7	9.0	10.3	11.2	9.7	9.7
Cariaco trench									
~1000 years	–	13.9	27.0	19.4	11.0	5.9	10.1	4.2	8.0
~5000 years	–	12.4	27.9	18.1	10.9	4.9	9.8	3.1	12.4
Alluvial fan deposits of La Jolla Canyon	–	8.8	13.9	11.3	45.4	7.5	10.7	2.9	–
Saint Nicholas Basin	–	13.7	25.2	23.3	15.5	6.6	13.1	2.6	–
Mohole Drillings									
0 m	–	13.6	18.4	9.7	21.9	14.2	13.4	8.8	–
138 m	–	20.7	18.9	38.8	5.4	3.6	12.6	traces	–
Black Sea (56 – 2150 m)[g]	–	18.5	18.2	29[b]	14.1	29[b]	19.6	–	–

The table was compiled from the data published by the following authors: Rittenberg et al. (1963), Degens et al. (1964, with amendments), Starikova and Yablokova (1972), Artemiev and Romankevich (1973), Mopper (1980)

[a] The dash in this column – for unfractionated extract
[b] Average of 14 probes
[c] Diatomaceous silt
[d] Glauconite sands
[e] Miopelagic red clays
[f] Terrigenous muds
[g] Average of 7 samples

Table 50. Free-sugar composition of various seawater samples (mol. %) (Mopper et al. 1980)

Samples	Rha	Fuc	Rib	Ara	Xyl	Fru	Man	Gal	Glu	Total (μM)	Ratio[b] Fru/Glu
1 Black Sea	tr[a]	0.8	2.7	3.3	1.6	42.4	2.4	0.9	41.5	0.51	1.02
1 Black Sea, hydrolyzed	5.3	4.5	2.6	7.0	7.1	10.7	12.8	8.3	41.7	0.73	0.25
2 Black Sea	tr	1.2	12.9	6.1	7.0	25.9	1.2	6.1	39.6	0.08	0.65
3 Baltic Sea	2.2	2.2	0.9	2.2	2.2	36.4	6.8	1.6	44.4	0.29	0.82
4 Baltic Sea	1.5	0.7	2.8	2.8	2.0	31.0	10.0	2.0	44.0	0.13	0.70
5 Baltic Sea	tr	tr	9.0	tr	tr	46.9	tr	tr	44.1	1.08	1.06
6 North Sea	1.3	1.5	2.8	2.9	1.9	38.9	1.3	2.2	47.2	1.42	0.82
7 North Sea	3.3	2.1	13.7	4.1	1.8	22.6	1.8	6.5	41.2	0.34	0.55
8 Sargasso Sea	tr	tr	tr	tr	4.6	36.0	1.4	10.8	47.1	0.49	0.76
9 Sargasso Sea	tr	tr	tr	tr	tr	27.5	tr	32.2	40.4	0.74	0.68
10 Harrington Sound	tr	tr	5.3	tr	4.0	42.1	4.1	tr	44.4	0.25	0.95
11 Pacific Ocean	tr	tr	1.2	tr	tr	37.5	tr	2.7	58.8	1.02	0.64

[a] tr = trace (<1 nM)

[b] The equilibrium ratio of Fru/Glu from isomerization experiments was found to be $0.3-0.4\ l^{-1}$

surface layer. It can be inferred that the polysaccharides that were formed from them (glucuronomannanes and glucuronogalactanes?), are the most resistant to biochemical and chemical degradation, both in solution and in suspension. However, these conclusions require further substantiation based on far more data. They will be probably be found to hold true only for the monosaccharide content of the total dissolved carbohydrates of the fine colloidal fraction and the truly dissolved OM outside high-production zones and not at the time of phytoplankton bloom.

The composition of free monosaccharides seems to be more diversified than that of the total dissolved carbohydrates (Table 50; Mopper et al. 1980). Numerous determinations of free sugars have demonstrated that their total concentration varies widely (0.080 μM in the deep-water area of the Black Sea to 1.418 μM in the Weser estuarine waters, the North Sea), glucose and fructose being predominant in all the samples analyzed. While the presence of glucose (the most common sugar in phytoplankton) is by no means surprising, the relative abundance of free fructose in sea water is a somewhat new occurrence. It may be the result of a complex of phenomena, such as isomerization of glucose to fructose, insufficiency of our knowledge of fructose content in various marine organisms, differences in the selective assimilation of sugars by heterotrophs and in the formation of complexes of low solubility.

7.3 Carbohydrates in the Suspended Matter

The content of the carbohydrates in the suspended material of the surface layer in seas and oceans varies between traces to several hundred micrograms per litre. According to our data, obtained by phenol-extraction from samples procured in the Pacific Ocean, the concentration of particulate CHC in the 0 – 10 m layer is 4.9 – 160 $\mu g\,l^{-1}$, and in the 0 – 100 m layer, 4.9 – 313.1 $\mu g\,l^{-1}$. The maximum concentrations of the particulate CHC are restricted to the nearshore areas of high productivity in the seas and oceans. The content of CHC here may vary from several $\mu g\,l^{-1}$ to thousands $\mu g\,l^{-1}$, for example in the 0 – 37 cm layer off Capetown from 14.5 to 1183.0 $\mu g\,l^{-1}$ (Barlow and Swart 1981).

A usually direct relationship with high correlation coefficient is apparent between the concentration of the particulate CHC and total POM content (a similar relationship was established for the CHC and POC of the eolian suspended matter, collected over the aquatory of the Pacific Ocean). As a consequence, the distribution patterns of the carbohydrates in the water and eolian suspensions exhibit the same zonality as the C_{org} in the respective suspensions. For instance, the CHC in the water suspension exhibit characteristically a circumcontinental, latitudinal and vertical zonality, and a direct correlation of chlorophyll "a" content and the biological productivity. As was shown above, such correlations are not typical of the dissolved CHC: in areas distant from the near-shore shallow waters, no relationship has been observed between the dissolved CHC and the phototrophic productivity. Evidently, the suspended carbohydrates are genetically more closely related to the OM of living organisms, whereas in the dissolved CHC this relationship is obscured, due to their greater separation in space and time.

The content of the particulate CHC decreases in the vertical profile of the water column. For instance, in the Pacific Ocean the mean values of CHC concentrations are 19.5 $\mu g\,l^{-1}$ in the 0–100 m layer, and 4.4 $\mu g\,l^{-1}$ below 1000 m. The content of the particulate CHC varies in the vertical profile, mainly within the upper 200 m (Figs. 53 and 54). Here the decrease in content is due to the breakdown of the WSP, which the algae, zooplankton and bacteria use as nutrients. In the Pacific Ocean, below 700 m, the concentration of the particulate CHC in the 0–100 m layer has a tendency to decrease in the order shelf → slope → ocean floor, falling from 34.8 to 18.6 to 13 $\mu g\,l^{-1}$, respectively. In the deeper waters of the Pacific Ocean, as far as the present data show, no recognizable variation is found in the CHC content.

In the POM of the surface and deep waters the CHC content varies mainly from 10 to 30%. In separation samples of suspensates from the Pacific and Indian Oceans (see Fig. 50), the CHC content averages 14.8% of OM, varying

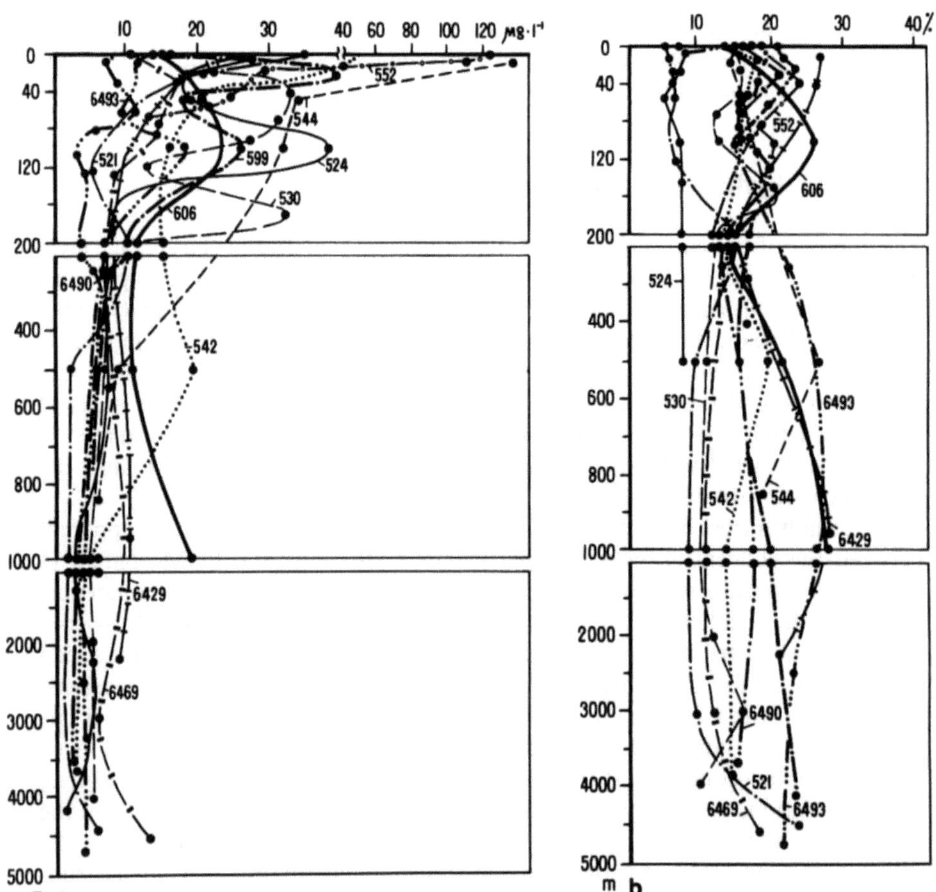

Fig. 53a, b. Distribution of particulate carbohydrates in the Pacific water column. **a** $\mu g\,l^{-1}$; **b** in OM (C_{CHC}/C-POM 100). *Figures near curves* show Station Nos

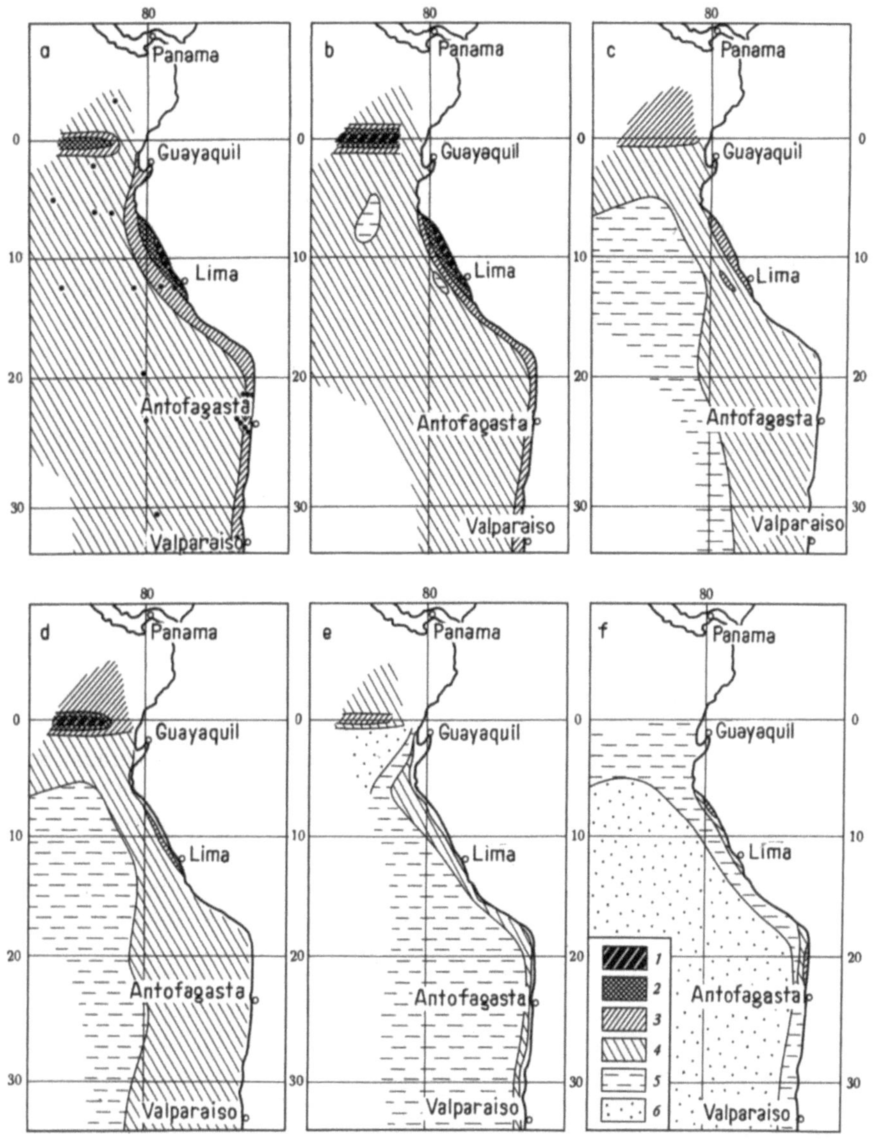

Fig. 54a–f. Distribution of particulate carbohydrates in the Southeastern Pacific waters, μg l^{-1}. *1* = >30; *2* = 30–20; *3* = 20–10; *4* = 10–5; *5* = 5–1; *6* = <1. Layers in m: **a** = 0; **b** = 10; **c** = 30; **d** = 50; **e** = 100; **f** bottom water layer (0.9 m above Ocean floor). *Dots* on map **a** show sampling sites

from 5.1 to 28.2%. The extreme values and average content of the carbohydrates in the POM of separation suspensates are practically the same as those in the diatom, peridinean and blue-green algae (7.3–31.3%). They are on average higher than in zooplankton (see Table 43).

The analysis of filtration samples (Whatman GF/C filtres) procured from the Pacific and Atlantic Oceans, showed that the variations in the CHC content of

OM in the vertical profile, observed at particular stations, are small. Furthermore, conspicuous differences are apparent in the mean values of CHC concentrations of the Pacific both spatially and at different depths: 0–100 m – 16.2%; 100–200 m – 15.8%, 200–1000 m – 16.6%, 1000 m – 17.8%. This suggests that the particulate carbohydrates are not extremely labile compounds. Their content decreases (in $\mu g\, l^{-1}$) in the water column, but their breakdown does not forestall the mineralization of the POM, as a whole. The preservation of carbohydrates in relatively consistent concentrations in the OM of the ocean water column can be attributed to the similarity in the biochemical composition of the initial OM in different pelagical areas of the ocean, and the general similarity in the OM transformation trends in the process of sedimentogenesis. The data on the content and distribution of CHC in the water column, also of the proteins, give support to the suggestion that the POM in the ocean is, according to its composition, a carbohydrate-proteinaceous complex.

The major constituents of the carbohydrates, both in the surface and deep water suspensions, are WISP, which account for 70–80% of the total carbohydrates. ASP and WSP are present in small quantities. In the probes of the surface suspension procured from the Northwestern Pacific, the ASS, WSP and WISP average 8.7, 11.8 and 76.2%, respectively, of the total carbohydrates (see Table 38). It is conceivable, that the proportion of the WISP fraction in the carbohydrate complex of the surface suspension is 1.5 times greater than in the phyto- and zooplankton. This may be ascribed to the fact that the particulate carbohydrates are incorporated in the organic detritus, where the more labile and assimilable reserve carbohydrates from the substrate for respiration of the dying algae, and also are utilized by consumers during the process of OM degradation.

According to the data available on the suspended matter from the Western Pacific, the amount of ASS + WSP in the 0–150 m layer approximates 13–30% of the total carbohydrates, and at depths between 200–500 m decreases to 5% and less (Handa et al. 1972). The relative content of the WISP fraction in the carbohydrates increases with depth, which results mainly from the more rapid breakdown of reserve products, such as glucane.

The monosaccharide composition of the particulate carbohydrates is apparently of much more diversity, as compared to the monosaccharides of the dissolved CHC (Table 49; Ittekkot and Degens 1982). According to the available data, they consist, apparently, of a mixture of homopolymers, and, probably, of heteropolysaccharides. Although the suspension contains a diverse set of monosaccharides, most obviously glucose predominates (>50–70%), as in the diatom plankton (35–74% of the total carbohydrates). As in the diatom plankton, most of the glucose in the alcohol- and water-soluble carbohydrates is contained in glucane. The instability of glucane explains the rapid breakdown of this polysaccharide in the upper water layers (Handa and Tominaga 1969). Among the polysaccharides in the suspended matter, fructose has been found lately in great quantities; earlier, it was not observed in phytoplankton, with the exception of fructan as food reserve material in Chlorophycae (Ittekkot and Degens 1982). Presumably the discovery of fructose is associated with the use of a softer OM hydrolysis method (1.8 N HCl, 100°C, 3.5 h, atmosphere N_2) against the previously applied (1 N H_2SO_4, 100°C, 10 h). As was shown, during the phytoplankton bloom the carbohydrates of suspended matter contain enhanced relative amounts of minor sugars (arabinose, fucose, ramnose), while the relationship of glucose (mainly storage material) and galactose (largely structural material) undergoes a change in favour of the latter (Ittekkot and Degens 1982).

The glucose content decreases in the order ASS – WSP – WISP. Seemingly, in the change to stable polystructural polysaccharides the composition of the carbohydrates becomes more complex. This is consistent with the materials on the carbohydrates in living organisms and certain structural elements of the latter.

After glucose, the second in importance constituent of the carbohydrate complex in the surface suspension, is sometimes fructose (during the phytoplankton bloom), but more often either galactose, or mannose, or xylose. The increase in their content in the sequence ASS – WSP – WISP, is similar to their increase in the diatom plankton. The composition of the suspended matter monosaccharides at depths between 500 – 700 m does not differ from that of the WISP in the surface layer.

In general, the carbohydrates of suspended matter in the deep waters appear to consist mainly of glucoronomannanes which are related to the polysaccharides in marine diatoms.

7.4 Carbohydrates in Bottom Sediments

Until the middle of the 1960s only limited information was available on the content and composition of carbohydrates in sea and ocean sediments (Shabarova 1950, 1955, Rodionova 1951, Plunkett 1957, Prashnowsky et al. 1961, Rittenberg et al. 1963, Degens et al. 1964, Degens 1967). Some real progress has been made since improvement of the analytical techniques, particularly since the introduction of the phenol method and its successful application to the determination of the total CHC content of different types of sediments (Romankevich and Artemiev 1969, Artemiev et al. 1971, Romankevich 1971, Romankevich and Urbanovich 1971, Artemiev and Romankevich 1973, Rodionova et al. 1973, Artemiev 1974, 1976, Bordovsky et al. 1974, Michaelis et al. 1980). However, owing to the difficulties in the extraction of the carbohydrates, which in the analytical procedures are subjected to profound changes, no information has as yet been obtained on the polymer composition of the carbohydrates in bottom sediments.

7.4.1 The Content and Composition of the Carbohydrates in the Sediments of the Surface Layer

According to the data of P. P. Shirshov Institute of Oceanology obtained from the Pacific, Atlantic and Indian Oceans, the concentration of carbohydrates in the surface layer of bottom sediments (0 – 5 cm) varies from 66 to 21,333 μg per g, that is more than 300-fold (Tables 51 and 52). The maximum concentrations occur characteristically in the shallow water pelitic sediments enriched in C_{org},

Table 51. Concentrations of carbohydrates in upper layer of the Pacific Ocean sediments

Facies type	Number of samples	Sandy-aleuritic sediments	Number of samples	Pelitic sediments
Terrigenous, of shelf areas	8	*939 – 9472 (3095)* 1.5 – 15.6(7.9)	5	*5826 – 10838 (8075)* 2.2 – 8.5(5.9)
Carbonate, of shelf areas	6	*956 – 6889 (4226)* 2.8 – 9.5(6.8)	4	*1396 – 5511 (2938)* 3.2 – 20.7(12.6)
Glauconite, of slope areas	11	*173 – 4799 (2344)* 2.8 – 7.3(4.6)	–	– –
Terrigenous, from upper slope areas (0 – 3000 m)	12	*190 – 4244 (2110)* 2.1 – 18.9(10.4)	8	*798 – 9433 (5501)* 0.9 – 15.3(9.1)
Terrigenous, from lower slope areas (>3000 m)	6	*224 – 2056 (984)* 1.5 – 11.0(6.4)	16	*155 – 8667 (1788)* 0.9 – 14.7(5.4)
Siliceous-diatomaceous, of continental slopes	–	– –	2	*1430 – 4497 (2963)* 4.8 – 9.3(7.1)
Pelagic siliceous-ethmodiscuscidic muds	–	– –	1	*800* 15.2
Pelagic calciferous (30 – 50% $CaCO_3$) foraminiferal	–	– –	6	*320 – 3200 (1874)* 3.3 – 27.0(14.8)
Pelagic calcareous-foraminiferal (>50% $CaCO_3$)	6	*831 – 1267 (1040)* 7.1 – 16.9(13.7)	28	*210 – 2311 (1280)* 2.6 – 24.3(15.0)
Pelagic carbonate algal-coral, bryozoan-shell	2	*1494 – 1544 (1519)* 4.0 – 9.4(6.7)	–	– –
Radiolarian muds	–	– –	12	*328 – 2583 (1551)* 3.6 – 22.6(15.2)
Miopelagic red clays	–	– –	15	*92 – 3556 (1074)* 0.9 – 29.6(12.0)
Eupelagic red clays	–	– –	24	*214 – 2600 (1115)* 3.5 – 29.4(16.7)
Pelagic volcanogenic	1	*703* 10.7	5	*294 – 1134 (472)* 5.3 – 11.6(7.8)
Average for sediments of:				
submarine continental margins	43	*173 – 9472 (2490)* 1.5 – 18.9(7.4)	35	*155 – 10838 (3733)* 0.9 – 20.7(7.2)
marginal part of ocean floor	3	*798 – 958 (881)* 5.0 – 11.5(9.0)	17	*66 – 3311 (1339)* 0.5 – 23.3(8.9)
central part of ocean floor	9	*703 – 1544 (1108)* 4.0 – 16.9(11.8)	86	*92 – 3556 (1300)* 0.9 – 29.6(15.4)

Italics = extreme values, in $\mu g\ g^{-1}$ of air-dry sediments; Roman = extreme values in OM ($C_{CHC}/C_{org} \times 100$); figures in parenthesis are mean values

Table 52. Concentrations of carbohydrates in upper layer of the Atlantic Ocean sediments

Type of sediments	Depth (m)	Number of samples	Carbohydrates
North Atlantic			
Sands, calcareous shell and foraminiferal	130 – 1025	4	*225 – 1488 (564)* 3.6 – 9.8 (5.9)
Aleurites, terrigenous	932 – 1600	4	*546 – 1270 (844)* 4.5 – 10.3 (7.4)
Pelites, terrigenous and volcano-clastic	395 – 3640	15[a]	*809 – 4610 (2084)* 6.0 – 24.8 (11.7)
Canary and Cape Verde Basins			
Aleurite, terrigenous	5117 – 5140	2	*74 – 149 (111)* –
Pelites, terrigenous	5090 – 5300	6	*66 – 262 (158)* –
Area of Falkland and South-Sandwich Islands			
Sands, terrigenous	1700	1	*642* 9.9
Aleurites, terrigenous	455 – 3942	6	*619 – 1534 (1033)* 6.6 – 19.8 (13.1)
Shelf and slope off Western Africa Humid zone			
Sands, terrigenous	45 – 223	7	*1035 – 4877 (3217)* 5.7 – 23.5 (15.2)
Aleurites, terrigenous	41 – 197	4	*4722 – 7366 (6150)* 9.2 – 16.7 (13.8)
Pelites, terrigenous	28 – 1495	7	*6600 – 10411 (8692)* 12.0 – 18.0 (14.1)
Arid zone			
Sands	62 – 360	23	*2222 – 15511 (6362)* 3.3 – 12.5 (8.4)
Aleurites	50 – 952	20	*2366 – 10377 (4820)* 6.2 – 13.5 (10.6)
Pelites	75 – 148	8	*7388 – 21333 (13577)* 5.8 – 9.2 (7.8)

Italics = extreme values in $\mu g\ g^{-1}$ of air-dry sediments; Roman = extreme values in OM ($C_{CHC}/C_{org} \times 100$); figures in parenthesis are mean values

[a] Content in OM, calculated from analysis of eight samples

which accumulate in the zones of high biological productivity; the minimal concentrations occur in the pelagic deposits. Between the CHC content and total C_{org} content a direct relationship of high correlation coefficient has been observed in various areas of the Pacific and Atlantic Oceans (Fig. 52). In general, the correlation coefficients between the CHC and C_{org} are lower in the sediments of the continental margin, which apparently is a consequence of the regional differences in the OM composition. In the pelagial areas, where the CHC and C_{org}

values are low, no correlation is obvious between them (d_1 and d_2 regions, on Fig. 52).

The distribution pattern of the carbohydrates in bottom sediments reveals the same regularities as the C_{org}, namely: a general decrease in the CHC content in the direction shelf-pelagic areas; large areas of direct correlation with the biological productivity; the pronouced reversed relationship between the CHC content and the range of particle dimensions in the sediments of the submarine continental margin.

The CHC content in the OM of the sediments varies mainly from 5% to 15%, the extreme values being 0.5 – 29.6 and 3.3 – 24.8% for the Pacific and Atlantic, respectively.

Three observations deserve special attention. First of all, the enhanced CHC content of the OM near the estuaries of certain rivers, which can be ascribed to the input of allochtonous OM, rich in carbonyl units. For example, the influence of the rivers on the CHC content in the OM is markedly reflected in the area of the submarine margin off western Africa (Fig. 55). In general, the sediments in the arid near-shore areas of the ocean are, seemingly, poorer in carbohydrates, whereas in the humid areas they are richer. Secondly, within the continental margin of the Pacific Ocean, the CHC content in the OM of the bottom sediments (7%, on the average) is less than in the particulate (15 – 18%, on the average) and dissolved OM (17 – 20%), whereas in the pelagic sediments it is practically the same (averages 15%). Thirdly, the CHC content in the OM increases in the order shelf → pelagic facies (Fig. 56). This is consistent with the Uspensky-Vassoyevich law, which in respect to the carbohydrates is well expressed in discrete areas.

These features of the distribution of the carbohydrates in the OM of the bottom sediments can be explained by the fact that the carbohydrates and the complexes in which they occur undergo more complete utilization in those areas where the benthic communities are of higher activity, for instance, within the shelf areas and slopes, beyond the zones of hydrogen sulphide contamination. The high resistance of some part of the CHC to mineralization (compared with the entire OM) leads to their preservation as the residue of the well-balanced OM synthesis – decay system in the area of the ocean floor.

The selective preservation of carbohydrates begins in the trophical chains of the zooplankton (this is supported by Harding's observations on the better preservation of carbohydrates, compared with proteins, during their passage through the gut of copepods; Harding 1974), but is best reflected only during the benthos attack and in the process of diagenesis of the sediments. Comparison of the data shows that the CHC content in the OM that is not digested by the detritus-feeders (the content from the lower end of the gut) is higher, than in the consumed food (sediments) and in the bodies of the same consumers (Artemiev 1976). This serves to suggest that the detritus-feeders may be the cause of the enhancement of the CHC content in the sediments. However, it should be noted, firstly, that there are seston-feeders although practically nothing is known as yet of their capability of transforming carbohydrates; secondly, during benthic attacks, a much greater portion of OM is destroyed than generated, thirdly, the resynthesized OM contains more labile carbohydrates, and, finally, the proportion

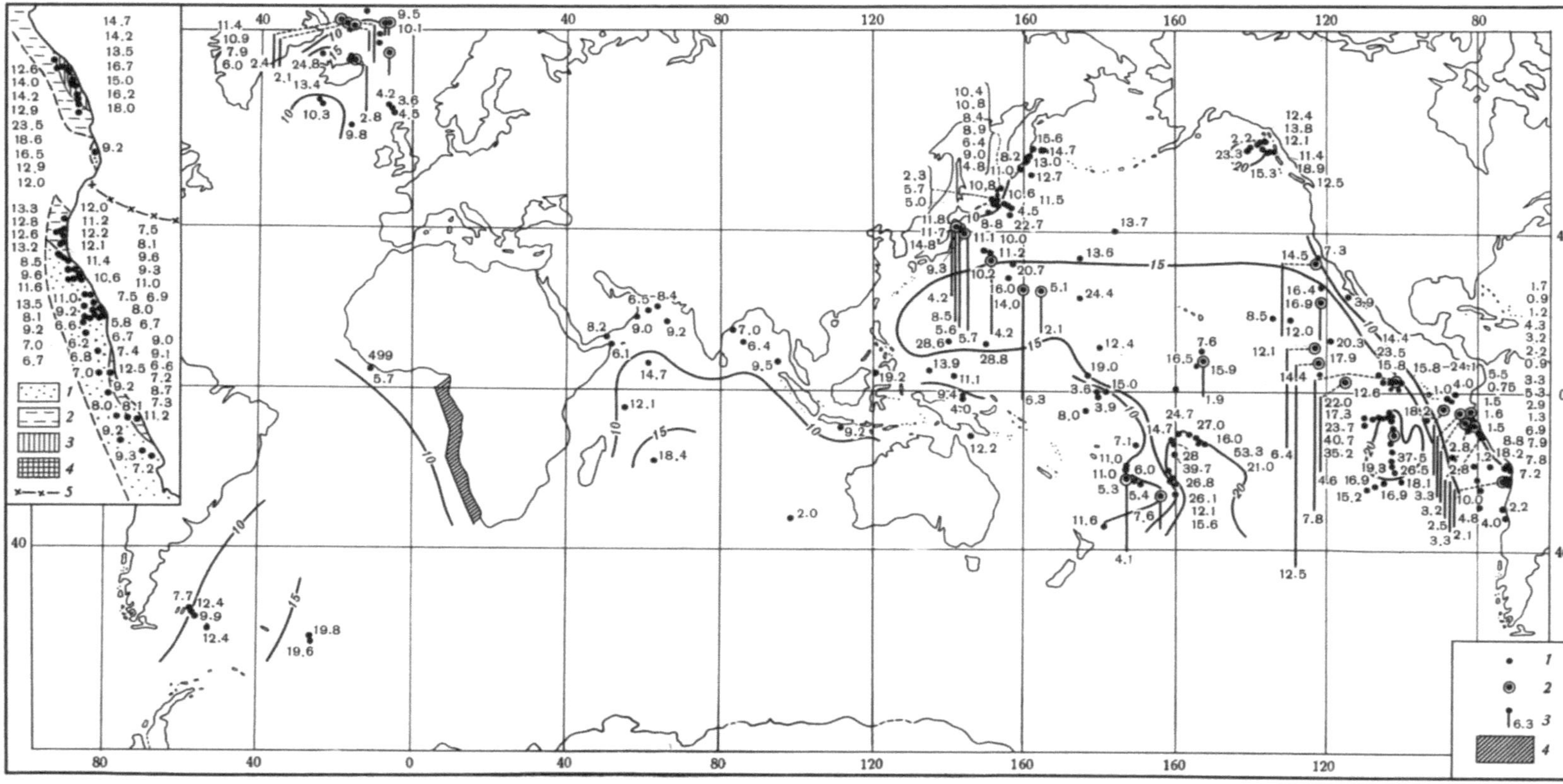

Fig. 55. Sampling sites of bottom sediments and interstitial waters, in which measurements of the carbohydrate content were made by uniform techniques in the Institute of Oceanology of the USSR Academy of Sciences.

1 sediments; *2* interstitial waters; *3* bottom sediment cores (length, m); *4* insert map. Figures and contour lines show CHC content of OM (C_{CHC}/C_{org} 100). *Insert map:* CHC content of OM in sediments of the submarine continental margin off Southwestern Africa, %: *1* = 5 – 10; *2* = 10 – 15; *3* = 15 – 20; *4* = 20 – 25; *5* = northern boundary of arid region

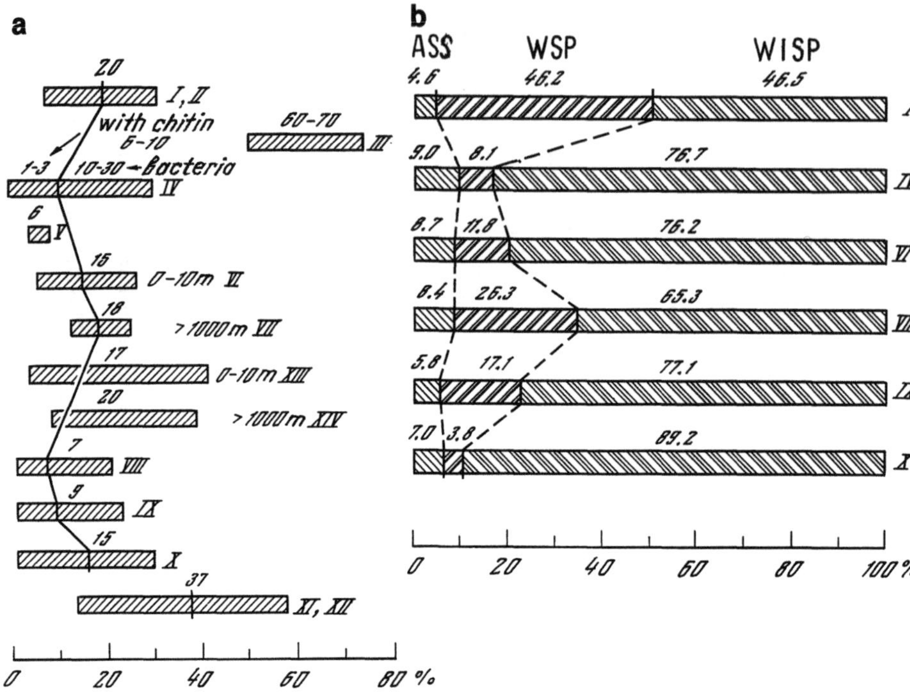

Fig. 56 a, b. Changes in the content of carbohydrates (**a** % of OM) and their group composition (**b** % of total CHC) in the process of sedimentogenesis. *XIV* = sea waters from depths below 1000 m. Other symbols as on Fig. 46

of CHC in this OM is relatively small. Conductive of these and other processes, is the lower content of carbohydrates in the OM of terrigenous sediments than in the POM (see Fig. 56).

From studies of the CHC composition it was found that in the surface layer of the Pacific Ocean sediments the contents of the alcohol-soluble, WSP and WISP fractions average 7.6%, 20.9% and 71.5%, respectively, of the total CHC content (Table 53; Artemiev and Romankevich 1973, Artemiev 1974).

The group composition of the CHC reveals a general trend to change from the terrigenous sediments of the continental margins towards the pelagic muds of the deep ocean basins, which is reflected in the decrease of the WSP content and increase of the WISP content, that is, of compounds resistant to mineralization, owing to the tight bonding of the carbohydrates with the mineral base of the sediments.

At the water-sediment interface and in the top layer of the sediments the monosugar spectrum undergoes reconstruction. The CHC monosaccharides in the bottom sediments, as in the phytoplankton and particulate matter, are represented by a set of several sugars. This is also typical of the individual groups of compounds that comprise the carbohydrates (see Table 49). For example, in the diatomaceous ooze from the shelf off south-western Africa, the alcohol-soluble fraction was found to contain glucose, galactose and traces of rhamnose; the WSP – glucose, galactose, mannose, rhamnose, xylose and unidentified sugars,

Table 53. Group composition of carbohydrates in sediments of the Pacific Ocean (in % of total)

Sampling area	ASS	WSP	WISP
Surface layer sediments			
Shelf areas off Peru and south-western Africa (terrigenous and diatomaceous muds)	*5.4–10.8* 7.8(4)	*12.4–32.8* 20.8(4)	*59.3–82.2* 71.4(4)
Slopes of the Kuril-Kamchatka, Japan, Peru-Chile Trenches (terrigenous and glauconite sediments)	*2.9–18.8* 8.7(8)	*6.4–52.7* 29.1(8)	*35.1–90.7* 62.2(8)
Marginal part of Pacific Ocean floor (hemipelagic terrigenous muds)	*1.8–7.9* 5.8(4)	*2.4–33.6* 17.1(4)	*59.3–95.8* 77.1(4)
Central areas of the Pacific Ocean floor (eupelagic and miopelagic red clays)	*5.1–9.0* 7.0(3)	*0.0–6.3* 3.8(4)	*84.7–95.9* 89.2(3)
Total	*1.8–18.8* 7.6(19)	*0.0–52.7* 20.9(19)	*35.1–95.9* 71.5(19)
Bottom sediments cores			
Japan Deep Trench			
0–5 cm	7.0	36.1	56.8
50–100 cm	6.7	23.9	69.4
370–430 cm	9.7	37.6	52.7
North-western Basin of Pacific Ocean			
0–5 cm	7.0	8.8	84.2
270–280 cm	8.4	0.0	91.6
610–630 cm	5.2	0.0	94.8

Italics = extreme values; Roman = average values, and those in parenthesis number of samples

in the ratios of 1:1.5:1.9:1.5:1.9:0.4. The same sugars that are present in the WSP fraction, were identified in the WISP (glucose, galactose, mannose, rhamnose, xylose, unidentified sugars), plus ribose, their ratios being 1:1:1.3:0.9:1.4:0.8:0.03. Arabinose, too, was found to be present in traces in the WSP and WISP fractions of the diatomaceous ooze.

The sugar spectrum of the carbohydrates in the bottom sediments is more diverse than that in sea waters, and apparently differs from that of the suspended matter in the higher content of rhamnose and xylose polymers and occasionally in the absence of glucose. As in the suspension no free sugars have been found in the bottom sediments. The above mentioned diatomaceous ooze from the shelf off South West Africa revealed, as in the suspension, an increase in the complexity of the carbohydrate composition in the order ASS – WISP fractions. This is further evidence of the genetic relationship between the WISP of the sediments, suspension, and the most stable structural elements of the phytoplankton and bacteria. However, it remains as yet unclear to what extent the pronounced reconstruction of the carbohydrate complex at the water-bottom sediment interface and within the upper sedimentary layer is due to the activities of the benthos, the different ratios in the sinking rate of the allochtonous and phytobenthos OM, where the products of decay are rich in xylenes and mannanes and more resistant to degradation. The microfloral activity is an essential factor in the transformation of the carbohydrates. Little is known, however, of the role of filter-feeders and detritus-feeders in the transformation of OM, and it would seem that we underrate the role of zoobenthos in these processes. The preferential accumulation of carbohydrates in the gut of deep water Holoturius, and the decrease in their content from the first to the third section of the gut (Artemiev 1976), are new facts that argue in favor of this conclusion.

7.4.2 Carbohydrates in Bottom Sediment Cores

During 1968 – 1982 investigations were performed on bottom sediments from the Pacific and Atlantic Oceans in the P. P. Shirshov Institute of Oceanology of the USSR Academy of Sciences; the CHC content was studied in 48 cores, 1 – 12.5 m in length (see Fig. 55) which penetrated sediments from the Recent to the Miocene. In several cores the group composition of the carbohydrates was studied.

The CHC distribution pattern in the sedimentary column exhibits the same regularities that are characteristic of the surface layer; namely, the decrease in the concentrations in the shelf – pelagial direction (comparison was made of sediments of the same age), and a direct correlation with the content of the pelitic fraction in the sediments of the continental margins. A sharp drop in the CHC concentrations, both on a dry sediment basis, and in the OM, is apparent only in the upper layers. Below 0.5 – 1 m, seldom more, the CHC content differs only slightly. As a whole, the sedimentary series within the studied cores, inherits the CHC concentrations that are typical of the contemporary sediments. The maximum concentrations occur in zones of upwellings and high biological productivity; the minimal concentrations in the pelagial areas of the ocean. The depth of the zone of intensive mineralization of the carbohydrates depends upon the rate of sedimentation, activities of the benthic population, and conjugated diagenetic processes of reconstruction of the sediments. It is of minimum depth in the red clays (a few tens of centimeters) and much greater in those areas where OM undergoes rapid burial and there is a deficiency in oxygen and/or sulphates. Below this zone the rate of breakdown of the carbohydrates is slow, and their content reveals a small difference during several million years.

In the Pacific and north Atlantic Oceans the CHC content in the OM of the sediments from the Recent to the Miocene, varies chiefly from 5 to 15%. The CHC content in OM decreases in the depth profile of the sediments of the continental margin; in the pelagic sediments the differences are small and within measurement error.

The Uspensky-Vassoyevich law (inverse correlation between the C_{org} and content of one or another group of compounds in the OM), which was established for bitumoids in sedimentary rocks, holds true for the CHC content in bottom sediments. This regularity is not characteristic of the particulate fraction. It is initiated during the diagenetic transformation of OM in the upper layer of sediments and is inherited by the series of sediments.

The differences in the CHC group composition within the sementary series, allow for the subdivision of the cores into three types. The first type includes those cores which have a thin upper oxidized layer, and below contain free hydrogen sulphide. All the cores of this type have characteristically the highest CHC content, and reveal an increase or tendency to increase of the WSP content with depth, which is obvious down to the lowest studied layer (8 m). The increase in depth of the WSP content in the carbohydrate complex can, seemingly, be attributed to the destruction of the WISP fraction and its change to water-soluble form, considering that in the depth profile, below the surface layer, the total CHC content is of relative consistency.

The cores of the second type reveal alternations of oxidized (brown) and anoxic (gray) layers of sediments in the depth profile. Free hydrogen sulphide is absent from the cores. The CHC content is less here than in the cores of the first type. The composition of the carbohydrates reveals a sharp change towards an increase in the WSP fraction in the vertical profile, which is reflected in the upper layer of sediments (0 – 80 cm), at the transition from the oxidized gray-brown muds to the anoxic gray sediments. This, probably, can be attributed, as in the cores of the first type, to the destruction of the WISP fraction, its change into labile compounds and release into the liquid phase of the sediments; proof of this is found in the increase of the CHC content in the vertical profile of the interstitial waters.

The third type includes cores of red abyssal clays, which are characterized by oxidation sedimentation conditions, and the marked predominance of WISP in the carbohydrate complex (91 – 96% of the total CHC).

The supposition that the carbohydrates are labile compounds and do not undergo fossilization may be unwarranted. Part of the carbohydrates contained in marine organisms and also the input from the land, being represented chiefly by structural water-insoluble heteropolysaccharides with a mixture of polymers, are preserved during sedimentogenesis in the form of intricate organo-mineral complexes. The fact that the CHC mean values of the OM are subjected only to small changes during the process of sedimentogenesis (see Fig. 56), coupled with the fact that the carbohydrates are preserved in marine sediments of 10 – 14 million years, and the results obtained from earlier investigations (the presence of carbohydrates in peat bogs, sapropels, kukersite shales, crude oils, Paleozoic and Precambrian rock deposits) all show that a portion of the carbohydrates is of high stability within a wide range of natural conditions.

7.4.3 Carbohydrate Compounds in Interstitial Waters

The content of CHC in the interstitial waters depends primarily on the lability of the carbohydrate complex of the solid phase, the intensity of formation of incomplete OM decay products, their solubility, the rates of mineralisation and coagulation from the solution, as well as diffusion into the near-bottom water. According to the data available, the concentration of CHC in the interstitial water varies from less than 1 to ~65 mg l^{-1} (Table 54; Poliakova 1974, Romankevich 1977), while their distribution in the upper layer and the mass of deposits follows the same pattern as the distribution of the CHC in sediments. The interstitial waters of reduced sediments contain more CHC than the oxidized ones.

Table 54. CHC and C_{org} content in the interstitial waters of sediments from the Pacific Ocean and Black Sea

Sampling site	Type of sediments[a]	CHC		C_{org} mg l^{-1}
		mg l^{-1}	% of OM	
Peruvian shelf	Strongly reduced	*12.3±0.2–64.5±1.1* 30.5 (4)	*30.9–54.4* 40.4 (4)	*12.2–47.4* 28.3 (4)
Black Sea	Strongly reduced, reduced	*2.0–56.2* 15.1 (59)	*7.3–89.5* 28 (60)	*6.5–77.4* 21.6 (59)
	Weakly reduced, weakly oxidized	*4.3–54.5* 20.5 (33)	*9.5–77.0* 30.8 (34)	*6.8–73.5* 26.6 (33)
Continental slope off Peru	Weakly reduced, weakly oxidized	*10.3±0.1–21.6±0.4* 17.6 (4)	*32.9–50.1* 40.3 (4)	*8.2–20.9* 18.9 (4)
Slope of Honshu Island	Weakly reduced, weakly oxidized	*4.4–9.8* 6.3 (5)	*13.5–38.8* 22.2 (5)	*10.1–13.9* 8.4
Shelf and slope off New Guinea	Weakly reduced, weakly oxidized	*2.2±0.2–6.9±0.1* 4.4 (4)	*17.1–28.1* 21.4 (4)	*4.4–13.9* 11.9
Peruvian basin	Weekly and strongly oxidized	*2.5±0.2–6.5±0.4* 4.5 (2)	*51.1–57.4* 54.2 (2)	*1.7–5.1* 3.4 (2)
North-western basin	Weekly and strongly oxidized	2.4	20.9	4.6
West Marianna basin	Weekly and strongly oxidized	*2.8–4.5* 3.5 (3)	*16.5–35.5* 23.2 (3)	*5.1–7.7* 6.4 (3)

Italics = extreme values, Roman = average values, in parenthesis = number of samples

[a] Strongly reduced sediments: Eh < −200 mV; reduced: Eh = −200 ÷ 0 mV; weakly reduced: Eh = 0 ÷ 200 mV; weakly oxidized: Eh = 200 ÷ 400 mV; strongly oxidized: Eh > 400 mV

With long-term reduction processes of the diagenesis the content of CHC in the interstitial water increases with the depth of penetration into the mass of sediments. This is caused by the formation of a large amounts of incomplete OM decay products, the diffusion of which into the near-bottom water proceeds at a slow rate. Thus the study of CHC in interstitial waters confirms the thesis that the conditions of a reducing environment favor the accumulation of incomplete OM decay products.

In the OM of interstitial waters in the surface sedimentary layer the content of CHC is usually higher (13.5–57.4%) than in the OM of the solid phase of sediments. The concentration of CHC in the interstitial water OM decreases vertically. This trend was little marked in cores that penetrated the Holocene sediments and the upper Pleistocene. With changes in the facial conditions the regular decrease of CHC as part of OM may become disturbed, as takes place in transition from the metalliferous clays to the coccolith silts.

The practically invariable content of CHC in the OM of interstitial waters in the mass of eupelagic clays reflects the balacing of the supply and expenditure of CHC even in the upper sedimentary layer, accompanied by a weakly marked attenuation of decay in vertical columns. The enhanced content of CHC in the OM of interstitial waters compared with the sediments stems from the fact that the stable metabolitic products, including the major part of CHC of interstitial and marine waters, tend to be long preserved in solutions.

8. Chemical Transformations of Lipids During Sedimentogenesis

Lipids constitute a group of hydrophobic organic compounds easily soluble in organic solvents. A characteristic feature of the chemical composition of lipids is the presence of higher alkyl radicals in the molecules. The sources of lipids in the seas and oceans are the auto- and heterotrophic organisms inhabiting the marine environment. The community of living organisms produces annually in the world oceans 7×10^9 tons of fatty substances (Bogorov 1971). A certain portion of these lipids reaches the aquatic environment with allochthonous OM; this comprises natural lipid compounds and the constituents of oil (Goldberg 1975, Cowell 1976, Nesterova and Simonov 1979, Nemirovskaya and Nesterova 1982).

All forms of lipids in the ocean, dissolved and colloidal (DL), particulate (PL), those contained in the bottom sediments and in interstitial water are interrelated and can be converted one into another. However, until recently most attention was devoted to the study of lipids contained in living organisms and in bottom sediments. The lipids of the suspended material beneath the surface layer were practically inaccessible to quantitative investigation, since this required precise methods of analysis. Furthermore, the results of such investigations were not easily comparable because they were obtained for different purposes and by different methods. Determination of the quantitative relationships between DL, PL, lipids contained in the bottom sediments and in interstitial water, and the estimation of the trend of their transformations in sedimentogenesis became possible after techniques had been elaborated for the extraction of the aforementioned forms of lipids in a little-transformed state and after development of a microtechnique for their quantitative and qualitative determination (Romankevich and Batrakov 1971, Beliayeva and Romankevich 1976, Beliayeva 1980a – c).

8.1 Lipids of Marine Organisms

According to the widely recognized biochemical classification, the lipids of marine organisms, in correspondence with their functions, are subdivided into intracellular (phospholipids, lipoproteins, lipopolysaccharides) and reserve substances, chiefly represented by neutral ethers of fatty acids and alcohols and by their derivatives. The content and composition, as well as the relationship between them, of the intracellular and reserve lipids in marine plants and zooplankton vary widely (Table 55), being dependent on the complex of ecological factors and the physiological state of individual organisms. In the case of phytoplankton the content of lipids may increase under unfavourable conditions of

Table 55. Lipid content in modern marine plants and animals (% of dry mass)

Object of research	Extreme values	Source
Modern marine plants		
Diatoms	5.0 – 38.8(12.0)	Clarke and Mazur (1941), Collyer and Fogg (1955), Low (1955), Fogg (1956), Lee et al. (1971), Barashkov (1972), Orcutt and Patterson (1975)
Peridineans	4.0 – 18.0(11.4)	Brandt and Raben (1920, cit. after Strickland 1965), Parsons et al. (1961), Vinogradova (1967)
Green	2.0 – 46.0(18.7)	Krogh et al. (1930), Ketchum and Redfild (1949), Collyer and Fogg (1955), Serenkov et al. (1957), Spoehr and Milner (1959)
Golden	4.6 – 11.6(8.1)	Parsons et al. (1961)
Blue-green	0.8 – 15.0(7.0)	Holm-Hansen (1968), Pakhomova (1969)
Red	0.4 – 3.6	Kanazawa et al. (1972), Hayashi et al. (1974)
Brown	0.5 – 4.8	Adam et al. (1971), Kizevetter (1973), Hayashi et al. (1974)
Modern marine animals		
Copepods	3.0 – 68.0(34.3)	Klem (1932), Nakai (1952), Kizevetter (1954), Littlepage (1964), Lee et al. (1971), Lee and Hirota (1973), Lee (1974), Sargent and Lee (1975)
Decapods	7.0 – 33.0(16.5)	Raymont et al. (1967, 1969), Culkin and Morris (1969), Lee and Hirota (1973), Lee (1974)
Euphausiids	7.0 – 35.0(19.0)	Nakai (1952), Littlepage (1964), Pierce et al. (1969), Raymont et al. (1971), Lee and Hirota (1973), Lee (1974), Sargent and Lee (1975)
Mysids	11.0 – 24.0(17.5)	Lee and Hirota (1973)
Amphipods	4.8 – 49.0(27.0)	Lee and Hirota (1973), Rakusa-Suszczewski and Dominas (1974), Lee (1974), Sargent and Lee (1975)
Chaetognathae	9.3 – 27.0(19.4)	Reeve et al. (1970), Lee and Hirota (1973)
Ctenophores	9.0 – 13.0(11.0)	Lee (1974), Sargent and Lee (1975)

Figures in parenthesis are average values

nutrition (for instance, with a deficiency of nitrogen in the environment) up to 46% (dry weight) (Ketchum and Redfield 1949, Collyer and Fogg 1955). This dependence also holds in the case of macrophytes (Mohsen et al. 1974). In the case of individual representatives of crustacean plankton in periods of good nourishment (shown by the presence of fatty drops) the content of lipids attains 68% (dry weight).

The dependence of the content of lipids in zooplankton upon hydrological and geographic conditions finds expression in the direct correlation with plankton biomass (lipids attaining a maximum in deep-water upwelling zones) and in the increasing fatness of the plankton with the greater latitude and depth of its habitation (Bogorov 1960, Littlepage 1964, Bogorov et al. 1966, Vinogradova 1967a, b, Vinogradov et al. 1970, Lee et al. 1971, Herring 1972, Morris 1972, Lee and Hirota 1973, Childress and Nygaard 1974). The relationship between lipid content and depth can be numerically expressed as follows: copepods living at depths of up to 250 m contain, on the average, 14% of lipids (dry weight); the migrational species – 14.8% and the deep-water species – 29%. The lipid content is, moreover, dependent on the age and sex of the animals and on the time of day, but is generally controlled by the total content of OM and the availability of food (Petipa 1964, 1981, Lee et al. 1970, 1971, Raymont et al. 1971).

Despite the large variations of the content of lipids in marine organisms, Bogorov (1967b) summarized the existing data and gave the average concentrations of fat in phyto- and zooplankton (see Table 41).

As regards the content and composition of lipids in marine bacteria which process almost all the OM in the ocean, very little information is available at present. Bacteria contain 10 – 15% of lipids on the dry weight basis, but in the case of individual acid-resistant strains growing on glycerol media, the lipid content may rise up to 40%. Most of the lipids are confined to bacterial cytoplasmic membranes in the form of a lipoproteinic complex and in the cellular walls of bacteria (Rose 1968, Kuznetsov 1970).

The composition of lipids in phyto- and zooplankton is dependent to a greater extent on physiological characteristics of the organisms and on ecological environmental factors. The lipids of plankton algae appear to be mainly the ethers of fatty acids and alcohols. For instance, in three diatom species phospho- and galactolipids account for 50 – 70% and triglycerides for 12 – 16% of the total lipids (Lee et al. 1971). Individual algae may accumulate significant quantities of fatty acids.

Among the lipids of phyto- and zooplankton, the fatty acids have been particularly well studied (e.g., Clarke and Mazur 1941, Lewis 1962, Koyama et al. 1963, Chuecas and Riley 1969, Pakhomova 1969, Culkin and Morris 1969, 1970, Ackman and Hooper 1970, Ackman et al. 1970, Harrington et al. 1970, Kenyon and Stanier 1970, Lee et al. 1971, Pugh 1971, Bottino 1972, Lee 1974, Akulin et al. 1975). The content of fatty acids in phytoplankton amounts to 1.5 – 12% of dry mass, as much as 62% (in diatoms) being present in a free state (Clarke and Mazur 1941). Fatty acids make up part of the reserve (neutral) and intracellular (polar) lipids. While the composition of fatty acids in intracellular lipids is constant and is determined by the parameters of biosynthesis and the membrane structure (Lee et al. 1971), that of reserve lipids, including fatty acids, is dependent on ecological factors (Gastaud 1973). This finds expression, for example, in a higher proportion of unsaturated fatty acids with greater latitude and depth of algal and zooplankton habitation (Lewis 1962, Patterson 1970, Benson and Lee 1975). In the phytoplankton, fatty acids are represented by polyunsaturated compounds (27 – 84% of total fatty acids) with 5 – 6 double bonds. The $C_{14}-C_{22}$ molecules are predominant among the unsaturated acids, while among the saturated ones $C_{14}-C_{18}$ occur most frequently.

In phytoplankton the content of unsaponifiable compounds is usually not very high (3 – 22%). These are principally hydrocarbons, as well as fatty alcohols and sterols. The phytoplankton hydrocarbons are principally represented by n-paraffins and olefins, isoparaffins and aromatic compounds being virtually absent, while pristane was discovered in trace quantities (Blumer et al. 1971). Sterols account on the average for 0.01 – 1.16% of dry algal weight (Voyshvillo et al. 1972), but in the composition of lipids they may constitute 10 – 17% (Lee et al. 1971).

In macrophytes the neutral and polar lipids are contained in approximately equal quantities, triglycerides accounting for 12%, hydrocarbons for 4%, and sterols for 10 – 25% of the total lipids (Laur and Liem 1970, Alam et al. 1971).

Zooplankton lipids are represented by phosphatides (4 – 86%), triglycerides (4 – 75%) and waxes (from traces to 86%) (Lee 1974). In contrast to phytoplankton, the main part of fatty acids in zooplankton exist in a combined state as a constituent part of different complex ethers (Lewis 1962, Iwata 1964, Bogorov et al. 1966). Recent research has demonstrated that the main reserve substances and energy sources in zooplankton are waxes (ethers of high-molecular fatty acids and higher fatty alcohols) the content of which may increase to attain 63% of total lipids in deep-water animals living in high latitudes (Lee et al. 1970, 1971, Lee and Hirota 1973). In the case of copepods playing such an important role in the transformation of phytoplanktonic OM, waxes account for 90% of total lipids (Benson and Lee 1975). At the same time, ostracodes, pteropodes, euphausiids, amphipods and decapods are principally synthesizing triglycerides.

The zooplankton fatty acids are characterized compositionally by a varying, though high degree of non-saturation (up to 84%). Among the unsaturated fatty acids, the $C_{16} - C_{22}$ molecules are predominant, and among the saturated ones, the $C_{14} - C_{18}$ molecules. Under natural condition the fatty acid and fatty alcohols of the zooplankton are obtained partly from the phytoplankton and partly from the synthesis of polynonsaturated molecules which may be absent in the food but are necessary to zooplankton (Koyama et al. 1963, Bottino 1972). In the unsaponifiable fraction, the content of which in zooplankton is highly variable (from 11% in amphipods to 50% in calanids), fatty acids make up 6 – 8% (euphausiids, mysids, amphipods) to 80% (copepods), hydrocarbons up to 1.7%, sterols from 2% (copepods) to 57 – 74% (Yamada 1972). The principal constituents of hydrocarbons in crustaceans are n-alkanes of $C_{14} - C_{33}$, accounting for 78% of the total hydrocarbons (Hamilton et al. 1975).

Thus during the transformation of the OM of primary producers by zooplankton, as well as by nekton, the content of lipids in the dry mass and in the OM increases, while the group and molecular composition of the lipids is subject to significant variations.

8.2 Sea Water Lipids

The first information on lipids compounds in sea water dates back to 1892 – 1894 when Natterer discovered stearic and palmitinic acids in the Mediterranean's coastal waters (Natterer 1892 – 1894, cited by Jeffrey 1970). However, methodo-

Table 56. The content of lipids in sea water in different areas of the Oceans

Sampling site	Depth, m	The content of lipids			Source
		$\mu g\,l^{-1}$	$\mu g\,C\,l^{-1}$	% of C_{org}	
Gulf of Mexico	8–2780	4580–13800	–	–	Jeffrey et al. (1964)
	–	400–8000	–	–	Jeffrey et al. (1966)
– coastal waters	0–100	170–1590	107–954	10.6–16.4	Jeffrey (1970)
– central part	10	180–310	120–160	14.6–19.9	Jeffrey (1970)
– western part	3200	180–200	120–130	13.5–21.0	Jeffrey (1970)
– north-eastern part	surface	60–160	–	4.0	Kenicutt and Jeffrey (1981a)
	near-bottom layer	61–116	–	4.0	Kenicutt and Jeffrey (1981a)
The Atlantic near Africa	Film 0–4500	200–1000 77–1130 (383)	– –	– –	Garret (1967), Saliot and Barbiert (1974)
Peruvian shelf	10	~333[a]	248	30.5	Beliayeva and Romankevich (1976)
South-eastern part of the Pacific	0–3800	~143–209[a]	107–157 (132)	11.4–28.4 (16.0)	Beliayeva and Romankevich (1976)
Barents sea	0–200	–	200–660 (400)	5.0–35.0 (21.3)	Beliayeva and Romankevich (1983)

Figures in parenthesis are average values

[a] In terms of lipid carbon the content of which in the matter is assumed to be 75%

logical difficulties involved in the isolation and chemical analysis of DL, which are due to their low concentrations in sea water, accounted for the fact that reliable quantitative data on the content of dissolved lipids were unavailable until the 1970s (Jeffrey 1970, Saliot and Barbier 1973, 1974, Beliayeva 1974, Beliayeva and Romankevich 1976, Table 56).

The content of dissolved lipids in different areas of the Pacific and Atlantic Oceans amounts to 60–1590 $\mu g\,l^{-1}$ or 4.0–30.5% of DOC in sea water (lipids extracted by chloroform at pH 2). Investigations performed in the southeastern part of the Pacific Ocean have revealed that a characteristic feature in the distribution of dissolved lipids is a comparatively narrow variability range of their concentrations. The available data are very few yet to enable us to assess the variability trend in the content of dissolved lipids as part of DOM. It can be noted that the southeastern Pacific Ocean at the depth of 3800 m (St. M-542) and 200 m (St. M-530) the content of dissolved lipids in the OM slightly higher (15.3 and 28.4%, respectively) than that in the surface layers at the respective stations (13.0 and 11.4%; Beliayeva and Romankevich 1976). Also the same was found later in the Barents sea (Beliayeva and Romankevich 1983). In the Gulf of Mexico, on the contrary, the surface waters contain about twice as much dissolved lipids as do the bottom waters, the lipid concentrations in sea water both at the surface and at the bottom varying from season to season (Kennicutt and Jeffrey 1981 a).

By using the improved methods of gas-liquid and thin-layer chromatography it became possible to establish the group composition of DL and the molecular composition of their constituent fractions.

However, because of the fact that the majority of available methods are suitable only for the study of narrow lipid fractions many of which cannot yet be isolated, the data obtained pertain mainly to hydrocarbons, fatty acids, and sterols (Table 57).

A complex investigation of the group composition of DL was carried out in the southeastern Pacific Ocean and in the Gulf of Mexico using a standard procedure (Jeffrey 1970, Beliayeva and Romakevich 1976). The chloroform extracts were separated either by the liquid method (the Gulf of Mexico) or by quantitative thin-layer chromatography (southeastern Pacific).

The concentration of hydrocarbons in sea water varies from 10 to 290 $\mu g\, l^{-1}$, and in the majority of samples hydrocarbons were found to account for a significant proportion of DL (6.5 – 61.5%). The main characteristic of the distribution of hydrocarbons is the increase of their concentrations in the surface water layer and, particularly, in the microlayer where the hydrocarbon content may attain 0.2 mg m^{-2} (Ledet and Laseter 1974), as well as in coastal productive waters (Zsolnay 1971, Saliot 1981). Owing to the continuous oil pollution of the ocean the composition of seawater hydrocarbons in most of the areas is of mixed biogenic and technogenic origin. The distinguishing features of biogenic seawater alkanes are the clearly expressed maxima in the low-molecular spectral region (n-C_{15}, n-C_{17}, n-C_{19}) corresponding to the predominant alkanes of marine organisms, and in their high-molecular part (n-C_{27} – n-C_{30}) marking the presence of hydrocarbons of a terrestrial origin, as well as the presence of pristane. The ratio of even and odd homologues in the alkanes series (Bray and Evans 1965) is close to one (Kennicutt and Jeffrey 1981 a), n-alkanes makes up about 10 – 15% of total hydrocarbons (Saliot and Barbier 1974). According to the available data, the composition of seawater hydrocarbons does not undergo great variations with depth (Iliffe and Calder 1974).

Combined fatty acids comprise several lipid fractions of a complex structure, which contain no heteroatoms: sterol ethers, fatty acid ethers, triglycerides. The content of these fractions was determined in the waters of the southeastern Pacific where they amount to 0.1 – 33 $\mu g\, C\, l^{-1}$ or 0 – 26.4% of dissolved lipids. The proportion of sterol ethers in the DL decreases with depth (Beliayeva and Romankevich 1976), which does not agree with the concept of stability of sterol ethers in the sea water (Gagosian 1975). Presumably the content of this fraction in the mass of water depends on the composition of sterols and fatty acids constituting the ethers, and on their resistance to hydrolysis. Combined sterols make up only about 30% of the total dissolved sterols in sea waters, cholesterol being of the most frequent occurrence among them (Gagosian 1975).

The concentration of fatty acid ethers in the filtered water of the southeastern Pacific does not exceed 10 $\mu g\, C\, l^{-1}$ at the surface and sharply declines with depth (Beliayeva and Romankevich 1976). In the Gulf of Mexico a much higher content of fatty acid ethers has been found – up to 122 $\mu g\, l^{-1}$ or 35% of the lipids,

Table 57. The composition of lipids in sea water and in suspended material in different areas of the oceans

Sampling area (depth)	Hydrocarbons	Sterol ethers	Fatty acid ethers	Triglycerides
Sea water				
Eastern Pacific	–	–	–	–
	–	–	–	–
20–2310 m	–	–	–	–
	–	–	–	–
Southeastern	*45–66*[b]	*1*[b]	*0.1–10*[b]	*14–33*[b]
Pacific	36.2–61.5	1.0	0.0–7.8	13.7–26.4
(15–3800 m)				
Gulf of Mexico	–	–	–	–
	–	–	–	–
– 10–1900 m	–	–	–	–
	–	–	–	–
– 8–2780 m[a]	*860–2390*	*60–330*	–	–
	8.0–52.2	1.4–7.0	–	–
– 0–3200 m	*290*	*30*	–	*160*[d]
	19.2	2.0	–	10.0
– north-eastern	*0.1–3.8*	–	*1.1–122.0*	–
part	1.0–4.0	–	3.0–35.0	
Coastal waters off	–	–	–	–
Western Bretagne	–	–	–	–
Sea of Japan	–	–	–	–
	–	–	–	–
Atlantic Ocean	*10–140(40)*	–	–	–
near Africa	7.0–20.0	–	–	–
– 0–4500 m	*10–137*	–	–	–
	6.5–40.0	–	–	–
North-western	–	–	–	–
Atlantic	–	–	–	–
Black Sea	–	–	–	–
(0–2000 m)	–	–	–	–
Mediterranean Sea	–	–	–	–
	–	–	–	–
Aquatic suspended material				
Eastern Pacific	–	–	–	–
(20–400 m)	–	–	–	–
Southeastern Pacific	*0–16(4)*[b]	*0.1–8(2)*[b]	*0–5(1)*[b]	*0.1–17(3)*[b]
0–50 m	0.0–38.2(16.6)	0.5–9.5(4.7)	0.0–38.2(16.6)	0.5–18.4(5.8)
50–500 m	*1.9(4)*[b]	*0–0.1(0.05)*[b]	*0–0.2(0.08)*[b]	*0–0.1(0.05)*[b]
	20.5–48.2(36.4)	0.0–3.6(1.0)	0.0–0.5(1.5)	0.0–2.7(0.8)
500–3800 m	*1.2(1)*[b]	*0–0.1(0)*[b]	*0–0.2(0)*[b]	*0.0–0.1(0)*[b]
	19.5–48.2(38.7)	0.0–3.5(0.9)	0.0–5.5(1.4)	0.0–2.7(0.7)
Sea of Japan	–	–	–	–
	–	–	–	–
Gulf of Mexico	–	–	–	–
(0–3200 m)	3.0	1.5	–	–
Gulf of Mexico	*0.6–7.9*	–	*0.6–17.5*	–
(0–3200 m)	4.9–14.5	–	5.8–23.9	–
Northern part of	–	–	–	–
the Atlantic Ocean	–	–	–	–
(0–0.2)				
Baltic Sea[e]	–	–	–	–
	1.2–3.3	0.6–1.6	Traces-6.1	2.3–4.6

Italics = extreme values in micrograms per litre; Roman = percentage of total lipids; ( ) = average values. Data are given for suspended material obtained by filtration

[a] Overestimates because of imperfections in the lipid isolation procedures

[b] Concentration given in µg C per litre

Fatty acids	Fatty alcohols	Sterols	Polar lipids	Source
10–120[a]	–	–	–	Williams (1961)
–	–	–	–	
1–9	–	–	–	Williams (1965)
–	–	–	–	
1–2[b]	*1–2*[b]	*6–10*[b]	*17–27*[b]	Beliayeva and
1.2–2.0	1.2–1.7	5.6–7.5	16.1–20.4	Romankevich (1976)
–	–	*10–135*	–	Matthews and
–	–	–	–	Smith (1968)
400–800[a]	–	–	–	Slowey et al.
–	–	–	–	(1962)
300–2900[c]	–	*260–3000*	*540–700*	Jeffrey et al.
–	–	5.7–30.0	5.7–12.7	(1964)
160[d]	*390*	*20*	*660*	Jeffrey (1970)
10.0	25.3	1.2	41.4	
–	–	–	*11.8–144.6*	Kennicutt and
–	–	–	55.0–96.0	Jeffrey (1981a)
10–40	–	–	–	Trequer et al.
–	–	–	–	(1972)
–	–	*1–4(3)*	–	Kanazawa and
–	–	–	–	Teshima (1971)
–	–	–	–	Barbier et al.
–	–	–	–	(1973)
1–10	–	*3–15*	–	Saliot and
–	–	0.0–8.7	–	Barbier (1974)
–	–	*0–1*	–	Gagosian (1975)
–	–	–	–	
177–342	–	–	–	Vitiuk (1967)
–	–	–	–	
51–153	–	–	–	Vitiuk (1970)
–	–	–	–	
1–3	–	–	–	Williams (1965)
–	–	–	–	
0.3–15(4)[b]	*0.1–22(5)*[b]	*0.2–25(4)*[b]	*2–49(24)*[b]	Beliayeva and
2.5–18.4(5.8)	0.5–22.2(7.7)	0.9–24.8(6.7)	23.2–75.0(44.2)	Romankevich (1976)
0.3–3(1.3)[b]	*0.2–3(2)*[b]	*0.1–2(1)*[b]	*0.4–9(4.2)*[b]	Beliayeva and
8.2–16.7(10.5)	8.2–14.0(11.7)	0.5–8.2(15.5)	11.2–50.0(30.7)	Romankevich (1976)
0.3–1(0.5)[b]	*0.3–2(1)*[b]		*0.1–1(0.6)*[b]	Beliayeva and
8.2–20.9(15.5)	11.8–52.1(22.9)		7.5–33.2(19.9)	Romankevich (1976)
–	–	*2.20(8)*	–	Kanazawa and
–	–	–	–	Teshima (1971)
–	–	–	–	Jeffrey (1970)
–	12.0	5.5	69.6	
–	–	–	*7.2–78.7*	Kennicutt and
–	–	–	49.4–84.9	Jeffrey (1981b)
4–26	–	–	–	Schultz and
–	–	–	–	Quinn (1972)
–	–	–	–	Bobyleva and
8.0–12.7	4.2–5.6	18.3–22.3	32.6–49.3	Beliayeva (1982)

[c] Total triglycerides, fatty acids, fatty alcolhols
[d] Total triglycerides, fatty acids
[e] Suspended material obtained by method of separation

varying from season to season (Kennicutt and Jeffrey 1981a). The fatty acid ethers consist mainly of methyl, ethyl and propyl ethers of fatty acids and are the conversion products of humic acids.

Following hydrocarbons and polar lipids, triglycerides[5] are quantitatively the third fraction of the DL. Their content in the seawater DL of the southeastern Pacific (14 – 33 μg C l^{-1} or about 14 – 26% of the total lipids) is higher than in the suspended material, while vertically this difference becomes even greater, presumably on account of the greater stability of dissolved triglycerides compared to the triglycerides of the suspended matter, as well as to other fractions of combined fatty acids comprised in the DL.

The content of free fatty acids in sea water varies from <1 to 342 μg l^{-1}. Maximum concentrations have been recorded in the waters of the Black and Mediterranean Seas (51 – 342 μg l^{-1}; Vitiuk 1967, 1970). These free fatty acids cover a wide spectrum of acids from C_{11} to C_{22} (Williams 1961, Ushakov et al. 1966, Garrett 1967, Schultz and Quinn 1972). Most of them are unsaturated compounds typical of plankton (Schultz and Quinn 1972, Saliot and Barbier 1974), but with increasing depth the content of saturated acids rises while the number of carbon atoms in their chains decreases (Slowey et al. 1962, Williams 1965, Vitiuk 1967). For instance, in the Black Sea saturated acids at the 0 and 2000 m levels account, respectively, for 50.8% and 55.1% of the total fatty acids, the concentration of acids with 16 and more carbon atoms within the same depth interval diminishes from 73.4 to 3.9% (Vitiuk 1967). The change in the length of the acids carbon chain, initiated by ultraviolet irradiation, apparently takes place at a high rate in the surface-film water (Garrett 1970).

The concentration of fatty alcohols and sterols in sea water varies widely: from 1 to 390 μg l^{-1} (1.2 – 25.3% of DL) for alcohols and from 0 to 135 g l^{-1} (0 – 8.7% of DL) for sterols. The composition of fatty alcohols in sea water is unknown, while sterols isolated from the water of the surface layers in the Sea of Japan and in the northwestern Pacific were found to be similar to the vegetative and animal ones. The predominant sterols are cholesterol (12 – 45%) and β-cytosterol (32 – 55% of the total sterols), campesterol being present in a lesser quantity (Kanazawa and Teshima 1971, Gagosian 1975).

Polar lipids (esters containing heteroatoms: phospho-, galacto-, and sphingolipids, as well as part of the pigments) occur in sea water in concentrations ranging from 20 to 660 μg l^{-1}, i.e. about 16 – 41% of the DL. Polar lipids were found to contain as much as 15.4% of phospholipids, 23.2% of pigments, and 2.8% of sphingolipids (Jeffrey 1970, Beliayeva and Romankevich 1976). Thus most comprehensive complex research, carried out by identical methods in two ocean areas, has revealed that the DL contain all the principal groups of neutral (low-polar) lipids the relationship of which, as well as of polar lipids in the areas studied, was not similar.

---

5 The triglyceride fraction comprises the ethers of fatty acids and glycerol, as well as of other polyols which are not separated by the employed method of thin-layer chromatography

8.3 Lipids in Suspended Material

8.3.1 The Quantitative Distribution of Particulate Lipids in the Ocean

The first data on the content and composition of lipids in the suspended material were obtained in the mid-1960s by means of the chemico-bitumenological technique used on separation samples (Bogdanov and Ovchinnikova 1965, Gorskaya and Glebovskaya 1966). The content of PL isolated by means of different solvents from the separation samples of the suspended material from the surface layer in the Pacific and Indian Oceans was found to vary from 3.4 to 31.3% of the OM (Table 58). In the Pacific Ocean the content of the chloroform (or chloroform-methanol) fraction of the lipids (β_{chl}) averages 19.5%. According to the materials obtained for the Pacific and Indian Oceans, the total lipids extractable after the acid hydrolysis with 10% HCl average about 21% of the OM.

The lipids in the filtration samples which had been taken at different ocean depths were investigated by means of the luminescent and histochemical methods (Bogdanov and Lisitzin 1968, Lisitzin et al. 1971). For quite a number of reasons these methods cannot yield a quantitative picture of the distribution of lipids. It is therefore not surprising that the results of the express luminescent method proved to be significantly overstated. At present the most reliable data on variations in the content and composition of PL from the surface to the seabed have been obtained in the southeastern Pacific, the northern part of the Indian Ocean and the northeastern part of the Gulf of Mexico. These data can be used in considering the behavior patterns of the PL (see Tables 57 and 59; Beliayeva and Romankevich 1976, Kennicutt and Jeffrey 1981b, Nemirovskaya and Romankevich 1982). The materials pertaining to the Baffin Bay are not typical, since they relate to the period of intensive phytoplankton bloom when the concentration of PL was higher than that of DL.

The content of PL and of their separate fractions in the southeastern Pacific (at the profile Peruvian shelf – Peruvian-Chilean trench and in the southeastern part of the northeastern basin), as well as in the Mozambique area, was determined directly from the carbon of the lipid compounds following their extraction from wet samples of the suspended material isolated on fibreglass filters CF/C by means of the mixture of chloroform and methanol (2:1) at normal temperature (Beliayeva 1980a–c).

The concentration of PL in these areas (2.4–99.6 $\mu g\,C\,l^{-1}$ or ~3.2–133 $\mu g\,l^{-1}$) was found to be directly dependent on POC. Therefore, the horizontal and vertical distribution of PL was determined by a complex of the same factors and was subject to the same regularities as was particulate carbon (circumcontinental and vertical zonalities). It was earlier shown that this was also typical of the protein-like and carbohydrate-like compounds.

The vertical gradient of the decrease of PL in the water column in the southeastern Pacific and in the Mozambique area was found to be the greater, the higher their concentration in the surface layer, which, in its turn, is directly correlated with the primary production. Vertically, the concentration of PL decreases, the decrease being particularly rapid in the upper 20–80 m, i.e. above

Table 58. The content of lipids in the separation suspensates (0 – 7 m layer) for different areas of the world ocean

Sampling site	Solvent[a]	Lipid content (%)		Source
		of dry mass	of OM	
Central Pacific	I	1.8 – 8.3	15.4 – 26.8 (22.2)	Bogdanov and Ovchinnikova (1965)
	II	4.2 – 9.5	20.5 – 30.7 (25.7)	Bogdanov and Ovchinnikova (1965)
Central and Southeastern Pacific	III	0.2 – 5.1	3.4 – 15.8 (12.7)	Gorskaya and Glebovskaya (1966)
	IV	0.2 – 6.4	3.8 – 18.4 (15.0)	Gorskaya and Gleboskaya (1966)
Northwestern Pacific	Ia	–	14.6 – 22.6 (17.3)	Romankevich (1977)
The Mediterranean Sea	Ia	2.8 – 7.0 (4.8)[b]	17.9 – 44.5 (31.8)[b]	Nemirovskaya et al. (1982)
The Red Sea	III	10.3	26.7	Gorskaya and Glebovskaya (1966)
	IV	12.1	31.3	Gorskaya and Glebovskaya (1966)
	Ia	4.6 – 5.8 (5.2)[b]	25.1 – 39.9 (32.5)[b]	Nemirovskaya et al. (1982)
Northern part of the Indian Ocean	III	4.7 – 7.7	14.7 – 22.3 (18.1)	Gorskaya and Glebovskaya (1966)
	IV	5.0 – 8.2	16.2 – 23.8 (19.7)	Gorskaya and Glebovskaya (1966)
	Ia	0.9 – 5.4 (3.0)[b]	10.0 – 32.1 (23.0)[b]	Nemirovskaya et al. (1982)
Mozambique area	Ia	2.7 – 5.3 (3.7)[b]	18.5 – 45.0 (27.4)	Nemirovskaya et al. (1982)
Bay of Bengal	Ia	2.8 – 3.9 (3.4)[b]	24.0	Nemirovskaya et al. (1982)
South-China Sea	Ia	1.4 – 3.9 (2.8)[b]	–	Nemirovskaya et al. (1982)
Antarctic area	III	1.3 – 3.6	9.7 – 23.2 (16.8)	Gorskaya and Glebovskaya (1966)
	IV	1.4 – 4.2	14.3 – 26.7 (20.3)	Romankevich (1977)
	I + Ia	–	(19.5)	Romankevich (1977)
Average	III	0.2 – 10.3	16.4	
	IV	0.2 – 12.1	20.8	

Figures in parenthesis are average values

[a] Solvents employed for the extraction of lipids: I = chloroform; Ia = mixture of chloroform and methanol (2:1); II = chloroform and a mixture of alcohol and benzene (in succession) 1. III = a mixture of alcohol and benzene (the benzene part of the extract was studied); IV – A mixture of alcohol and benzene (1:1). Lipids extracted by means of solvents III and IV were obtained after treatment of dry (t^0 40 – 80 °C) samples of suspended material with 10% HCl; lipids extracted by means of solvents I, Ia, II were obtained from natural dry (t^0 40 – 60 °C) samples without the preliminary acid hydrolysis

[b] Investigations were performed on the wet samples just obtained. Extraction was preceded by freezing (t^0 18 – 20 °C) and methanol treatment

Table 59. The content of lipids in the filtration samples of suspended material for different areas of the world ocean

Sampling site	Depth, m	Lipid content			Source
		$\mu g\,l^{-1}$	$\mu g\,C\,l^{-1}$	% of C_{org}	
Tropical Pacific	0–200	2.2–48.3 (19.4)	–	2.1–27.5 (11.4)	Bogdanov and Agatova (1972)
	>200	–	–	1.9–49.2 (22.3)	Bogdanov and Agatova (1972)
Baffin Bay	0	3540	–	–	Jeffrey (1970)
Southeastern Pacific	0–15	–	7.2–99.6 (50.5)	26.2–64.6 (46.4)	Beliayeva and Romankevich (1976)
	0.50	–	2.4–99.6 (31.6)	11.6–64.6 (39.0)	Beliayeva and Romankevich (1976)
	>50–500	–	2.5–23.5 (8.9)	26.0–57.6 (43.4)	Beliayeva and Romankevich (1976)
	>500–4000	–	3.5–6.5 (4.3)	35.8–46.7 (31.5)	Beliayeva and Romankevich (1976)
Mozambique area	0	–	13–48 (27.5)	25.4–37.2 (35.9)	Nemirovskaya et al. (1982)
	25–120	–	11–26 (17.8)	–	Nemirovskaya et al. (1982)
	200–500	–	6–7	–	Nemirovskaya et al. (1982)
Gulf of Mexico	surface	12–70	–	20	Kennicutt and Jeffrey (1981c)
	bottom layer	9–52	–	20	Kennicutt and Jeffrey (1981b)
Barents Sea	0–270	–	6–53 (23.0)	8–63 (27.2)	Beliayeva and Romankevich (1983)

Figures in parenthesis are average values

the pycnocline. Here there is a great deal of labile OM and it decays rapidly. For the OM as a whole the decomposition rate is proportional to its concentration (Skopintsev 1966).

In the suspended material collected off Peru the layers with the maximum concentrations of PL and ATP coincide, but vertically the content of living OM (ATP × 250) is subject to a more rapid decrease. This is indicative of the presence of stable PL constituents selectively preserved in the particulate organic detritus which is principally the product of the bacterial processing of OM[6].

The hydrological characteristics of water masses affect the distribution of PL. This is manifested in a certain increase of the concentration of PL near the bottom and at discrete levels in the water column. The irregular distribution of PL in the southern part of the Pacific northeastern basin may be the reflection of the structural complexity of the water masses in this area. In the mass of water

6 Organic detritus is assumed to comprise all the OM determined from C_{org} with the exception of phyto- and microzooplanktonic OM

the latitudinal zonality in the distribution of PL seems to be preserved (Lisitzin et al. 1971), but this needs to be verified by using more reliable methods.

The content of PL in the OM of the surface suspended material in the southeastern Pacific varies from 26.2 to 64.6%, which exceeds their content in the OM from dry separation samples. No unambiguous explanation for this discrepancy has been given so far. Supposedly, this is due, firstly, to the fact that diatom algae are predominant in the phytoplankton of the Peruvian area (up to 95%), which are prominent among the autotrophic organisms for their high lipid content; secondly, to the neoformation and selective preservation of lipids during the intensive utilization of POM by the zooplankton and phytophagous fishes; thirdly, to the possible underestimation of the content of PL in the separation samples because of the loss of fine OM fractions, as well as the entrance of a part of the lipids into the non-extractable complexes during the drying of the separation suspensates; fourthly, to the unproportionally high loss, compared to the lipids, of part of the water-soluble OM during the collection of suspended material on the filters.

In the intermediate layers the content of PL in the OM is observed to decrease due to the decay of lipids and their utilization by the microorganisms, but in the deep-lying layers where nearly all lipids are confined in the detritus the selective preservation of PL leads once again to their higher content in the OM. However, these data call for additional confirmation based on more abundant material.

Investigations carried out in the northern part of the Indian Ocean have revealed that the content of PL declines with the growing distance from the mouths of the Limpopo and Zambezi rivers into the central part of the Mozambique Channel by a factor of 5 – 7, and in the 0 – 500 m depth interval by a factor of 2 – 3.

At the water-atmosphere interface an increase of both the DL and PL and of their individual component groups takes place in the water microlayer (Hunter and Liss 1981, Nemirovskaya and Romankevich 1982). In the vicinity of the Seychelles the PL in the water microlayer accounted for 75% of the total lipids. In the north Atlantic (Maroccan Deep) and in the Mediterranean the ratio of PL and DL was reversed (1:3) (Nemirovskaya and Romankevich 1982).

8.3.2 The Composition of Particulate Lipids and Its Variations in the Mass of Ocean Waters

In the southeastern Pacific the lipids of the living OM account in the surface water layer for about 30 – 79% and in deep waters for less than 1% of the total (Fig. 57). The lipids that are part of the organic detritus generally sharply predominate in the mass of ocean waters (more than 90% of the total PL; Beliayeva and Romankevich 1976). Later similar data were also obtained for the Arabian Sea (Boussuge et al. 1981).

In the suspended material of the southeastern Pacific there have been quantitatively determined eight groups of lipid compounds containing the main classes of the substances: hydrocarbons, sterol ethers + waxes, fatty acid ethers, triglycerides, free fatty acids, fatty alcohols, sterols and polar lipids (see Table 57).

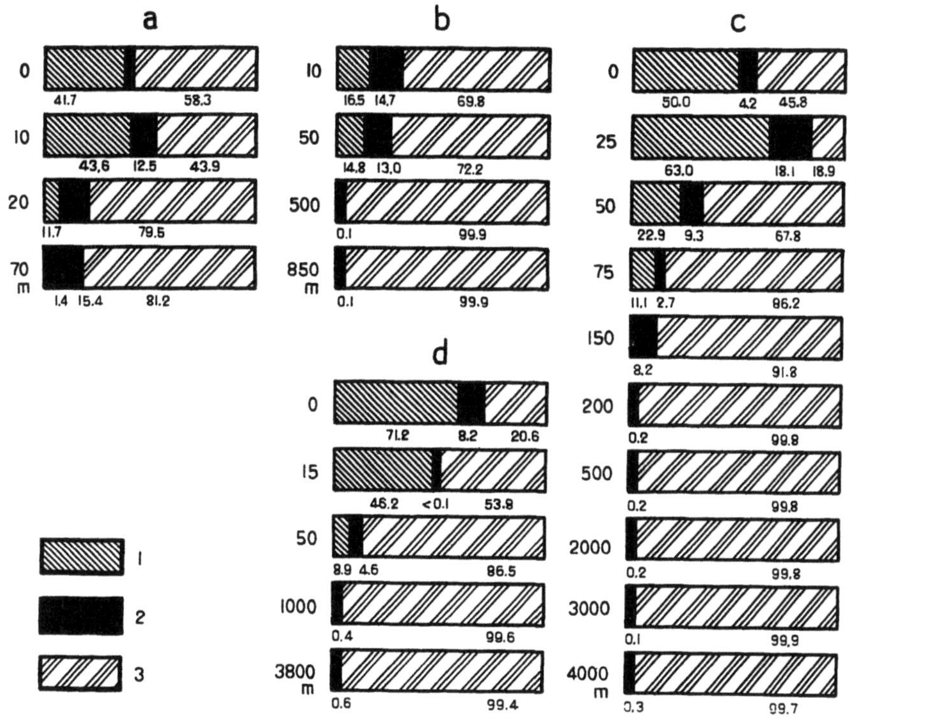

Fig. 57a–d. The genetic composition of particulate lipids, % of the total lipids. Station nos: **a** M-252; **b** M-544, **c** M-542; **d** M-530. Lipids: *1* phytoplankton; *2* microzooplankton; *3* detritus (Beliayeva and Romankevich 1976)

Hydrocarbons are one of the highly conservative lipid fractions. With depth, the content of hydrocarbons in PL approximately doubles. This takes place due to a sharp decrease of the other classes of PL. As a result, the relationship between the content of PL in the OM and hydrocarbons is characterized by a high correlation coefficient ($r = +0.784$). The content of hydrocarbons in the PL is connected with the concentration of living OM and polar lipids in the suspended matter by a reverse dependence ($r = -0.573$, $r = -0.850$) which reflects the stability and selective preservation of hydrocarbons in the hydrolysis. This gives rise to the fact that the hydrocarbon content in numerous samples of the suspended material is higher than in the diatom plankton (12.1–48.2% and 2–16%, respectively). According to Kennicutt and Jeffrey (1981c), the composition of particulate alkanes in the Gulf of Mexico corresponds to that of seawater alkanes, but the olefines and cycloalkanes were also found in the suspended material.

Combined fatty acids comprise sterol ethers plus waxes, fatty acid ethers, triglycerides and are a minor component of the PL. In the suspended material, they are presumably an unstable group of compounds. In the vicinity of the Peruvian-Chilean trough (St. M-544) they were recorded only in the 0–50 m layer, while at the depth of 100 m only trace concentrations were found. Under the conditions of alternate upwelling and downwelling waters, fixed fatty acids may be present at great depths. In the Gulf of Mexico the concentration of fatty acid ethers was found to decline with distance from the coast. A higher content of fatty acid ethers in February as compared to November, which may be associated with the effect of the river runoff, was also recorded (Kennicutt and Jeffrey 1981b). In the southeastern Pacific the sterol ethers, fatty acid ethers and triglycerides are not interrelated by a correlative dependence with the content of living matter in the suspended material. The content of fatty acid ethers and triglycerides in the diatoms amounting to 12–17% may be taken as evidence of their utilization in the trophic chains during the minor secondary synthesis by the heterotrophs. Sterol ethers have not been discovered at all in the diatom algae (Lee et al. 1971).

Free fatty acids in the suspended material in the southeastern Pacific are contained in approximately the same concentrations (0.3–15 $\mu g\,C\,l^{-1}$ or ~0.4–19 $\mu g\,l^{-1}$; Beliayeva and Romankevich 1976) as in its eastern part (1–30, on the average 3 $\mu g\,l^{-1}$; Williams 1965) and in the northern Atlantic (4–26 $\mu g\,l^{-1}$; Schultz and Quinn 1972). Their content in the PL of the southeastern Pacific varies from 2.5 to 20.9% and increases with depth. A distinct minimum of fatty acids in the PL of the Peruvian area corresponds to the maximum content of living OM and polar lipids there. Together with the negative correlation between fatty acids and the sum of complex lipids ($r = -0.500$), this furnishes evidence to the effect that their increasing content in deeper layers is due to the decay of the lipids of a complex structure.

Compositionally, the fatty acids contained in the suspended material are strongly varying in a vertical direction (in contrast to hydrocarbons). With depth, the amount of saturation increases and the length of carbon chains decreases. As a result of the bacterial effect, the quantity of fatty acids with an odd number of carbon atoms, which are not typical for plants and animals but are characteristic of bacteria, is observed to increase with depth (Slowey et al. 1962, Vitiuk 1967, Garrett 1970).

Fatty alcohols and sterols in the suspended material of the southeastern Pacific constitute 0.5–22.2% and 0.5–24.8% of the PL. At certain stations the content of fatty alcohols and sterols constituting the PL is higher than in the filtered water and reveals an increase of average values with depth. The vertical distribution of fatty alcohols and sterols resembles that of fatty acids, while their content is in direct correlation with the concentration of PL in the OM and in reverse correlation with the sum of complex lipids which are likely to be their source in the suspended material. The composition of fatty alcohols contained in the suspended material is unknown, and the composition of sterols was found to contain around 28 individual compounds (Kanazawa and Teshima 1971, Saliot and Barbier 1974, Gagosian et al. 1983).

Polar lipids of the suspended material constitute a highly dynamic fraction. Their average content in the PL of the southeastern Pacific (37.6%) and the variability of their concentrations (7.5–75.0%) are greater than in the case of DL (16.1–20.4%, on average 18.2%), but less than in the diatom plankton (50–57%), as recorded by Lee et al. (1971) which is predominant in this area (Beliayeva and Romankevich 1976). The concentrations of the polar lipids ($\mu g\,l^{-1}$) and their content in the PL were found to decrease with distance from the Peruvian shelf towards the open part of the ocean and from the surface to the seabed. Polar lipids are the only group of PL that manifest a distinct direct correlation with the content of living matter in the suspended material ($r = +0.581$) and a reverse correlation with the content of PL in the OM ($r = -0.834$). These dependences reflect on overall trend involving an increase of the content of PL in the OM contained in the suspended material due to the enrichment of lipids with more stable compounds.

Little is known about the composition of polar lipids contained in the suspended material. In the southeastern Pacific they were found to contain phospholipids (42.1%), pigments (25.5%) and sphingolipids (2.0%). In the Gulf of Mexico the polar lipids also included substituted phenols and a complex mixture of substituted ketones (Kennicutt and Jeffrey 1981c).

8.3.3 Comparative Characteristics of Particulate and Dissolved Lipids

By using a standard procedure for the extraction of lipids from the suspended material and sea water, as from the interstitial water and bottom sediments (by means of chloroform at pH 2 from sea and interstitial water and by means of a mixture of chloroform–methanol (2:1) from wet samples of suspended material and bottom sediments at 25 °C) and a newly-developed technique for the determination of micro-concentrations of lipids (Beliayeva 1980a–c), it was possible to compare the distribution and composition of different forms of lipids in the area off Peru (Table 60, see also Table 57; Beliayeva and Romankevich 1976).

The distribution of PL and DL in sea water is different. In transition from the productive waters of the Peruvian shelf to the open part of the ocean the concentration of PL decreases more than 10-fold whereas that of DL only 2-fold. The concentration of DL changes even less with depth: at the depth of 200 and 3800 m it is only 8–15% less than in the surface layer, whereas the corresponding variation for PL amounts to about 50%.

The content of PL in the surface water layers is directly correlated with primary production; no such correlation is observed in the case of DL. The maximum variability gradient in the concentration of PL occurs in the upper water layers, since in the case of the 80-fold difference in depths the content of PL decreases by about 50%.

In absolute figures the amount of dissolved plus colloidal form of lipids exceeds that of the particulate one, thus reflecting the overall relationship of the DOC and POC masses. In the shelf-pelagial direction the ratio of DL to PL drops from 38.5 to 4.8 and decreases still further lower in the water column (down to 3.2–4.0 assuming the DL = 100). The main reason for the variation of the ratio of DL to PL in the vertical direction proves to be the enzymatic hydrolysis, which in the first place affects the lipids at the interfaces. As a result of this, the low-molecular fragments of biopolymers go over from the suspended material into the solution and shift the balance towards the DL. Part of DL in the mass of water goes over once again into the particulate form, the balance being thus dynamic. Owing to a lower sorptional capacity of low-molecular compounds and

Table 60. Distribution of different forms of lipids in the water column and in the silts over the Peruvian shelf and in the Peruvian-Chilean trough (weighted average concentrations)

Station no. (depth of seabed)	Lipid concentration, μg C l^{-1} [a]					
	suspended matter		sea water		sediment, 0–10 cm	interstitial water, 0–10 cm
	0–10 m	0–bottom	0–10 m	0–bottom		
552 (78 m)	96.8 (28.0) [b]	36.3 (12.8) [b]	248.0	248	3800	1070
542 (6250 m)	12.3 (8.9) [b]	4.5 (3.8) [b]	126.0	113.0	945	1040

[a] For sediments in micrograms of carbon per gram of dry sediment

[b] Figures in parenthesis are ratios PL/DL + PL, expressed as a percentage

[c] The accepted value is that for the 0–10 m layer which is permissible for a low depth at the station site and homogeneous water

the reduced surface of the suspended material (per litre of water) there is a relative accumulation of DL as compared to PL in the water mass.

The data obtained at the same stations in the area of the Peruvian-Chilean trough have shown that the PL contain the following classes of compounds (average values, in %): hydrocarbons 23.2; sterol ethers 3.3; fatty acid ethers 3.2; triglycerides 3.8; fatty acids 11.7; fatty alcohols + sterols 15.7; polar lipids 37.6; DL comprise (in %): hydrocarbons 48.8; sterol ethers <1; fatty acid ethers 3.9; triglycerides 20.0; fatty acids 1.5; fatty alcohols 1.5; sterols 6.5; polar lipids 18.2. Thus the DL contain more hydrocarbons and triglycerides, but less free fatty acids and polar lipids. The larger content of such conservative compounds as hydrocarbons and decay-resistant fats in the DL is due to a significantly longer residence time of the DL in sea water, to the selective preservation of stable compounds, as well as to the differences in the sorptional capacity of separate lipid fractions. Hydrocarbons may selectively accumulate in the solution as a result of the low sorptional capacity. This is presumably characteristic of triglycerides too. In fact, hydrocarbons prove to be the only group among the DL, the concentration of which increases with depth, whereas the content of hydrocarbons in the PL decreases.

In contrast to hydrocarbons, the sorptional capacity of fatty acids is high, and for this reason they most commonly occur in the particulate form. The concentration of fatty acids as part of PL is 5–10 times as high as in the DL. It was shown that the overwhelming part of fatty acids (under experimental conditions as much as 98.7 ± 1.4%, Meyers and Quinn 1971) may adsorb from sea water at the particles of calcium carbonate. The capacity of fatty acids and lipids extracted from diatom algae to aggregate and form particles retained on the 0.8 μm filter under irradiation with ultraviolet light was noted by Williams (1965) and later experimentally demonstrated by Wheeler (1972).

Sterol ethers, fatty acid ethers, triglycerides and polar lipids are the particularly labile fractions of PL. Their decrease with depth is accompanied by an overall increase, though not at all of the horizons, of the content of hydro-

Table 60 (continued)

Lipid content in OM, %						Eh of int. water, mV
suspended matter		sea water		sediment 0–10 m	interstitial water 0–10 cm	
0–10 m	0–bottom	0–10 m	0–bottom			
56.3	52.4	30.5	30.5[c]	7.7	8.8	−227
48.1	46.1	13.0	14.6	2.1	5.4	from +48 to −22

carbons, fatty acids and, possibly, fatty alcohols and sterols in the PL. Generally, the PL happen to be a more labile constituent of the sea water as compared to the DL. Their concentration and composition are both varying in moving away from the coastal areas towards the open ocean and from the surface to the seabed. The results of a comparative investigation of the composition of DL and PL in the Gulf of Mexico have furnished evidence concerning the greater complexity of the PL containing, in contrast to the DL, olefines, substituted benzenes, quinones, ketones, phenols, indans, aromatic compounds and certain other constituents which are absent in the composition of the DL (Kennicutt and Jeffrey 1981b). In contrast to the PL the composition of the DL appears to be little dependent on the primary production at a given place and features a more monotonous distribution along the vertical; the DL appear to be conservative constituents of the seawater OM.

In the marine OM the concentration of the DL and PL at the respective stations varies but little with depth due to the absence of the advance mineralization or to the selective preservation of lipids. A sharp reduction in the content of lipids of the OM occurs in transition from the suspended material (46.1% – 52.4%) to the uppermost sedimentary layer (2.1% – 7.7%). On the shelf, during the passing of the suspended material into the sediments, the content of lipids in the organic matter composition declines by a factor of seven, and at the bottom of the trough by a factor of 22. It is only natural to look for the causes behind such differences in the characteristic features of these sediments; these include the presence of free H_2S in shelf sediments and a low benthonic biomass. The effect of these factors, as well as the supply of large masses of little-transformed OM to the shelf sediments, account for a higher content of lipids in the shallow-water muds as is described in the next section.

8.4 Lipids in Bottom Sediments

Lipids (bitumoids) in the bottom sediments were formerly looked upon as the most likely predecessors of hydrocarbons and other compounds present in the composition of oil (Engler 1888, Stadnikov 1937). Their regional distribution and composition have been intensively studied from the late 1920s, but the most complex research has been carried out since the 1960s. Among the general works based on an abundance of experimental material and summing up a certain research stage the following should be noted: Smith (1954), Weber et al. (1956), Romankevich (1962, 1976, 1977), Bordovsky (1964, 1974), Degens (1965), Weber (1966, 1973), Rodionova (1967), Uspensky (1970), Vassoyevich (1973), Degens and Mopper (1976), Tissot and Welte (1978).

8.4.1 Regularities in the Distribution of Lipids in Bottom Sediments

The content of lipids (bitumoids) in different types of bottom sediments, identified according to the lithologic-mineralogic, physico-chemical and paleontological indices and the sedimentation rates within the confines of three largest structural-geomorphological ocean zones (submarine margin, marginal and central ocean floor), is given in Tables 61 and 62. The analysis of this material permits us to discuss the basic regularities in the distribution of lipids in bottom sediments.

The general dependence which determines the distribution of lipids in oceanic sediments is a direct correlation with the total OM (Bordovsky 1964, 1974, Rodionova et al. 1973, Romankevich 1974, 1977). The relationships with other parameters – depth, the content of the <0.01 mm fraction, median diameter (Md), $CaCO_3$, $SiO_{2(amorph.)}$ – are generally either absent or expressed indistinctly (for instance, a reverse correlation between the lipids in the bottom sediments and $CaCO_3$). The result of the direct relationship between lipids and OM is that both are maximally concentrated in the peripheral oceanic zone. This regularity in the distribution of lipids can be clearly seen on the distribution charts of luminescent lipids in the upper sedimentary layer of the Pacific and Indian Oceans (Figs. 58 and 59). High concentrations of luminescent lipids as well as of lipids (bitumoids) A_{chl} and A + C are found to coincide and they occur in the divergence and upwelling zones which are the areas of high biologic productivity and of the C_{org} content significantly superior to the Clarke level. All types of the sediments there (for instance, those off South West Africa, Peru and Japan – terrigenous, diatomaceous and carbonaceous, of different granulometric composition) are characterized by a higher lipid content, both with respect to sediment and in the OM, in comparison with the average values for the ocean (Rodionova et al. 1973, Romankevich and Baturin 1974, Romankevich 1977, Gershanovich and Koniukhov 1980, Coastal upwelling; Its Sediment Record 1981). The facies of the diatom silts deposited in the arid region on the South West African shelf ($A_{chl} = 0.680 - 1.376\%$, $\beta_{chl} = 12 - 19.6\%$, on average 14.8%) is particularly noted for the high lipid content. This is also due to the influence of the initial material rich in lipids (diatom suspension). The varying fat content of the initial material, for instance of the diatom algae and phytobenthos, affects the content of lipids in the sediments and determines their increased content in the sapropel OM of the rocks (e.g., Veber 1966, Vassoyevich 1973).

A low lipid content is peculiar to different types of ocean floor sediments: siliceous diatom silts of the ocean basins, carbonacous foraminiferal and radio-

Table 61. The content of lipids in the sediments of the structural-geomorphological zones of the Pacific, Indian, and Atlantic Oceans (% of dry sediment)

Structural-geomorphological zones and facial types of sediments	Number of samples	Lipids[a]			Total C_{org}	<0.01 mm fraction
		benzene part of alcohol-benzene extract		average for total lipids		
		extreme value	average			
Submarine margin	24	0.010 – 0.308	0.079	0.122	0.78	27.3
– terrigenous sandy-aleuritic of the shelf	5	0.042 – 0.308	0.142	0.215	–	–
– terrigenous-iceberg sandy aleuritic of the shelf	4	0.018 – 0.039	0.028	0.039	0.27	20.0
– terrigenous weakly-siliceous of the slopes	7	0.036 – 0.201	0.119	0.194	1.26	58.4
– carbonaceous (more than 50% $CaCO_3$) and shelly sands, broyozoanforaminiferal of the shelf	7	0.010 – 0.024	0.026	0.034	0.41	13.3
– carbonaceous algal-coral	1	–	0.069	0.115	0.55	4.8
Marginal parts of the ocean floor						
– hemi-pelagic terrigenous	2	0.030 – 0.031	0.030	0.036	0.60	72.8
Central parts of the ocean floor	14	0.014 – 0.050	0.027	0.039	0.39	70.0
– Antarctic siliceous diatomaceous	3	0.022 – 0.038	0.028	0.047	0.40	65.7
– carbonaceous foraminiferal	6	0.014 – 0.032	0.025	0.033	0.36	67.2
– miopelagic red clays	3	0.034 – 0.050	0.040	0.053	0.47	72.4
– eupelagic red clays	2	0.014 – 0.017	0.016	0.022	0.38	76.8
Average for the ocean	40	0.010 – 0.308	0.059	0.089	0.64	44.5

[a] Lipids (bitumoids) were extracted with a mixture of alcohol and benzene (1:1) in Soxhlet apparatuses from the samples dried at 40 – 70 °C after their decalcination by 10% HCl and washing away of water-soluble compounds

larian-foraminiferal deposits and deep-water red clays. Particularly low values are typical of eupelagic red clays – the sediments deposited at an extremely slow rate in low-productivity areas. On the average, they contain less lipids in the OM (see Table 62). However, some individual varieties of red clays are noted for the high β_{chl} values, as was first pointed out by Gorskaya (Weber et al. 1960) (Gorskaya's law).

Table 62. The group composition of the organic matter in bottom sediments and in ferro-manganese nodules of the Pacific, Atlantic and Indian Oceans

Facial types of sediments	In the OM composition, % lipids							C_{org}, % of dry mass	<0.01 mm fraction, %	Eh, mV
	β_{chl}	benzene part of alcohol-benzene extract	$A_{alc\text{-}benz}$	A+C sum	easily-hydrolyzable OM	Humic acids (γ)	ROM			
Bottom sediments										
Submarine margin sediments, 80 samples	6.3	7.2	6.6	10.1	16.8	30.1	36.4	2.18	41.2	−97
Including										
– terrigenous and terrigenous-glauconite of the shelf	5.7	8.9	10.5	12.1	20.8	23.5	31.4	3.02	36.4	−154
– terrigenous-iceberg of the Antarctic shelf	–	7.6	–	9.7	–	20.6	–	0.28	20.0	−112
– diatomaceous of the South West Africa's shelf	14.8	–	–	17.8	17.3	44.8	24.3	5.33	71.8	−204
– terrigenous and weakly-siliceous	2.4	6.6	5.4	7.3	20.0	30.9	41.2	1.41	54.2	−6
– average for terrigenous sediments (56 samples)	4.3	7.6	6.4	9.5	19.3	28.7	39.4	1.91	44.9	−60
– carbonaceous sediments of the shelf (>50% $CaCO_3$, 19 samples)	6.2	5.9	7.0	8.8	6.9	30.8	45.6	2.14	19.2	−134
Sediments of the marginal part of the ocean floor (12 samples)	2.3	3.5	3.4	6.2	15.6	17.1	67.6	0.61	65.1	+366
Sediments of the central part of the ocean floor (26 samples)	0.9	5.0	3.3	6.6	19.1	5.7	73.8	0.32	76.9	+442
Including										
– Antarctic siliceous-diatomaceous	–	5.7	–	8.4	–	5.6	–	0.40	66.2	+440
– carbonaceous-foraminiferal	–	4.7	1.5	5.9	–	9.8	–	0.42	68.8	+414
– miopelagic red clays	0.84	6.0	3.4	7.7	–	7.0	–	0.43	80.4	+454
– eupelagic red clays	1.01	2.9	3.7	4.0	19.1	1.3	73.8	0.24	82.4	+474

Table 62 (continued)

Facial types of sediments	In the OM composition, %									
	lipids									
	β_{chl}	benzene part of alcohol-benzene extract	$A_{alc\text{-}benz}$	A+C sum	easily-hydrolyzable OM	Humic acids (γ)	ROM	C_{org}, % of dry mass	<0.01 mm fraction, %	Eh, mV
Average for the sediments of the world ocean (118 samples)	5.4	6.3	5.2	9.0	16.8	22.8	43.4	1.61	51.9	+70
Ferro-manganese nodules										
Ore body (3 samples)	–	–	2.2	–	–	0.5	–	0.11	–	–
Nuclei (2 samples)	–	–	3.0	–	–	0.2	–	0.19	–	–

In sedimentary rocks it has been found that the higher the carbonate content of the sediments, the greater the quantity of lipids extractable by chloroform (Vassoyevich 1973). This regularity is not manifested in Recent and Upper Quaternary oceanic sediments. The carbonaceous sediments (>50% $CaCO_3$) of the shelves, slopes and ocean floor rises contain less lipids than the terrigenous sediments, but in the percentage composition of the OM the former contain as much or even more lipids. This is also typical of the sediments of some seas (Bordovsky 1969). A notable feature of the highly-productive shelves and upper slopes is the presence of the facies of carbonaceous foraminiferal and shell-fragmentary sandy-aleuritic sediments with a greatly increased C_{org} and lipid content. For instance, the OM of such sediments off South West Africa contains as much as 12–13% of A+C lipids while their β_{chl} content amounts to 9–11%, this value being only slightly lower than in the shallow-water diatom silts which are rich in lipids (β_{chl} 12.0–19.6, on the average 14.8; Romankevich and Baturin 1974). Thus the available data indicate that the magnitude of bioproduction, the quantity of particulate carbon arriving at the seabed and its burial rates, plus the environment arising largely under the effect of these factors and the gaseous regime of the sediment, determine the concentrations of lipids and their content in the OM. The relationship between the OM and its oxidizers (saprophites, metal oxides) is mainly controlled by the same factors.

In oceanic sediments the maximum content of lipids in the OM is characteristic of extreme geochemical environments: of shallow-water reduced facies with a high content of C_{org} (from 3–5% to 20%), free H_2S (sediments in the upwelling zones) and of the eupelagic red clays with a minimum content of C_{org} (0.04–0.56%, on average 0.25%) and no indications of the reduction of Mn^{4+}. Under such conditions β_{chl} attains 12–20% (Beliayeva 1974, Romankevich and

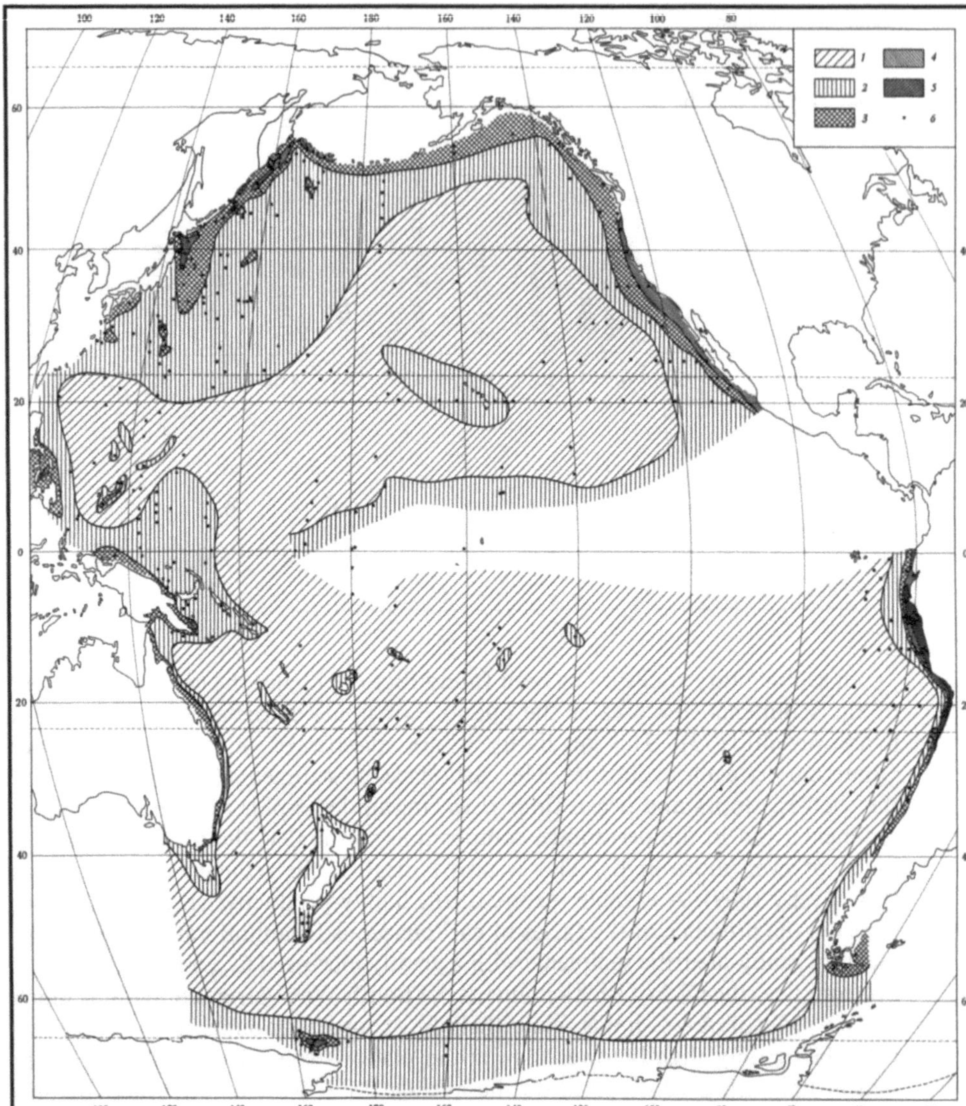

Fig. 58. Distribution of luminescent lipids in the Pacific sediments, μg per g of dry sediment: *1* = <25; *2* = 25 – 100; *3* = 100 – 200; *4* = 200 – 500; *5* = >500. Lipids extracted with a mixture of alcohol and benzene (1 : 2) at 20° – 25 °C. The content determined by the method of the luminescent-chromatographic analysis. *Dots* are sampling sites (Florovskaya 1957)

Baturin 1974). However, in the red clays or in other sediments similar to them such high oxidation values are rare, whereas in the reduced sediments of high-productivity zones they are quite common.

Generally the Uspensky-Vassoyevich law (a reverse relationship between the content of lipids in the organic matter and C_{org}) in the case of Recent sediments

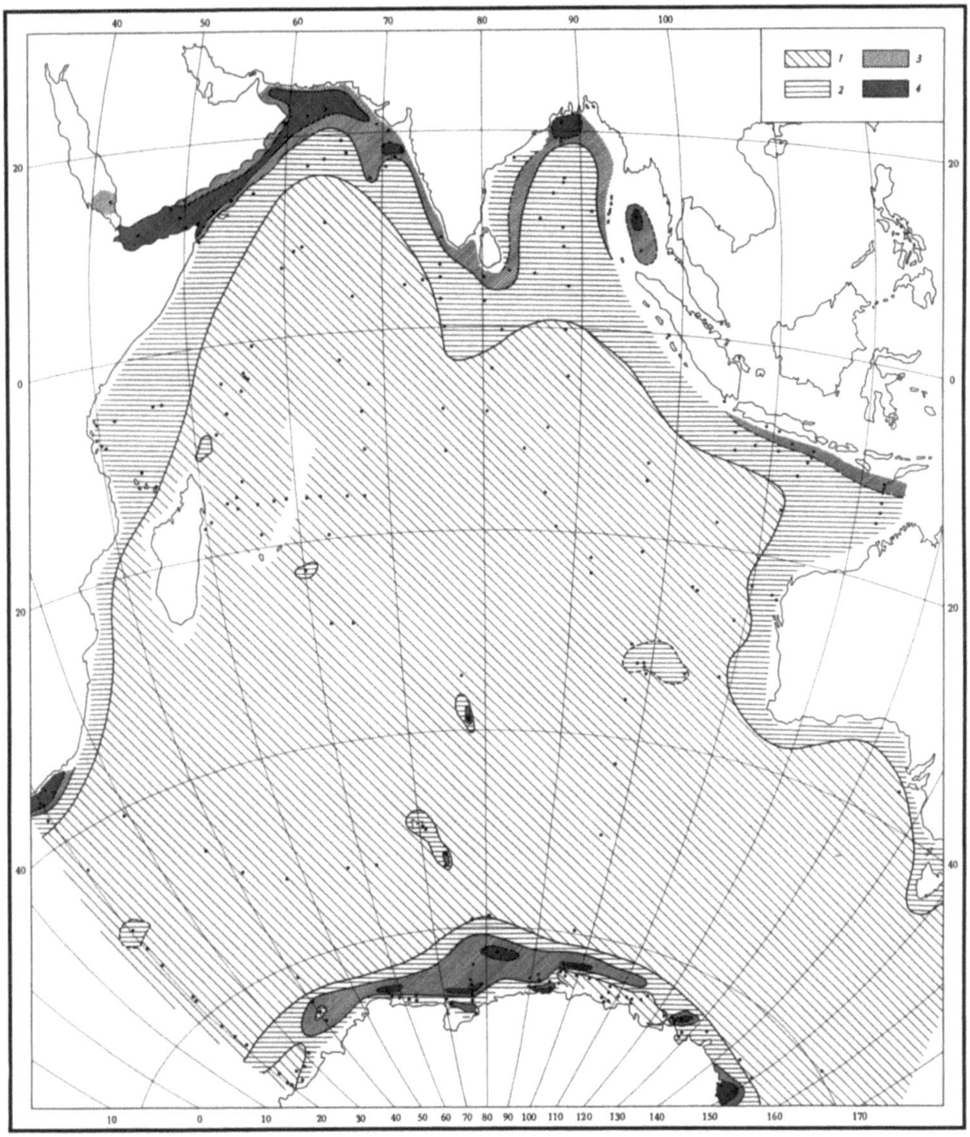

Fig. 59. Distribution of luminescent lipids in the sediments of the Indian Ocean, μg per g of dry sediment: *1* = <25; *2* = 25 – 100; *3* = 100 – 200; *4* = 200 – 500. Lipids extracted with a mixture of alcohol and benzene (1 : 2) at 20° – 25 °C. The content determined by the method of the luminescent-chromatographic analysis. Dots are sampling sites (Florovskaya 1957)

does not manifest itself very distinctly. For instance, it is not manifested for $\beta_{chl} - C_{org}$ on the Peruvian shelf – Peruvian basin, Japan's slope – northwestern basin profiles. In either case this relationship approaches a straight line. With respect to average data, the content of A_{chl} and A + C in the OM is observed to decrease in the series: the submarine oceanic margin – marginal part of the ocean floor – central part of the ocean floor, but for the lipids of the bottom sediments this decrease is smaller (for A + C 100:62:66) than for the total C_{org} (100:200:10). This can be taken as a manifestation of the Uspensky-Vassoyevich law arising in the case of a selective preservation of some lipophilic compounds when the time of OM oxidation increases. At the later stages of OM transformation this law is more distinctly expressed, particularly at low concentrations of the OM. In the cores of deep-water red clays where a high positive Eh is preserved millions of years, there is a clearly observed increase of lipids from 2.2 to 9.3% (the benzene part of A + C) with the decrease of C_{org} from 0.45 to 0.06% (Weber et al. 1960, Romankevich 1970, 1977). Thus, the lesser the content of the OM and the longer the transformation period it has been subject to, the more distinctly manifested is the Uspensky-Vassoyevich law. The principal role is played here by the geological time, other factors being of secondary importance. For instance, this relationship is more distinct in the sea water with a conservative composition of the OM and lipids. For the oceanic suspended material with a residence time about 100 times as short as that of dissolved carbon, the Uspensky-Vassoyevich law is practically not applicable.

8.4.2 Lipids in the Interstitial Water

Quantitative data on the content of lipids in the interstitial water are very scarce. In the reduced (Eh between −22 and −227 mV) and oxidized (Eh from 420 to 470 mV) sediments off Peru their concentration is 0.90 – 1.07 mg C l^{-1} (about 1.2 – 1.4 mg l^{-1}), that is 4 – 9 times as high as in the sea water at the same stations, but in the composition of the interstitial water OM the lipid content is only 9 – 18%, that is about one third that of sea water. This is associated with a less conservative composition of the interstitial water OM than sea water (Beliayeva and Romankevich 1976).

The evidence obtained in the eastern part of tropical Atlantic rendered support to the relationship between the lipid content in the sea and in the interstitial water discovered in the southeastern Pacific (Boussuge et al. 1979). The distinguishing features of the composition of lipids in the interstitial water are: a predominance of even n-alkanes over the odd ones, an increase in the content of unsaturated fatty acids with depth and the presence of certain aromatic polycyclic hydrocarbons. It seems probable that some individual components of the interstitial water lipids and the bottom sediments solid phase are quantitatively correlated in the same way as those of the sea water and the suspended material. Thus, for example, fatty acids are principally associated with solid sedimentary particles where their content is 50 – 1000 times as high as in the interstitial water. At the same time the concentration of free fatty acids in the interstitial water is 3 – 79 times as high as in the superlying sea water (Saliot and Boussuge 1978). These facts, along with the presence of unsaturated and odd fatty acids in the interstitial water, point to a high biological activity in the sediments.

The bulk of lipids is contained in the solid phase of the sediments, lipids in the interstitial water accounting only for 0.1 – 0.3% of their total content in the sediments (Beliayeva and Romankevich 1976).

8.4.3 The Composition of Lipids in Bottom Sediments

8.4.3.1 The Composition of Lipids Extracted from Wet Sediments at Normal Temperature

In studying variations in lipid composition during sedimentogenesis in the system: plankton – suspended material – sediments, the methods of bitumenologic analysis involving high-temperature extraction in the Soxhlet apparatuses proved to be inapplicable due to the transformation of labile compounds which make up a significant proportion of particulate lipids and lipids of bottom sediments. In 1971 the method of soft extraction was proposed; this permitted the isolation of lipids from wet samples of the sediments in a comparatively little transformed state (extraction at 20–25 °C by using the 2:1 mixture of chloroform and methanol) and the separation of lipids in bottom sediments into groups of compounds by means of thin-layer chromatography (Romankevich and Batrakov 1971). Later this method was modified by substituting thin-layer chromatography with the liquid column chromatography with the subsequent gravimetric determination of the various fractions of lipids from the bottom sediments. Liquid chromatography on silicagel (Hirsch and Ahrens 1958) has enabled the separation of lipids from bottom sediments to be carried out in a single stage dividing them into the same fractions as in the case of the quantitative thin-layer chromatography of DL and PL.

The composition of lipids in bottom sediments was determined in this manner in samples of the terrigenous, carbonaceous sediments and deep-water red clays taken from the shelf, slope and ocean floor in the Pacific and Atlantic Oceans and Baltic Sea (Romankevich and Batrakov 1971, Beliayeva 1974, Beliayeva and Romankevich 1976, Bobyleva and Beliayeva 1982). The results (Table 63), though revealing certain differences, demonstrate a common tendency of the compositional variation of lipids from the highly productive shelf facies to the low-production pelagial facies: in terrigenous sediments, carbonaceous (including shallow-water) sediments and deep-water red clays, the minimum and maximum, as well as the average content of hydrocarbons and fatty acids are observed to increase both relative to the total lipids and to the OM composition.

The content of sterols and fatty alcohols is not high and reveals a reverse correlation which is typical of zooplankton (Yamada 1972). It can be inferred that this is the result of the effect of benthos on lipids in the bottom sediments (Beliayeva 1974).

The content of polar lipids in the upper sedimentary layer decreases with the change-over from the shallow-water to the deep-water facies and from the terrigenous sediments of the submarine margin (averaging about 64%) to the carbonaceous (including the shallow-water algal-coral) and pelagic (averaging about 40%) sediments and further to the red clays (averaging about 18%). In the mass of the sediments of different types polar lipids continue to decrease, whereas the proportion of low-polar lipids rises. The composition of polar lipids is yet unknown. A very low phosphorus content in this fraction ($0.0-6.6\times10^{-3}$% of polar lipids) leads to believe that phospholipids constitute a small proportion of

Table 63. The composition of lipids contained in the bottom sediments of Pacific and Atlantic Oceans, isolated from samples of natural moisture at 25 °C with a mixture of chlorofom and methanol (2:1)

Level (m)	Number of samples	Hydrocarbons	Fatty acid ethers	Triglycerides	Fatty acids	Fatty alcohols	Sterols	Polar lipids	Hydrocarbons
		% of low-polar lipids[a]						% of total lipids	
Terrigenous sediments									
0–0.1	10	16.3–75.3 (33.3)	0.0–4.2 (0.8)	3.6–20.3 (11.5)	11.1–45.3 (30.3)	0.0–42.3 (15.5)	2.8–28.8 (11.9)	51.1–74.8 (63.9)	4.8–29.9 (12.4)
3.5–8.0	4	42.2–44.2 (43.3)	Not found	5.8–12.5 (8.8)	28.4–44.7 (36.3)	4.1–19.2 (8.7)	4.2–8.2 (6.2)	57.6–63.8 (61.2)	16.0–17.9 (14.3)
Carbonaceous sediments									
0–0.1	3	38.0–41.8 (39.7)	Not found	0.0–12.3 (5.3)	33.2–60.6 (47.0)	0.0–1.7 (0.6)	0.0–12.7 (7.4)	24.1–51.5 (39.6)	18.4–30.3 (24.1)
Red clays									
0–0.1	3	23.4–44.0 (30.7)	Not found	0.0–2.8 (0.9)	46.5–75.2 (57.0)	0.0–30.1 (10.0)	Not found	13.2–23.9 (17.7)	17.9–38.2 (25.6)
2.4–6.5	2	40.8–41.7 (41.2)	Not found	1.1–1.2 (1.1)	54.9–55.1 (55.0)	2.0–3.1 (2.5)	Not found	6.1–6.6 (6.3)	38.3–39.0 (38.6)
Average for the 0–0.1 m layer		34.0	Traces	8.4	38.4	11.9	0.9	50.7	17.1

Table 63 (continued)

Level (m)	Number of samples	Fatty acid ethers	Triglycerides	Fatty acids	Fatty alcohols	Sterols	Hydrocarbons	Fatty acids	Hydrocarbons	Fatty acids
		% of total lipids					% of OM		% of dry sediment	
Terrigenous sediments										
0 – 0.1	10	0.1 – 1.0 (0.2)	0.9 – 9.1 (4.1)	4.4 – 20.4 (11.1)	0.0 – 12.6 (4.9)	1.1 – 9.9 (4.4)	0.25 – 1.20 (0.60)	0.50 – 0.78 (0.59)	0.0012 – 0.0243 (0.0106)	0.0018 – 0.0328 (0.0121)
3.5 – 8.0	4	Not found	2.1 – 5.0 (3.5)	11.4 – 16.7 (14.1)	1.5 – 7.7 (3.5)	1.8 – 3.0 (2.4)	0.88 – 1.07 (0.98)	0.49 – 0.87 (0.69)	0.0113 – 0.0158 (0.0136)	0.0084 – 0.0147 (0.0116)
Carbonaceous sediments										
0 – 0.1	3	Not found	0.0 – 7.0 (2.9)	18.9 – 45.9 (29.2)	0.0 – 0.8 (0.3)	0.0 – 7.2 (3.9)	0.42 – 1.54 (0.97)	0.52 – 1.42 (1.05)	0.0042 – 0.0135 (0.0080)	0.0064 – 0.0108 (0.0083)
Red clays										
0 – 0.1	3	Not found	0.0 – 2.4 (0.8)	35.5 – 63.2 (47.2)	0.0 – 23.0 (8.8)	Not found	1.35 – 2.43 (1.78)	1.26 – 7.39 (3.91)	0.0042 – 0.0087 (0.0065)	0.0047 – 0.0265 (0.0147)
2.5 – 6.5	2	Not found	1.0 – 1.1 (1.0)	51.5 – 51.6 (51.5)	1.9 – 2.9 (2.4)	Not found	2.91 – 3.04 (2.97)	3.26 – 3.34 (3.30)	0.0020 – 0.0027 (0.0023)	0.0027 – 0.0030 (0.0028)
Average for the 0 – 0.1 m layer		Traces	3.3	23.3	4.7	3.6	0.91	1.35	0.0093	0.0119

Figures in parenthesis are average values

[a] Low-polar lipids are the difference between the total lipids and the polar fraction

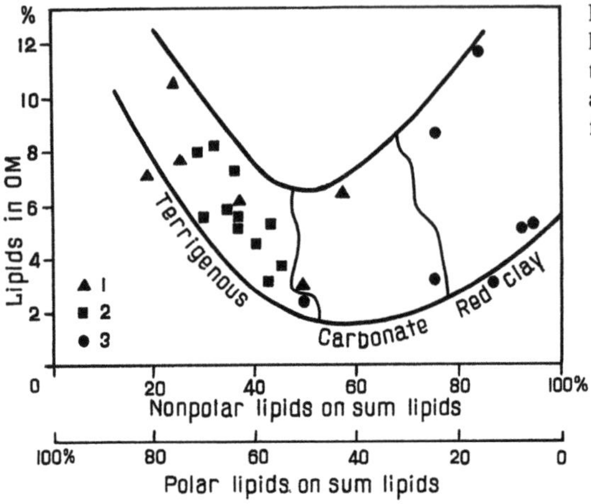

Fig. 60. The content of the low-polar lipids versus the total content of lipids in the OM: *1* shelf sediments; *2* continental and islandic slope sediments; *3* ocean floor sediments

the polar lipids, their content, moreover, declining from the coastal to the pelagic sediments (Beliayeva 1974).

The content of polar and low-polar lipids is in a specific relation with the concentration of the lipids of the bottom sediments as part of the OM (Fig. 60). This relationship indicates that the maximum lipid concentration in the terrigenous shelf sediments is due to the polar lipids, and in the red clays to the low-polar lipids. The relationship reflects the general trend of OM transformation from the shelf to the pelagic facies and from the terrigenous to the carbonaceous sediments, namely: an increase of the more stable low-polar lipids represented mainly by hydrocarbons and fatty acids.

The facial conditions of sediment accumulation influence the composition of lipids in bottom sediments. For instance, the terrigenous sediments of the Peruvian shelf, accumulating in an environment which is sharply reductional as regards certain metals, contain more fatty alcohols and less fatty acids compared to the less reduced sediments of the continental slope and the oxidized muds of the ocean floor, though in all these cases the OM is principally of the planktonogenic origin. In an oxidation environment the reactive compounds of the type of fatty acid ethers and fatty alcohols undergo oxidation, while hydrocarbons and fatty acids are selectively preserved, being more inert.

Not only the environment, but differences in the source of OM will influence the quantity of labile compounds in the class of lipids. For instance, the coral and biogenic-fragmentary carbonaceous sediments taken respectively at the Longan atoll (Ninigo Ils.) and in the Langemak Bay (New Guinea), in contrast to the terrigenous sediments of the Peruvian shelf with planktogenic OM contain more triglycerides and no fatty alcohols.

The general trend in the compositional variability of bottom-sediment lipids in the direction submarine margin – ocean floor is a reduction in the content of polar lipids with the parallel increase of low-polar lipids, including hydrocarbons and fatty acids. This regularity remains valid for sedimentary masses of different composition and different physico-chemical conditions of formation (see Table 63).

The study of the isotopic composition of polar lipids, hydrocarbons and fatty acids contained in the bottom sediments has revealed that in the direction shelf – ocean floor sediments, as well as in the sedimentary mass, the lipids in the bottom sediments grow isotopically more light-weight due to an increasing proportion of hydrocarbons and fatty acids, for which the values of $\delta^{13}C$ are smaller than in the case of polar lipids (Shirinsky et al. 1974, Shadsky et al. 1980, 1982, Galimov 1981).

Thus the relative content of lipids, including low-polar lipids, hydrocarbons, fatty acids, fatty alcohols and sterols, as well as the isotopic composition of the fractions, may serve a good indicator of the extent of OM transformation at the different stages of diagenesis. It should be borne in mind that these inferences relate to that part of the lipids which is extractable from the sediments without hydrolysis and is conventionally classed among free lipids. Their transformation proceeds more rapidly than that of combined lipids (Van Vleet and Quinn 1979).

In considering the origin of oil, an important question is that of establishing the facies in which the neoformation of total hydrocarbons takes place. It is to be emphasized that the single fact of the simultaneous increase of hydrocarbons in the composition of OM and in the dry sediment mass, all other conditions being equal, provides a sufficient foundation for a positive answer to this question, since an increase (and a significant one) of hydrocarbons in the OM only takes place in conditions of their selective preservation in the sediments. A case in point are eupelagic red clays where the coefficient μ_3 attains 2.4 – 3.0%. Among the sediments studied by the aforementioned method the neoformation of hydrocarbons was recorded by us only in the mass of terrigenous sediments with an intensive sulphate reduction and a sharply reductional environment (the Japan trough's slope and bottom).

8.4.3.2 Some Results of Chemico-Bitumenologic Research

Extensive data are now available on the composition of lipids contained in the bottom sediments of seas and oceans, obtained by the chemico-bitumenologic method (Handbook on the analyses of bitumens and dispersed organic matter in rock deposits) (1966). The present section deals principally with the data on the composition of lipids in the bottom sediments of the Pacific, Indian and Atlantic Oceans obtained in recent years in joint research carried out by the Institute of Oceanology, USSR Academy of Sciences, and some petroleum institutions.

The Elementary Composition of Lipids (bitumoids A_{chl}) in the sediments of the southeastern and northwestern Pacific is characterized by the following (%): C 67.66 – 81.11; H 9.04 – 13.03; S + N + O 9,85 – 22.30; atomic H/C ratios 1.33 – 1.91; O/C 0.06 – 0.24 (Fig. 61); for the surface layer the range of variations is only slightly smaller: C 67.66 – 75.88; H 9.11 – 11.63; S + N + O 12.61 – 22.30; the H/C ratio 1.54 – 1.91; O/C 0.09 – 0.24 (Rodionova et al. 1973, Romankevich 1977).

The elementary composition of the benzene portion of the alcohol-benzene extract of lipids in the bottom sediments of the Pacific, Indian, and Atlantic

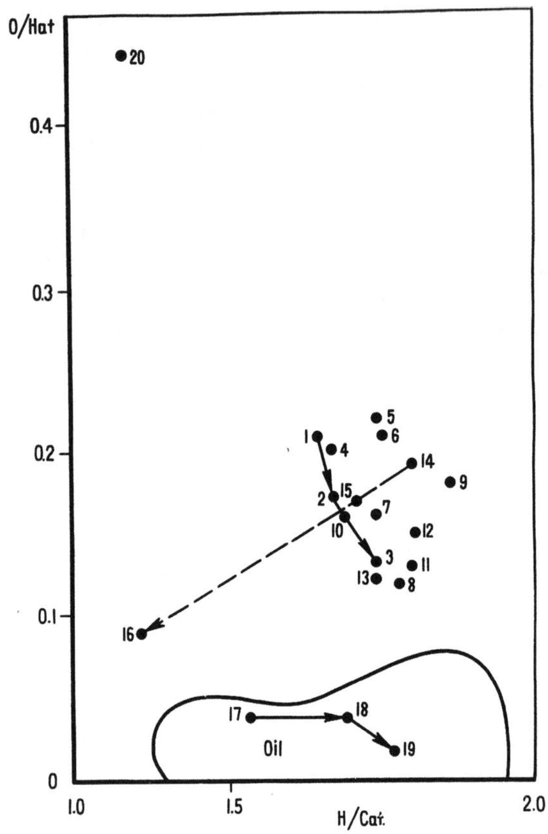

Fig. 61. The elementary composition of the lipids (bitumoid A_{chl}) in the Pacific sediments (*1–16*), of their oil fraction (*17–19*) and of the lipids-bitumoids C_{chl} (*20*). Terrigenous sediments of the shelf (*1*), the continental slope (*2*), and red clays (*3*) west of Peru. Terrigenous muds west of the Honshu i. (*4–13*). Eupelagic red clays in the North-West basin (*14–16*). *Solid arrow* is the variability trend of the elementary composition of lipids in the sequence: shelf facies – pelagic facies; *dotted arrow* is the composition of lipids in the sequence: shelf facies – pelagic facies; *dotted arrow* is the compositional variability of the lipids contained in deep-water red clays in the diagenesis

Oceans differs from A_{chl} in a lower content of C 61.68–76.30 (an average for 38 samples being 67.80%), H 8.20–11.75 (average 9.7%), S 0.02–2.05 (average 0.67%), O+N 11.14–29.79 (average 21.8%), H/C 1.53–1.85 (average 1.71%) and a larger O/C ratio 0.110–0.362 (average 0.25%) (Table 64, Fig. 62) (Weber et al. 1960, Gorskaya and Gleboskaya 1966, Romankevich 1977). In transition from the suspended matter to the sediments, the H/C ratio in the composition of lipids decreases and the O/C ratio increases. This is in agreement with the conclusion that there is a higher content of polar lipids in the sediments than in the suspended material. The field which represents sediments of different genetic types and different structural ocean zones (see Fig. 62) is elongated along the ordinate, thus pointing to significant differences between the facies with respect to O/C and small differences with respect to H/C. Its upper part is occupied by the hemipelagic terrigenous and pelagic foraminiferal silts, and the lower part by the terrigenous sediments of the Bering Sea and the Antarctic shelf, the shallow-water carbonaceous shell and coral sediments. The latter combination of facies is characterized by a relatively lower content of hetero-elements with an approximately equal H/C ratio.

Table 64. The elementary composition of the benzene portion of the alcohol-benzene extract from the marine organisms, aquatic suspended material and bottom sediments of the surface layer in the Pacific, Indian, and Atlantic Oceans (% of the total)

Object of study	Sampling site	C	H	S	O+N	H/C (atomic)
Marine organisms						
Diatom plankton	Caspian Sea	65.76	10.07	0.92	23.25	1.83
Zostera	Caspian Sea	69.77	8.81	0.37	21.05	1.51
Cladophora	Caspian Sea	72.30	10.66	0.85	16.19	1.77
Clopepod seston	Black Sea	67.43	10.20	–	22.37	1.81
Calanus cristatus	Kuril-Kamchatka Basin	77.16	11.52	11.32		1.79
Hymenodora glacialis (deep-water decapod)	Pacific Ocean	78.87	11.65	9.48		1.77
Zooplankton	Indian Ocean	66.85	9.79	23.36		1.75
Aquatic suspended material						
Aquatic suspended material	Tropical part of the Pacific Ocean					
	0–7 m[a]	67.57	10.80	0.26	21.37	1.92
	50–75 m	72.40	10.57	–	17.03	1.75
	500 m	71.37	11.25	–	17.38	1.89
	1000–3000 m	69.24	10.90	–	19.86	1.89
	Antarctic area[a]	72.72	11.16	0.38	15.74	1.84
	Tropical part of the Indian Ocean[a]	68.53	10.90	0.33	20.24	1.92
Average for suspended material of the surface layer	–	70.30	10.86	–	18.84	1.86
Superficial bottom sediments						
Bottom sediments	Submarine margin	69.19	9.84	0.90	20.02	1.70
Including						
– terrigenous	Shelves	68.39	9.60	1.36	20.64	1.68
– terrigenous-iceberg	Shelves	70.65	10.37	0.84	18.13	1.76
– terrigenous weakly-siliceous	Continental slopes	66.70	9.22	0.83	23.24	1.65
– shell, byrozoan-foraminiferal sands (>50% $CaCO_3$)	Shelves	71.33	10.32	0.69	17.65	1.73
– algal-coral sediments	Shelves	70.26	10.04	0.59	18.48	1.70
Hemipelagic terrigenous silts	Marginal ocean floor	62.03	8.64	0.17	29.15	1.67

Table 64 (continued)

Object of study	Sampling site	C	H	S	O+N	H/C (atomic)
Pelagic sediments	Central ocean floor	67.28	9.80	0.43	22.50	1.74
– including Antarctic's siliceous diatomaceous		69.76	10.25	0.21	19.77	1.76
– foraminiferal		65.19	9.32	0.40	25.16	1.71
– miopelagic red clays		67.82	10.05	0.69	21.42	1.77
– eupelagic red clays		69.13	10.18	0.47	20.21	1.77
Average for oceanic sediments (38 samples)		67.80	9.71	0.67	21.80	1.71
Bottom sediments	Caspian Sea	69.9	10.02	20.08		1.72
	Bering Sea	72.70	10.20	17.10		1.68

The table has been compiled after data by the following authors: Weber et al. (1960), Gorskaya and Glebovskaya (1966), Vinogradov et al. (1970), Lisitzin et al. (1971), Romankevich (1977)

[a] Average value from five samples

From the diagram for lipids (bitumoids A_{chl}, see Fig. 61) it can be inferred that in transition from the terrigenous reduced sediments of the Peruvian shelf to the compositionally similar sediments of the continental slope and the red clays of the ocean floor the O/C ratio tends to decrease and the H/C ratio to increase. A similar variability trend in the elementary composition of lipids is characteristic of the mass of terrigenous sediments on Japan's slope (Rodionova et al. 1973). This points to the growing degree of reduction of bitumoids A_{chl} and agrees with an earlier conclusion about the increase in this direction of low-polar lipids, hydrocarbons and fatty acids.

The differences in the elementary composition of lipids (bitumoids A_{chl} and the benzene portion of the alcohol-benzene extract) in the direction from the shelf facies to the pelagial are chiefly manifested in the decrease of heteroelements. A remarkable role in this respect is played by the sulphate-reducing bacteria. Their vital activity has brought about, on the one hand, the mineralization of the OM and the formation of metastable products of its incomplete decomposition and condensation, e.g. resins, asphaltenes, and other polar lipids which are poorly preserved as such in the sharply oxidized conditions, and, on the other, the formation of hydrogen sulphide and the reduction of certain elements. The progress of these processes was least intense in the pelagic facies, being accompanied by a partial condensation of OM and finding expression, as concerns the average data, in a higher H/C ratio in the lipids. In the mass of deep-water red clays the H/C and O/C ratios in the lipids have lower values, and they are observed to undergo carbonisation.

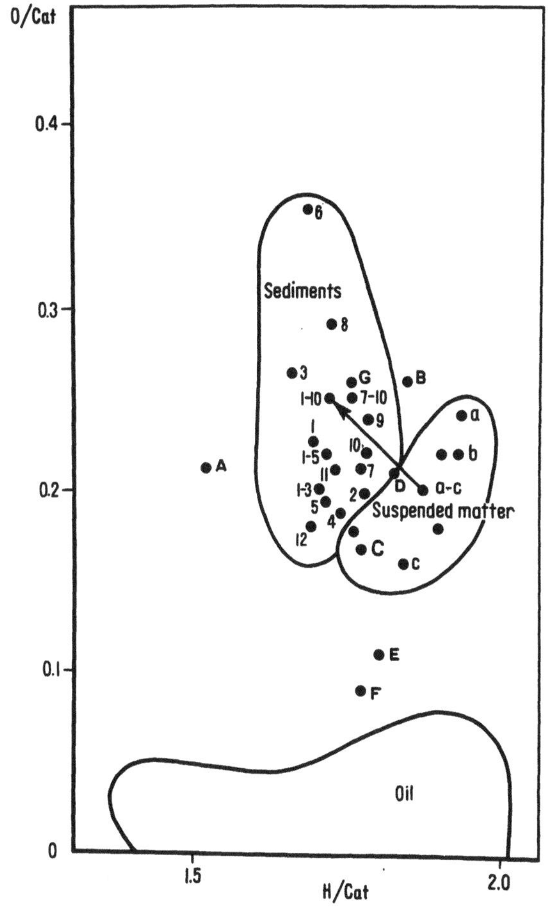

Fig. 62. The elementary composition of the lipids (the benzene portion of the alcohol-benzene extract) contained in marine organisms, suspended material and bottom sediments. *Marine organisms:* diatom plankton (*A*); Zostera (*B*) and Cladofora (*C*) of the Caspian Sea, copepod seston of the Black Sea (*D*) and of the Kuril-Kamchatka area (*E*); deep-water decapods (*F*) and zooplankton of the Indian Ocean (*G*). *Aquatic suspended material of the surface layer (a–c):* tropical areas of the Pacific (*a*), Indian Ocean (*b*); Antarctic (*c*). *Bottom sediments of the various facial types in the Pacific, Indian, and Atlantic Oceans* (*1–10*): submarine margin (*1–5*), marginal parts of the ocean floor (*6*), central parts of the ocean floor (*7–10*). The sediments of the Caspian Sea (*11*), Bering Sea (*12*). Arrow shows the variation of the elementary composition

The Group Composition. In the surface layer of the sediments in the Pacific, Indian, and Atlantic Oceans the group composition of the benzene portion of the alcohol-benzene extract varies within the following limits: oils 12.4–59.9% (on average 32.8%), benzene resing 8.2–33.3% (on average 18.7%), alcohol-benzene resins 0.4–26.6% (on average 15.5%), asphaltogene acids and asphaltenes 13.6–64.3% (on average 34.8%). Despite significant variability limits, no great differences are observed between the facial types of oceanic sediments when averages are considered. The differences find expression in a varying ratio of the oils and asphaltogenic acids (for instance, a maximum of oils and a minimum of asphaltogene acids in the iceberg sediments and the reverse ratios between these two parameters in the coral sediments). Both fractions of the resins are subject to only small variations. The suspended material (diatomaceous and peridinean) contains more oils (on average 45.7%) and less asphaltogene acids (24.7%), than the sediments. This is in line with the earlier established fact of the larger content of low-polar lipids in the suspended material.

The study of the lipids (bitumoids A_{chl}) in the coastal upwelling zones with little-transformed OM has revealed that the content of oils here is lower, while that of the asphaltogene acids, which are greatly predominant over asphaltenes, is higher than in the pelagic sediments. That means that once again the observed overall trend of OM transformation from the shelf facies to the ocean floor facies consists in a decrease of polar lipids and an increase of low-polar lipids. Evidence of the same effect is found in the large content of unsaponifiable compounds in the lipids of the deep-sea sediments in the Peruvian area (36–47%) compared with the shelf sediments (1–18%). The lipids (in bitumoid A_{chl}) of the Peruvian area sediments there comprise about 15–23% fatty acids in the composition of which C_{12}, C_{14}, C_{16} and C_{18} molecules were identified.

Research carried out in different ocean areas has revealed that molecules with an even number of carbon atoms are predominant in normal fatty acids of the sediments, while the branched acids, on the contrary, contain more molecules with an odd number of carbon atoms (Leo and Parker 1966, Peake et al. 1974). The synthesis of branched acids with an odd number of carbon atoms is the prerogative of the bacterial flora so that from the quantity and composition of these compounds in the sediments it is possible to assess the intensity of the bacterial effect on the initial OM. The composition of fatty acids in the sediments points to their biogenic origin.

Hydrocarbons. In the surface sedimentary layer of the Pacific, Indian, and Atlantic Oceans the average content of hydrocarbons amounts to about 21% in oils, ca. 8% in the benzene portion of the alcohol-benzene extract, ca. 0.3% in organic matter, 0.0028% in dry sediment (Table 65). All of the investigated sediments featured the predominance of the methane-naphthane hydrocarbons over the aromatic ones. In the sediments of the Pacific, Indian, and Atlantic oceans the methane-naphthane hydrocarbons account, on average, for about 87% and aromatic hydrocarbons for about 13% of the total hydrocarbons. In the sediments of the Japan's islandic slope it was found that naphthenic hydrocarbons (about 74%) were predominant over the methane ones, thus indicating the participation of non lipidic OM in the formation of hydrocarbons (Shishenina et al. 1974).

On the evidence of other materials (Smirnov 1976), the content of hydrocarbons extracted from dry samples by using the mixture of 70% of benzene, 15% of ethanol and 15% of acetone amounts on average to about 1% of C_{org}. According to these data, the hydrocarbons of recent marine and oceanic sediments have the following composition: napthenic hydrocarbons ($\geqslant$50%) with the $C_{22}-C_{30}$ maximum and, on the average, with two rings to a molecule and four alkyl substitutes; paraffins (<30%), basically of the normal $C_{18}-C_{36}$ structure with the $C_{24}-C_{28}$ maximum; isoparaffins (about 2%), principally 2- and 3-methyl substituted with the same number of atoms in the molecules and the $C_{25}-C_{29}$ maximum; isoprenoid alkanes ($\leqslant$0.2%) $C_{18}-C_{25}$; aromatic hydrocarbons (about 15%) predominantly mono- and bicyclic with 3–4 alkyl substitutes and polycyclic non-linearly-condensed (Smirnov 1976). The regular variations in the content of hydrocarbons in the sediments correlate with the OM distribution and manifest themselves in the circumcontinental and latitudinal local-

Table 65. The content and composition of hydrocarbons in aquatic suspended material and in bottom sediments

Object of study	Content of hydrocarbons (%)				Hydrocarbons (% of the total)	
	in oils	in benzene portion of alcohol-benzene extract	in organic matter (μ_3)	in dry sediment	methane-naphthenic	aromatic
Aquatic suspended material						
Diatom suspension (Antarctic area)	–	17.77	–	–	77.8	22.2
Suspended material with predominance of blue-green algae (Red Sea)	–	53.30	–	–	97.9	2.3
Bottom sediments in the Pacific, Indian and Atlantic Oceans						
Submarine margin sediments	20.16	6.18	0.30	0.0030	85.16	14.86
Including						
– terrigenous of the shelf	11.00	2.75	0.16	0.0035	77.9	22.3
– terrigenous iceberg of the shelf	32.59	11.45	0.68	0.0030	86.0	14.0
– terrigenous weakly siliceous of the slopes	18.17	5.42	0.25	0.0037	91.8	8.2
– shell, bryozoan-foraminiferal (>50% $CaCO_3$) sands of the shelf	20.79	5.63	0.16	0.0019	87.2	12.7
– coral sediments	18.80	5.77	0.14	0.0002	72.6	27.4
Terrigenous sediments of marginal part of the ocean floor	23.92	8.38	0.22	0.0025	91.1	8.8
Sediments of central regions	24.51	9.80	0.40	0.0026	87.7	12.1
Including						
– siliceous diatomaceous of the Antarctic	26.10	11.60	0.58	0.0037	88.4	11.6
– carbonaceous foraminiferal	16.95	5.15	0.28	0.0014	–	–
– miopelagic red clays	30.39	12.70	0.62	0.0042	83.4	16.6
– eupelagic red clays	33.46	14.00	0.42	0.0024	88.3	11.7
Total for the sediments (135 samples)	21.41	7.75	0.33	0.0028	86.72	13.20

ization of their masses. In passing from the highly productive oceanic zones to the regions of low bioproductivity there takes place an increase of the content of n-paraffins which are represented here by higher-molecular compounds ($>C_{22}$), accompanied by a decrease in the proportion of cyclic and isoparaffinic hydrocarbons. In Smirnov's opinion (1980) the process of hydrocarbon biosynthesis in the world ocean at the sedimentogenesis and early diagenesis stage involves the breaking off of the functional groups and the subsequent recombination of hydrocarbon radicals. The initial compounds in this case are fatty acids, amino acids with non-polar R-groups, aromatic amino acids and sterols.

The comparison of the composition of hydrocarbons in the sediments and in the principal OM producers in the seas and oceans reveals that there is generally no correlation between them. The hydrocarbons in phytoplankton are represented by n-paraffins and olefines, isoalkanes and aromatic compounds being practically absent and pristane being present in the amount from traces to about 3%. The transformation of the OM at the water – bottom interface is responsible for the lack of a direct relationship between the composition of hydrocarbons in the bottom sediments, in sea water, in suspended material and interstitial water (Saliot and Tissier 1977, 1978, Dastillung and Corbet 1978, Tissier and Dastillung 1978).

The distribution of n-alkanes in the sediments is generally characterized by the predominance of odd and high-molecular homologues of largely terrigenous origin. The alkanes n-C_{15} and n-C_{17} corresponding to the planktogenic origin of the hydrocarbons in the bottom sediments are frequently encountered in the shallow-water coastal sediments, since their residence time in the mass of water is too short to enable them to avoid significant transformations in the sedimentogenesis (Saliot 1981). The principal isoprenoidal hydrocarbons in the sediments are pristane and phytane with ratio between them varying from 0.85 to 1.51 (Saliot 1981) and capable of serving as an indicator of the reduction-oxidation conditions in the sediment (Didyk et al. 1978). Aromatic hydrocarbons in the bottom sediments are, as a rule, represented by non-alkylated polycyclic compounds with 3 – 5 rings. Their mean composition is 12% of phenanthrene, 16% of fluoranthene, 15% of pyrene, 23% of $C_{18}H_{12}$ and 35% of $C_{20}H_{12}$ (Laflamme and Hites 1978).

Owing to the oil pollution of the ocean, increased concentrations of hydrocarbons of all classes in the upper layers of bottom sediments are observed in the vicinity of industrial centres (Laflamme and Hites 1978, Barrick et al. 1980).

The structural and group composition of the hydrocarbons in recent sediments is in many respects similar to that of the scattered OM in ancient rocks and oils. However, there are certain differences which find expression, for instance, in the fact that molecules with less than C_{13} which constitute the basis of oils are practically absent from bottom sediments, while the content of isoparaffins and, within them, of isoprenoidal alkanes is lower by approximately an order of magnitude.

8.5 Transformation of Lipids in Sedimentogenesis

The difficulties involved in the comparative characterization of the composition of suspended lipids and of the lipids contained in bottom sediments are chiefly due to the fact that various fractions of lipids enter into complexes which cannot be dissolved in organic solvents. Any method of lipid extraction from such objects as suspended material and sediments is a compromise between the desire to isolate the group of compounds as fully as possible and to minimize the artefact processes. The subsequent discussion is entirely based on the single method of soft extraction of lipids from freshly obtained wet samples (Beliayeva 1980a – c). The yield of lipids with this kind of extraction proved to be unexpectedly high: for surface samples 78 – 121% and for the sediments at the depth of 4 – 10 m 44 – 70%, compared with the results obtained with lipid (bitumoids A + C) extraction on a complete basis (Handbook on the Analysis of Bitumens ... 1966). Presumably, sample drying interferes with lipid extraction and leads to the formation of non-extractable complexes. Therefore, the processing of natural wet sediments with methanol and thereupon their successive extraction with a mixture of chloroform-methanol (2:1, t^0 25 °C), chloroform, a mixture of ethanol and benzene (in the Soxhlet apparatuses) results in a perceptible increase in the yield of lipids, including hydrocarbons (by a factor of 1.5 – 5, according to the type of bottom sediments).

An overall pattern of distribution of lipids in the OM of oceanic objects points to a decrease of their content in passing from the living OM to POM and further to the bottom sediments (Fig. 63). However, in certain areas with a high productivity of diatom algae, for instance, off Peru and South West Africa, the suspended material and sediments may contain a significantly larger quantity of lipids in the OM than the producers and their consumers.

In the southeastern Pacific, where the complex research of DL, PL, lipids of bottom sediments and interstitial water was carried out, the primary source of lipids is the diatom plankton. A comparison of the composition of lipids of diatom algae (Lee et al. 1971) and those contained in the suspended material from the 0 – 50 m layer makes apparent certain differences: the surface suspended matter contains less polar lipids, fatty acid ethers with glycerol and single-atom alcohols and slightly more hydrocarbons in the composition of lipids. Furthermore, sterol ethers and free fatty alcohols, which are absent in diatoms, have been discovered in the surface suspended material, thus testifying to the decomposition and re-synthesis of lipids proceeding in the upper water layers (see Fig. 63).

During sedimentation the PL display a decrease in the content of labile complex compounds (polar lipids, sterol ethers, fatty acid ethers, as well as triglycerides) and an increase of the proportion of the stable constituents of the class of lipids (hydrocarbons, fatty acids, fatty alcohols + sterols). This process is the result of the decay of lipids and their re-synthesis by the organisms from the non-lipid OM, as well as of the sorption of DL and their transfer into the suspended material from the solution of the surface-active molecules. The increased content of fatty acids and fatty alcohols in the composition of PL of deep-water suspended material also points to the fact that the reactions of the enzymatic

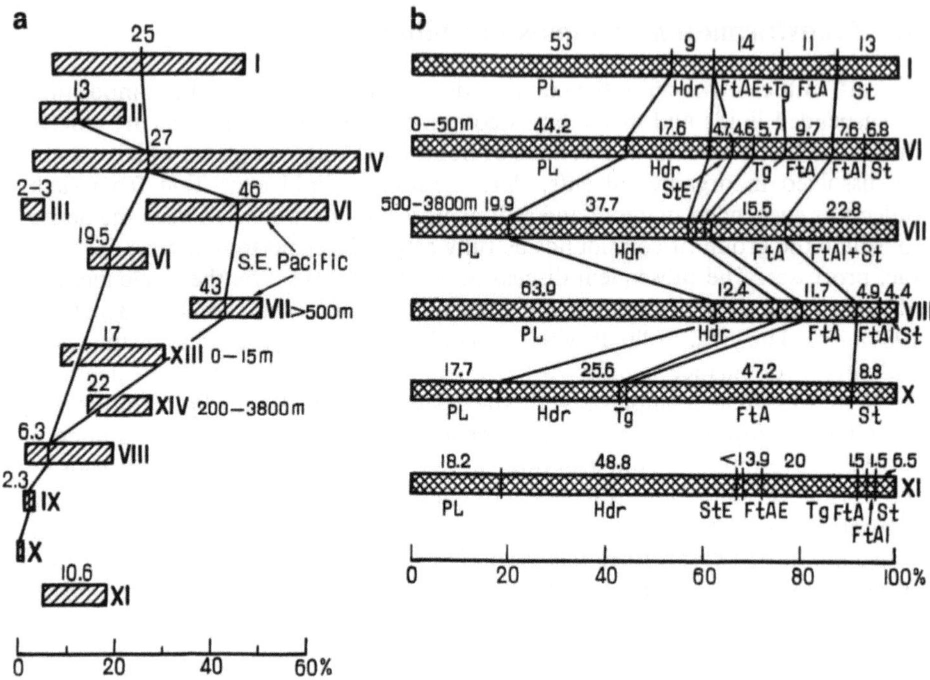

Fig. 63 a, b. Variation of the content of lipids (**a** % of OM) and of their group composition (**b** % of the total) during the sedimentogenesis. Symbols are as in Fig. 46

and bacterial hydrolysis of complex lipids in the mass of water do not proceed as far as the complete oxidation of these compounds. At the same time, the sedimentation of suspended material is accompanied by a decrease of the concentrations ($\mu g\ l^{-1}$) of all fractions of PL, thus indicating utilization of both the labile and the relatively stable compounds.

There is a direct relationship between the decrease of the concentrations of individual fractions of PL with depth and their initial concentration in the 0–50 m layer. The vertical variability gradient of PL is therefore determined not only by the length of the water column but also by the location of the station. As a result, the maximum variability gradient of the concentrations of PL and their individual fractions in the southeastern Pacific is characteristic for the Peruvian shelf where its value (3.64) is some thousand times larger than the respective gradients on the continental slope (0.9×10^{-3}), over the Peruvian-Chilean trough (4.59×10^{-3}) and on the ocean floor (0.75×10^{-3}).

As the suspended material passes into the sediment the composition of lipids undergoes restructuring as a result of catalytic reactions with and without the participation of the enzymes of the living organisms. The content of polar lipids in the sediments is 2.6–9.4 times as high as in the suspended matter (Table 66). The major part of the sedimentary polar lipids is the product of condensation; this difference is therefore more pronounced for deep-sea sediments which have a great age, than for shallow-water sediments.

Table 66. Comparative characteristics of the lipid composition of suspended material and bottom sediments (%) (0 – 10 cm layer)[a]

Lipid fraction	Content of lipid fractions (%)					
	of low-polar	of total lipids	of OM	of low-polar	of total lipids	of OM
	Peruvian shelf, St. M-552, seabed depth 78 m			Peruvian-Chilean trough, St. M-542, seabed depth 6250 m		
Polar						
– suspended material	–	28.4	16.4	–	7.5	3.5
– sediment	–	74.8	5.6	–	70.9	3.9
– Δ	–	+2.6	–3.0	–	+9.4	+1.1
Low-polar						
– suspended material	100	71.6	41.2	100	92.5	43.2
– sediment	100	25.2	1.9	100	29.1	1.6
– Δ	–	–2.8	–21.7	–	–3.2	–27.0
Hydrocarbons						
– suspended material	53.6	38.4	22.1	21.1	19.5	9.1
– sediment	19.1	4.8	0.37	16.3	4.7	0.26
– Δ	–2.8	–8.0	–60	–1.2	–4.1	–35
Sterol ethers						
– suspended material	Not found	Not found	Not found	Not found	Not found	Not found
– sediment	Not found	Not found	Not found	Not found	Not found	Not found
Fatty acid ethers						
– suspended material	Not found	Not found	Not found	Not found	Not found	Not found
– sediment	0.4	0.1	0.01	Not found	Not found	Not found
Triglycerides						
– suspended material	Not found	Not found	Not found	Not found	Not found	Not found
– sediment	3.6	0.9	0.07	8.1	2.4	0.13
Fatty acids						
– suspended material	15.4	11.0	6.3	22.1	20.9	9.8
– sediment	25.7	6.5	0.50	31.9	9.3	0.51
– Δ	+1.7	–1.7	–12.6	+1.4	–2.2	–19.2
Fatty alcohols						
– suspended material	19.6	14.0	8.0	56.4[b]	52.2[b]	24.4[b]

Table 66 (continued)

Lipid fraction	Content of lipid fractions (%)					
	of low-polar	of total lipids	of OM	of low-polar	of total lipids	of OM
	Peruvian shelf, St. M-552, seabed depth 78 m			Peruvian-Chilean trough, St. M-542, seabed depth 6250 m		
– sediment	42.3	10.7	0.8	43.7[b]	12.7[b]	0.78[b]
– Δ	+2.2	–1.3	–10	–1.3[b]	–4.1[b]	–31.3[b]
Sterols						
– suspended material	11.4	8.2	4.7			
– sediment	8.9	2.2	0.2			
– Δ	–1.3	–3.7	–23.5			

Δ is the coefficient indicating how many times the content of one or other fraction in the sediments is higher (+) or lower (–) as compared to the suspended material

[a] The values for the suspended material correspond to the lowermost of the studied horizons (St. M-442, 70 m; St. M-542, 3800 m)

[b] Fatty alcohols plus sterols

The low-polar lipids contained in sediments differ in composition from the PL due to the selective preservation of some individual fractions. Thus in the reduction environment on the Peruvian shelf the low-polar lipids in the sediments are observed to accumulate more fatty acids and fatty alcohols than those in the suspended material. In the oxidation environment at the bottom of the Peruvian-Chilean trough the content of fatty alcohols in low-polar lipids decreases in comparison to the suspended matter, while the content of fatty acids is enhanced, the latter being of greater stability in the diagenesis. The content of fatty acids and fatty alcohols in the total lipids in the sediments of both types, compared with the suspended material, diminishes, which is principally due to the altered relationship between the polar and low-polar compounds.

The appearance in the upper sedimentary layers of such compounds as fatty acid ethers and triglycerides, which mostly undergo disintegration in the 0–100 m layer when in suspension, is presumably due to their resynthesis by the living OM.

The long transformation period of the OM in the sediments is responsible not only for the appearance of polar condensation products but also for a part of the compounds going over into insoluble complexes. This is apparently the cause of the sharp decrease of the content of hydrocarbons in transition from the suspended material to the sediments, where this fraction was identified in the composition of lipids-bitumoids "C", as well as in the humic and fulvic acids. As regards the chemosorption of hydrocarbons in the sediments, an important role of the catalysts is played by the clay minerals (Galwey 1972, Meyers and Quinn 1973). The decomposition of lipids and their condensation in the upper water layers gain a new momentum in the sediments and are accompanied by a change of the relationship between the lipid components and other OM classes. Accordingly, the content of lipids and of all their constituent fractions in the sedimentary OM is less than in the suspended material (see Fig. 63).

Comparing the composition of lipids in the phytoplankton, suspended material, sea water and sediments reveals the fact that their transformation both in the mass of water and at the stage of the early diagenesis is directed under all environments (oxidation and reduction) towards the selective preservation of low-polar compounds and, among them, mostly of hydrocarbons and fatty acids.

The different facial conditions in the mass of water and at the water-bottom interface determine the masses of lipids arriving on the bottom, as well as their differences in composition. With respect to increasing stability, the natural compounds of the lipid class make up the following sequence: fixed fatty acids < polar lipids < fatty acids < fatty alcohols < sterols < hydrocarbons.

9. Specific Organic Compounds

In the oceanic waters and in the bottom sediments there is a wide occurrence of complex polyfunctional compounds which are generated in the process of OM decay from compounds originally included in living matter. By analogy with specific soil substances these natural geopolymers are called humic acids (HA), fulvic acids (FA) and humins. From the analytical viewpoint, humic acids are the organic matter which is extractable from the sediments or from the suspended material by treatment with aqueous alkaline solutions, mostly following the preliminary soxhlet extraction of lipids and treatment of the sediment with 10% HCl. Those known as fulvic acids (FA) comprise the acid-soluble organic matter contained in the alkaline extract. Humins, or ROM are the organic compounds which remain in the sample being treated in succession with organic solvents, HCl and KOH solutions and which are not to be extracted from the sediments unless the mineral matrix is destroyed. The living, normally functioning organisms do not contain FA and HA. Such substances may be formed in some quantity as the metabolic products of exchange in the organisms or they may be the result of the alkaline hydrolysis of native OM.

9.1 Humic Substances of Sea Water

Sea water humic substances exert a diversified influence on the physico-chemical, biological and biogeochemical properties of water. They largely determine the solubility of substances, sorption and precipitation of numerous elements, the optical properties of water, the damping of capillary waves, the foam-formation process and the trophic and sensor relations of marine organisms. The composition of the dissolved humic substances affects the value of a planetary constant such as the total mass of DOC in the world ocean.

Despite the various important functions of the humic substances in sea water, information on their composition is scant owing to the great difficulties involved in their separation in unaltered form, as well as of their investigation (Stuermer and Harvey 1974, 1978, Gagosian and Stuermer 1977, Gagosian 1980, Højerslev 1980). Most of the information was obtained in the Sargasso Sea and off Pier-Vineyard Sound where, from 200 – 2000 l of water taken at depths of 0 and 1500 m, the humic substances were separated by absorption on Amberlite XAD-2 resin with subsequent elution with alkaline solution. Its yield made up 4 – 23% of DOM. Humic substances are complex chemical compounds of variable composition with a wide range of functional groups and molecular weights. A molecule

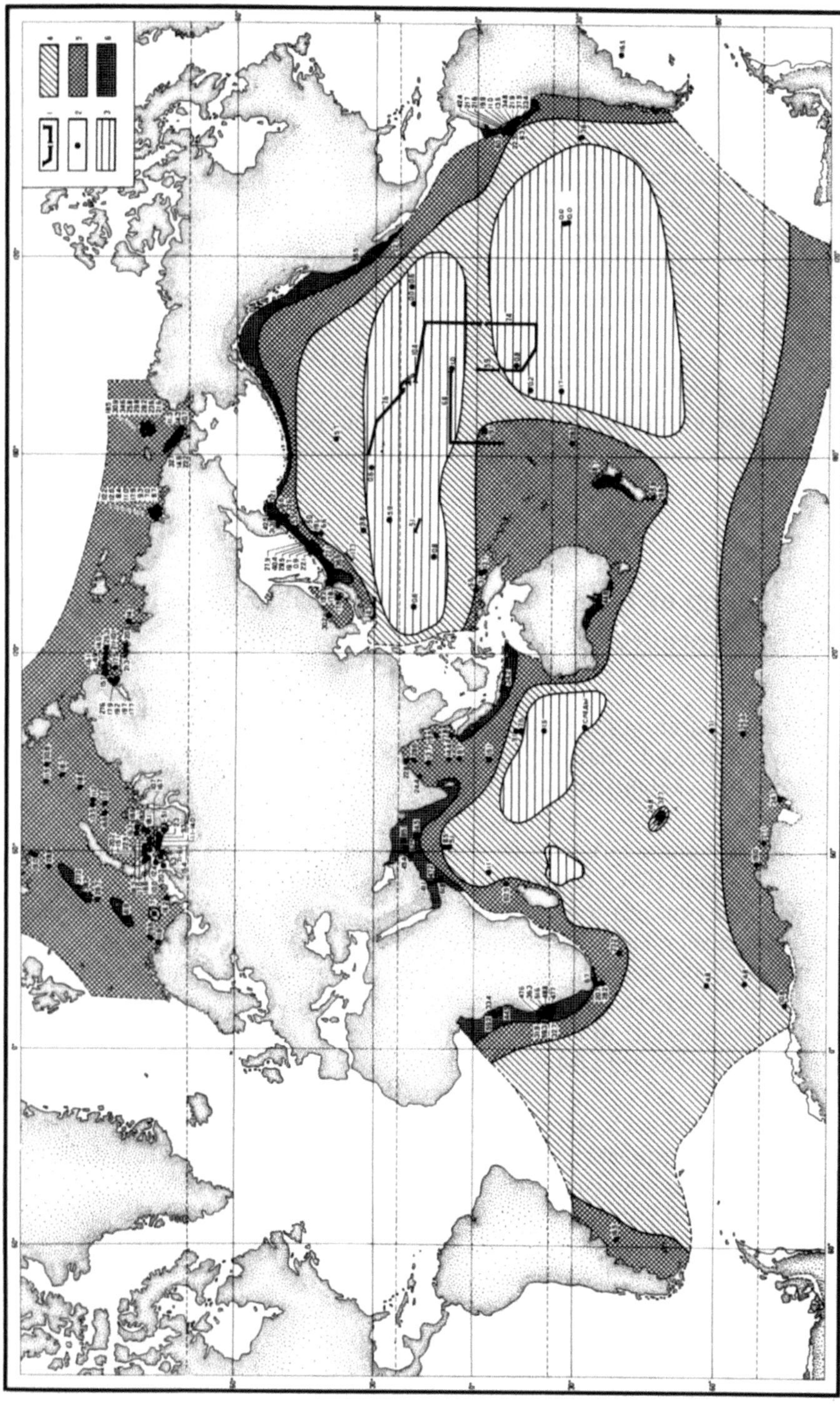

Fig. 64. Distribution of humic acids in suspended matter and oceanic bottom sediments, % of OM: *1* separation samples of suspended material in the surface layer; *2* bottom sediment samples; *3* = <2; *4* = 2 – 10; *5* = 10 – 30; *6* = >30

of a humic substance may include, in different proportions, the remnants of such biopolymers as peptides, saccharides, lipids, pigments, phenols etc.

The principal characteristic of humic substances of sea water is the predominance of aliphatic structures, poor phenol content, low degree of aromaticity, relatively low molecular weight (700 – >5000, 55 – 73% <700), high nitrogen content, low UV – VIS light extinction coefficients (higher in offshore waters than in the Sargasso Sea), the presence of amidic bonds (absorption band at 1560 cm^{-1} of fulvic acids).

The genesis of humic substances and the 'yellow stuff' apparently occurs, as was suggested by Kalle (1966) and Nissenbaum (1974), during the condensation of carbohydrates with amino-acids with subsequent rearrangement, cyclization and decarboxylation of the resulting polyfunctional melanoidin complexes having a brown colour. The aliphatic character of the bulk of humic substances of sea water and the presence of long-chain aliphatic structures in HA and FA suggests the incorporation in their composition of the remnants of lipid biopolymeres (Stuermer and Harvey 1978, Gagosian 1980).

9.2 Distribution of Humic Acids and Residual Organic Matter Humins in Suspended Matter and in Bottom Sediments

An analysis of separation samples of suspended matter from the Pacific Ocean's surface layer revealed that it contained HA + FA carbon amounting to 0.53 – 1.27% (dry weight) or 4.3 – 10.4% (average 6.2%) of the total POC (Table 67).

In Baltic suspensates HA + FA carbon makes up 5.6 – 12.2% (on the average 8.5%) from the total POC. In contrast to the analyzed pelagic samples from the Pacific, Baltic suspensates contain some 43 – 47% of residual C_{org} in corporated into POC (Daniushevskaya and Romankevich 1982).

It is not yet quite clear whether the degree of humification of suspended OM increases in deep waters. Isolated determinations fail to indicate this. It was furthermore not possible to establish any clear-cut differences in the degree of humification of suspended OM between the different areas of the Pacific which were explored.

Humic and fulvic acids were found to be contained in all samples of recent marine and oceanic sediments. Their content varies over a very wide range, from 0.003 to 9.16%. They were also encountered to some extent in the mass of Quaternary, Neogene and Paleogene terrigenous deposits. Their distribution there is, as a rule, symbatic to that of C_{org} (Bordovsky 1964, Romankevich 1977). Humic acids were found to occur in perceptible quantities in sedimentary rocks down to the stage of late protocatagenesis (Sub.Bit.A). In older rocks, in the catagenesis stage, they pass into a state which is insoluble in alkali, their content usually being not above 1% of OM (Rodionova and Pentina 1969, Chetverikova et al. 1974, Chetverikova and Pentina 1982).

Table 67. Content of Humic Acids and ROM in Aquatic Suspended Matter and Bottom Marine and Oceanic Sediments (% of OM)

Object of study, region	Humic acids	ROM	Source
Suspensates of the Pacific Ocean			
Coccolith-diatom suspension of Central Pacific	*5.9–10.4* (7.6)5	– –	Bogdanov and Lisitzin (1968)
Diatom suspensions, east of the Kurils	*4.3–5.3* (4.9)3	Not found	Romankevich (1977)
Peridinium suspensions, western tropical part	*4.3–5.1* (4.7)2	Traces	Romankevich (1974b)
The Baltic Sea			
Riga Bay	5.6	46.3	Daniushevskaya and Romankevich (1982)
Gdansk Bay	8.6	43.2	Daniushevskaya and Romankevich (1982)
West of Kurisch Gulf	12.2	47.3	Daniushevskaya and Romankevich (1982)
Gotland trough area	7.4	45.9	Daniushevskaya and Romankevich (1982)
Bottom marine and ocean sediments			
Terrigenous facies:			
of delta cones	*9.3–40.3* (21.3)	– –	Weber (1973)
of lagoons and semiclosed sea bays	*2.0–6.0* (4.5)	– –	Weber (1973)
of sublittoral areas	*2.7–2.7* (2.7)	– –	Weber (1973)
of inland and marginal seas	*9.8–21.9* (15.9)	– –	Weber (1973)
of icebergs	*22.9–29.3* (27.1)	– –	Weber (1973)
Carbonaceous facies:			
of sea gulfs and straits	*6.4–32.4* (18.4)	– –	Weber (1973)
of coral reefs	*1.2–10.2* (5.7)	– –	Weber (1973)
of shallow-water ocean areas	*22.2–32.7* (27.5)	– –	Weber (1973)
Black Sea sediments:			
recent	(12.1; 11.0)[a]	(60.7; 57.9)[a]	Kuprin and Potapova (1973, 1978)
calamitic	(13.2; 12.7)[a]	(55.5; 52.7)[a]	Kuprin and Potapova (1973, 1978)
Bugaz-Vityaz	(12.5; 11.9)[a]	(63.0; 62.7)[a]	Kuprin and Potapova (1973, 1978)
neoeuxine	(10.2; 12.6)[a]	(63.7; 60.4)[a]	Kuprin and Potapova (1973, 1978)
Caspian Sea	(10.7; 12.8)[a]	(64.6; 71.9)[a]	Kuprin and Potapova (1973, 1978)
Mediterranean Sea	(4.9)	(56.0)	
White Sea	(15.8; 21.2)[b]	(53.9; 56.6)[b]	

Table 67 (continued)

Object of study, region	Humic acids	ROM	Source
Californian borderland	(10.6 – 13.8)	–	Cit. Nissenbaum and Kaplan (1972)
Saanich Inlet	44 – 68	–	Cit. Nissenbaum and Kaplan (1972)
South-West African shelf:			
silicon-clay sediments	*27 – 60* (43) 8	*28.6 – 43.0* (37) 4	Romankevich and Baturin (1974), Korzhikova et al. (1979), Koniukhov et al. (1978)
carboniferous sediments	*31.5 – 51.6* (40.8) 5	*13.8 – 37.0* (24.2) 5	Romankevich and Baturin (1974)
Southern Baltic	*29 – 62* (43) 13	– –	Bojanowski and Pemkowiak (1980)
South-West Baltic:			
postglacial sediments	*11.2 – 34.1* (22.4) 12	*60.7 – 85.6* (73.2) 12	Daniushevskaya and Romankevich (1982)
Driass sediments	*2.8 – 12.1* (7.4) 5	*72.5 – 88.8* (79.7) 8	Daniushevskaya and Romankevich (1982)
Barentz Sea	*8.7 – 41.9* (23.6) 36	– –	Daniushevskaya et al. (1978, 1980), Romankevich et al. (1982)
Kara Sea	*5.1 – 30.8* (15.0) 30	– –	Daniushevskaya et al. (1978, 1980), Romankevich et al. (1982)
Laptev Sea	*4.9 – 27.6* (17.7) 16	– –	Daniushevskaya et al. (1978, 1980), Romankevich et al. (1982)
East-Siberian Sea	*7.0 – 17.9* (10.9) 8	– –	Daniushevskaya et al. (1978, 1980), Romankevich et al. (1982)
Chuckchee Sea	*10.7 – 34.6* (22.8) 14	– –	Daniushevskaya et al. (1978, 1980), Romankevich et al. (1982)
Peruvian upwelling area	*21.6 – 60.2* (37.3) 7	*20 – 60* 40 (7)	this volume, Koniukhov et al. (1978), Korzhikova et al. (1979), Vasilevskaya and Maksimov (1982)
Amur Gulf sediments	30.17	–	Vasilevskaya and Maksimov (1982)
East China Sea muds	9.94	–	Vasilevskaya and Maksimov (1982)
Japan Sea muds	17.93	–	Vasilevskaya and Maksimov (1982)
Muds east of Honshu i.	17.33	–	Vasilevskaya and Maksimov (1982)

Table 67 (continued)

Object of study, region	Humic acids	ROM	Source
Bering Sea terrigenous sediments	*16.8–39.8* (26)	– –	Bordovsky (1964, 1974)
Kuril-Kamchatka	*6.1–55.5* (30.5) 71	– –	Romankevich (1962)
Terrigenous muds of the Japanese deep-water trough:			
0–0.1 m	*17.8–36.9* (24.6) 3	*35.9–57.1* (49.3)	Romankevich (1974, 1976)
4–8 m	*31.3–39.8* (36.2) 4	*39.9–51.8* (44.8)	Romankevich (1974, 1976)
Pacific eupelagic red clays	*0.8–5.9* (3.3) 2	*72.8–74.9* (73.8)	Romankevich (1977)
World ocean (without seas):	*Traces-85.1* (22.8) 95	*13.8–83.6* (43.4)	Romankevich (1977)
submarine margin sediments	*6.8–85.2* (30.1) 63	*13.8–57.1* (36.4)	Romankevich (1977)
hemipelagic sediments of the ocean floor margin	*5.6–32.6* (17.1) 7	*57.6–83.6* (67.6)	Romankevich (1977)
sediments of central areas of the ocean floor	*Traces-23.2* (5.7) 25	*72.8–74.9* (73.8)	Romankevich (1977)

Italics = extreme values, Roman = numbers of samples, bracketed figures are average values. Values obtained for the surface layer (0–10–20 cm) are not generally specified

[a] The first figure is an average for the shelf, the second one an average for the continental slope
[b] The first figure is an average for open sea, the second one for bays

A direct correlation is observed between the contents of C_{org} and the carbon of humic acids in oceanic sediments, which continues into the mass of Quaternary terrigenous sediments. Eupelagic clays below 3–6 m were found not to contain humic or fulvic acids, which indicates a much deeper transformation of their OM content.

Maximum quantities of humic and fulvic acids and maximum humification of OM are characteristic features of reduced sediments (diatomaceous, terrigenous, carbonaceous (benthogenic) of the shelf, slopes and bottom of deep-water troughs) widespread both in humid and in arid zones of high biological productivity. A common property of such sediments is the high content of OM, the decay (utilization) of which at early deposition stages favours the neoformation of humic and fulvic acids as metabolic products of the exchange in the organisms and their subsequent condensation, and at later stages contributes to their preservation. The other characteristics of these sediments are relatively high sedimentation rates and a large content of little-consolidated OM. However, we have not yet sufficient data to be able to extend the inference about the conserva-

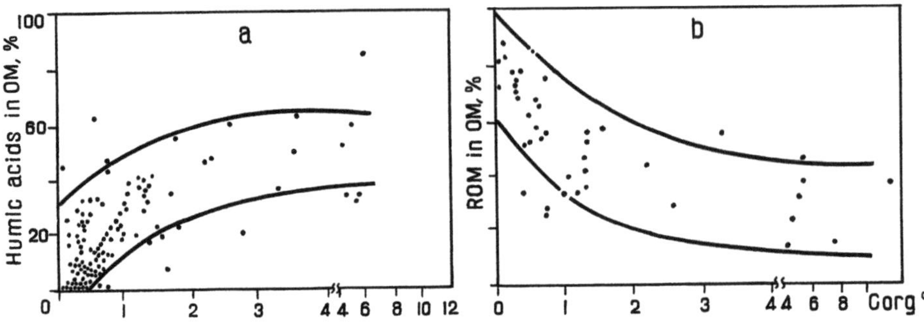

Fig. 65a, b. The content of humic acids (**a**) and ROM (**b**) in organic matter from the Pacific, Indian and Atlantic ocean sediments versus total C_{org}

tional effect of OM on the kinetics of mineralization and transition of humic and fulvic acids into an insoluble state. The deceleration of their decay in the mass of reduced sediments seems highly probable, due to a deficit of free and fixed oxygen and inhibited enzymatic destruction.

According to average data for the ocean, humification decreases in the series: submarine marginal facies (about 30%), marginal ocean floor facies (hemi-pelagic, about 17%) and central ocean floor facies (about 6%). In this series the maximum and minimum limits of humic and fulvic acid content in OM decline regularly. A minimum degree of humification is what characterizes eupelagic red clays (averagely 1–2%). Comparison of the maps showing the distribution of HA (% of OM) and that of C_{org} in sedimentary organic matter (cf. Fig. 64, 30) reveals their general similarity.

The decrease in the content of HA+FA in transition from the submarine marginal to pelagic facies is accompanied by a rising content of ROM. This proves that some part of HA resists mineralization and continues to be insoluble in alkali, being more strongly connected with the clay minerals of sediments (Gadel 1980, Goncharova and Kolosov 1980, Hedges 1980). The content of ROM in marine and oceanic sediments varies from 13 to 85%, the average values for the submarine margin, marginal and central parts of the ocean floor being, correspondingly, ~36, ~68, and ~74%. Whereas in the case of humic acids the Uspensky-Vassoyevich law (established for bituminoids and being the inverse correlation between C_{org} in bottom sediments and the relationship $C_{bit} \cdot C_{org}^{-1}$) does not manifest itself, it is well-defined as regards ROM the lower the degree of conversion of OM, the larger the quantity of HA and the lower the quantity of ROM contained in it (Fig. 65). The sum of HA, FA and ROM makes up the domineering part of OM in Recent and Quaternary sediments (60–90% on the average). All basic quantitative changes in the group composition of OM are attributable to varying relations precisely among these groups.

The relatively high endurance of HA to enzymatic hydrolysis is a firmly established fact. Nevertheless, HA may possibly be the source of energy for sulphate breathing of bacteria (e.g., Romankevich 1976). Studies of the growth of microorganisms on poor agar with the addition of 0.1% of HA which had been isolated from bottom sediments in the Kuril-Kamchatka trough showed that the microorganisms from deep ocean waters and sediments utilized HA whereas those from the surface water layers would now grow on such a substrate. Obviously, the former were capable of developing the corresponding enzymatic systems capable of utilizing HA during the experimental period (12–14 days, 24 °C, Mitskevich 1975).

Our special research of the uppermost film of bottom sediments with a thickness of 1 mm or less has demonstrated that the content of HA + FA and ROM in the OM of these sediments is 10 to 15 times higher than in surface or deep-water suspended matter. This means that the bulk of HA and FA are formed and partially change over into not easily extractable complexes at the water-bottom interface, as well as in the topmost layer of the sediments. Research undertaken in the Atlantic, northwest Pacific, in the Arabian Sea and in the Bay of Bengal has shown that the humification of OM is proceeding slowly in the sediment cores of the submarine margin and marginal part of the ocean floor. The fact that HA increase and ROM practically correspondingly decrease in the layers where biogenic sulphate reduction proceeds intensively, does not detract from the general law governing the distribution of HA in the sediments (Romankevich 1976, Emelianov and Romankevich 1979).

Inside the zone of active biochemical transformation of OM (and even in areas with low sedimentation rates this zone is not confined to a layer of a few centimetres but rather to several meters from the surface of the bottom) the absolute concentrations and the content of HA and FA in OM are observed to decrease, this being usually accompanied by a decline in the portion of easily hydrolyzable compounds and diminishing or invariable content of ROM. All this takes place against the background of the overall decline of OM and reflects the instability of some part of HA, FA and ROM during diagenesis. This is explained by two factors: partial bioutilization of HA and FA and their transition into ROM. The effect of the latter becomes particularly manifest in the older rocks at the long-flame stage of OM carbonization in the zone of mezokatagenesis (e.g., Rodionova and Pentina 1969, Chetverikova et al. 1972, Guseva and Leifman 1979, Daniushevskaya and Matveyeva 1979).

9.3 Elementary Composition

The elementary composition of HA and FA in aquatic suspended material remains unknown. Humic acids from the muds of the Kuril-Kamchatka trough contain (%): C 55.43, H 5.65, N 3.90, O + S 35.01. This differs but little from the average values for HA contained in the surface layer of marine and ocean sediments (%); C 55.8, H 5.7, N 4.0, O + S (usually determined as a difference) 34.7 (Table 68). Various facial types of surface layer sediments with planktogenic OM are observed to differ in the degree of oxidation and condensation of humic complexes (Fig. 66). However, the H/C ratio, its increase possibly providing evidence of condensation, aromatization, a larger number of double bonds, does not show any definite variability trend in transition from reduced to oxidized facies.

Table 68. Elementary composition of humic acids in marine and ocean sediments

Sampling site, layer (m)	Elementary composition, %					Source
	C	H	N	S+O[a]	H/C (atomic)	
Kuril-Kamchatka trough (terrigenous muds)	55.43	5.66	3.90	35.01	1.22	Romankevich (1977)
Bering Sea						
surface layer	54.11	6.48	3.00	36.41	1.44	Bordovsky (1964, 1974)
2.2 m	56.54	6.26	2.15	35.05	1.32	Bordovsky (1964, 1974)
4.4 m	57.45	5.86	1.81	34.88	1.22	Bordovsky (1964, 1974)
13.4 m	59.89	5.96	2.06	32.09	1.19	Bordovsky (1964, 1974)
Indian Ocean	51.45	6.68	3.90	37.97	1.56	Bordovsky (1964, 1974)
Antarctics	53.03	6.56	4.84	35.57	1.48	Bordovsky (1964, 1974)
Taman Peninsula water reservoirs	55.66	6.32	5.55	32.47	1.36	Gorskaya (1960)
Black Sea (muds)						
recent	58.35	5.12	3.73	32.80	1.05	Rodionova and Pentina (1969)
neoeuxinic	58.60	5.61	2.91	32.88	1.15	Rodionova and Pentina (1969)
Black Sea avan-delta sediments	52.73	5.52	–	–	1.26	Scherbakov et al. (1978)
recent	54.59	6.07	2.45	36.20	1.34	Kuprin and Potapova (1978)
calamitic	57.26	6.56	2.65	33.66	1.38	Kuprin and Potapova (1978)
bugaz-vityaz	58.48	6.10	1.95	33.46	1.25	Kuprin and Potapova (1978)
neoeuxinic	58.64	5.17	2.45	33.73	1.06	Kuprin and Potapova (1978)
White Sea						
lacustrine	57.51	6.56	1.26	34.67	1.40	Kuprin and Potapova (1978)
shelf	52.26	5.38	3.85	39.42	1.23	Kuprin and Potapova (1978)
slope	56.28	5.59	2.38	35.75	1.19	Kuprin and Potapova (1978)
Mediterranean Sea (sapropel oozes)	57.84	6.02	2.31	33.83	1.25	Kurpin and Potapova (1978)
Caspian Sea						
avandelta	59.17	4.85	3.82	32.16	0.98	Gorskaya (1960)
open part	55.10	6.18	5.95	32.77	1.35	Bordosky (1974)
open part	58.85	6.44	4.30	2.54; 27.87	1.31	Pentina (1973)
Zhdanov bank						
0–10 m	54.42	5.52	2.07	2.82; 35.15	1.21	Petina (1973)

Table 68 (continued)

Sampling site, layer (m)	Elementary composition, %					Source
	C	H	N	S+O[a]	H/C (atomic)	
15–20 m	56.56	5.46	1.60	2.81; 33.57	1.16	Pentina (1973)
45–50 m	56.34	5.09	2.31	2.32; 33.94	1.08	Pentina (1973)
Californian borderland	53.75	6.27	4.87	1.37; 33.74	1.39	Nissenbaum and Kaplan (1972)
Gulf of California	48.89	5.12	–	–	1.25	Nissenbaum and Kaplan (1972)
West of Oregon	55.92	5.01	–	–	1.07	Nissenbaum and Kaplan (1972)
Nova Scotia shelf	52.8	6.6	5.6	–	1.50	Nissenbaum and Kaplan (1972)
Saanich Inlet						
upper layer	56.7	5.9	2.2	5.9	1.25	Brown et al. (1972, cit. after Nissenbaum and Kaplan 1972)
34 m	57.8	5.4	1.5	3.0; 32.3	1.12	Brown et al. (1972, cit. after Nissenbaum and Kaplan 1972)
Southern Baltic	53.84	6.67	5.93	33.56	1.49	Bojanowski and Pemkowiak (1980)
Peruvian upwelling area	56.25	5.95	6.09	31.71	1.27	Vasilevskaya and Maksimov (1982)
	52.41	6.38	3.64	37.57	1.46	Koniukhov et al. (1978)
Walvis Bay	48.72	6.54	4.06	40.68	1.62	Koniukhov et al. (1978)
Cariaco trough						
oxidized sediments	55.7	5.8	4.8	–	1.25	Rashid and King (1970, cit. after Nissenbaum and Kaplan 1972)
reduced sediments	59.8	5.5	4.2	–	1.10	Rashid and King (1970, cit. after Nissenbaum and Kaplan 1972)
Off Japan	52.15	5.23	4.69	37.93	1.19	Ishiwatari (1973)
Sagami Bay	50.57	5.21	4.83	39.39	1.23	Ishiwatari (1973)
Amur Gulf	58.11	4.88	4.52	2.45; 29.90	1.00	Vasilevskaya et al. (1977)
East China Sea	56.28	6.32	4.81	0.76; 31.83	1.35	Vasilevskaya et al. (1977)
Sea of Japan	59.76	4.67	5.07	0.53; 29.56	0.94	Vasilevskaya et al. (1977)
East of Honshu i.	59.66	4.98	6.24	0.72; 27.66	1.00	Vasilevskaya et al. (1977)
Average for surface layer sediments	55.75	5.72	4.04	34.72	1.24	

[a] Where two figures in a column are separated by semi-colon the first represents sulphur content and the second oxygen content

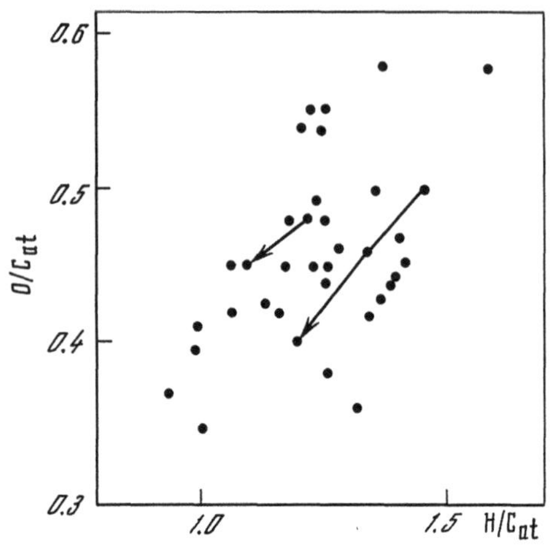

Fig. 66. The elementary composition of humic acids contained in Recent and Quaternary marine and oceanic sediments. *Dots* are average values for different regions; *arrows* are transformation trends in diagenesis

In terms of average values the elementary composition of HA in marine sediments little differs from that of river or lake waters (average, %): C 56 – 60, H 4.9 – 5.7, N 4.4 – 5.1, O 32 – 34 (Semenov 1971). Littoral, sublittoral, and deep-water sediments in various oceanic areas showed the carbon and hydrogen content in humic acids to be very similar to that of soils, peats, lake and river sediments (e.g., Nissenbaum and Kaplan 1972). Accordingly, it cannot be assumed that the H/C ratio is a reliable criterion of either autochthonous or allochthonous genesis of HA.

The main differences between HA contained in oceanic bottom sediments and those found in various land formations (soils, peats, river and lake sediments) lie generally in the higher content of nitrogen and sulphur and heavier isotopic composition ($\delta^{13}C$ usually varies from – 20 to – 23‰), as compared to HA of the soils (usually – 25 to – 26‰, Nissenbaum and Kaplan 1972, Nissenbaum 1974, Gadel 1980, Vasilevskaya and Maksimov 1982). Furthermore, they contain a large quantity of non-hydrolyzable 1, 2 and 6 N HCl nitrogen (66 – 75% of the HA nitrogen), which distinguishes them sharply from HA present in soils and peats where the content of hydrolyzable nitrogen may well attain 80% (Vasilevskaya et al. 1977).

The nitrogen content in the HA of the sedimental surface layer varies from 2.2% to 6.2% (average 4.4%). HA in the oxidized sediments of the Japanese trough, as evidenced by the H/C ratio, contain about as much nitrogen as the reduced sediments on the shelf of South West Africa. The nitrogen content of HA and FA is declining in the mass of the sediments. A significant proportion of N_{org} contained in the surface layer of terrigenous (53 – 67%), diatomaceous (34 – 54%) and carbonaceous (44 – 59%) sediments is included in HA and FA, the larger part of it contained in FA. As far as eupelagic clays are concerned, nitrogen occurs only to a minor degree, in HA (about 2%) and in FA (about 10%), being mainly included in the ROM. In the mass of the sediments these differences between the extreme groups of facies become still more pronounced (Romankevich 1976, 1977).

HA in the older sediments of the existing basins (the Neoeuxinic of the Black Sea, the Khvalynic of the Caspian, the Pleistocene of the Bering Sea) are different from those contained in surface-layer marine sediments in that their nitrogen content is almost halved (2% and 4%, respectively) and they show a tendency towards higher C and lower H/C with immersion depth. This indicates that the process of condensation in diagenesis is accompanied by an advance escape of nitrogen from humic complexes. This agrees with the change, during diagenesis, of the total sum of nitrogen-containing compounds which is observable both in marine conditions and on dry land and is the result of lower stability of the bulk of nitrogen-containing compounds in decomposition chains. The nitrogen content in the mass of the sediments could be somewhat influenced by differences in the supply and conservation of organic matter attributable to lower algae (planktonogenic component) and to dry-land plants (humus component) of which the latter is more deficient in nitrogen. This can be illustrated by the sediments of the Zhdanov bank (see Table 68) containing mixed sapropel-humus OM (Pentina 1973).

Investigations of the oxidized and reduced sediments in the surface layer of Nova Scotia's shelf (depth 198–274 m) and of the Cariaco trough (depth 808–1390 m) undertaken by Rashid and King (1970) revealed that, compared to HA, FA contained less C, H, N and, correspondingly, more O (FA (%): C 38.4–48.7, H 4.3–6.6, N2 2.6–5.2, O 41.8–54.5, C/H 7.00–8.93; HA (%): C 50.2–55.7, H 5.6–7.0, N 3.6–5.6, O 32.4–40.6, C/H 7.90–9.49). Similar differences between HA and FA are common to river and lake waters and soils (Semenov 1971, Orlov 1974). The fact that the C, H, O contents in FA are subject to greater fluctuations than HA indicates that compositionally they are

Table 69. Elementary composition of ROM in river, marine and oceanic sediments

Type of sediments	C	H	N	O+S	H/C (atomic)	N/C (wgt.%)
Indian Ocean						
Recent terrigenous muds	42.4–58.4 (51.0)	4.5–6.8 (5.5)	1.4–3.1 (2.1)	34.6–50.4 (41.1)	1.0–1.6 (1.3)	2.9–5.9 (4.0)
Caspian Sea						
Recent terrigenous muds	55.4–66.9	5.0–5.1	1.4–1.8	26.3–38.1	0.9–1.1	2.5–2.7
Black Sea						
Recent muds	64.66	5.27		30.07	0.98	–
Calamitic	65.11	6.52		28.36	1.20	–
Neoeuxinic	64.97	6.70		28.46	1.24	–
Shelf sediments	63–68	5.7–5.8		26–30	1.00–1.09	–
Continental slope sediments	62–66	5.8–7.6		27–32	1.13–1.44	
Danube avandelta sediments	66.4	4.86		29.74	0.87	–
Rivers of the Black-sea basin						
Bottom sediments	63.15	5.62		31.23	1.07	–

Bracketed figures are average values

more variable; in fact, the very definition of FA implies this. Indeed, a FA molecule contains more hydrophilic and protogenic groups, than does a HA molecule. These groups are retained in the dissolved state for a wide pH range. The transition of FA to low-molecular uncoloured compounds is smooth, and on the side of high-molecular compounds they are limited by HA.

Determinations of the elementary composition of "insoluble" OM (ROM) of the sediments in the Indian Ocean and in the Caspian Sea brought out significant differences between the samples as is shown in Table 69 (Bordovsky 1974, Kuprin and Potapova 1978, Scherbakov et al. 1978).

ROM from sapropel muds of the Black Sea is distinguished by the maximum content of hydrogen and the highest H/C_{at} value. Difractometric studies and X-ray structural analysis of Black Sea sediments have demonstrated a decrease in the amount of aromatic structures of ROM in the series: avandelta sediments – shelf sediments – continental slope sediments. The aromaticity coefficients for this series are 0.2, 0.1 and 0.03 respectively (Scherbakov et al. 1978).

In general, when compared to HA, ROM contain a greater quantity of carbon and a lower quantity of nitrogen and other heteroatoms. The similarity of average data on the elementary composition of both HA and ROM is an indication of their genetic affinity.

9.4 Molecular Weight Distribution

The application of gel-chromatography and ultrafiltration techniques to the study of the molecular-weight distribution of HA in various facial-genetic types of marine sediments furnished new evidence of their polymeric nature.

Molecular-mass content of HA of bottom sediments is highly variable: from $<5\times10^3$ to $>200\times10^3$. In the major part of samples taken in very different facial conditions (Peruvian area sediments, Walvis Bay, East China Sea, Sea of Japan, Red Sea, off New Guinea, pelagic regions of the northwest basin of the Pacific) there was an increased content of high-molecular fractions (mol m $>100\times10^3$). The molecular mass of HA in bottom-sediment cores was found to decrease with immersion. It was established by means of gel-filtration through sephadex, that FA contained in the sediments of Nova Scotia's shelf had lower molecular mass than HA (from <700 to 10 thousands for FA and <700 to >200 thousands for HA). The proportion of the >200 thousands molecular mass fraction in HA from these sediments amounted to 26–49%. This is also characteristic of FA forming in the course of decay of *Fucus* and *Laminaria* and of the latter's excretion products, as well as of FA present in river waters (e.g., Semenov 1971, Rashid and Prakash 1972). From this it follows that the transition from FA to HA is accompanied by the growing scale of molecules.

In highly oxidized eupelagic oceanic clays with consolidated OM (ROM 75–88%) the $FA \leftrightharpoons HA$ equilibrium is displaced towards the low-molecular FA with augmented content of oxygen functions (Romankevich 1977). Thus, humification at the high stages of OM conversion involves not only the condensation of HA and their transition into ROM but also a partial destruction of macromolecules into fragments: $FA \rightleftarrows HA \rightleftarrows ROM$.

9.5 Amino Acid Composition. The Content of Carbohydrates and Functional Groups

The amino-acid composition of HA in terrigenous muds taken from the inner slope of the Kuril-Kamchatka deep-sea trough (4260 m depth) and from the San Diego trough located within the California borderland area turned out to be quite similar (Table 70). Both samples contained increased amounts of acidic amino acids and, among the neutral ones, leucines and alanines. The group composition of HA revealed certain differences. For example, although similar content of acidic and aromatic amino acids in HA was observed in both samples, HA from sediments of the Kuril-Kamchatka trough contained less neutral (39.1%) and more basic amino acids (23.5%) compared to the San Diego trough sediments (55.0% and 11.4%, respectively).

Of the total amount of amino acids that could be extracted from the sediments, 10–23% were included in HA. Nitrogen of the amino acids which were part of HA from the sediments in the Kuril-Kamchatka trough accounted for about 25% of total nitrogen in amino acids of the sediments. The remaining 75% of the amino acid nitrogen were apparently included in FA and ROM. Of the total carbon of HA from the Kuril-Kamchatka trough sediments, amino acids account for 8.6%.

A set of amino acids observed present in the sediments and in HA was identical. The relationship was, however, different: HA were observed to be higher in acidic and lower in neutral amino acids.

In HA from terrigenous, diatomaceous and carbonaceous sediments there were present about 4–17% and in FA about 4–9% of carbohydrates (the total averaging about 15%). Between 33 and 95% (average 50%) of carbohydrates are represented by uronic acids (Romankevich 1977, Vasilevskaya et al. 1977, Vasilevskaya and Maksimov 1982).

For the mass of terrigenous deposits of submarine margins with little consolidated OM the carbohydrate content of HA and FA has remained practically the same during the past 0.1–0.2 million years. In the mass of eupelagic clays the content of carbohydrates has been rising in the past approximately one million years, which reflects the high stability of geopolymers formed with the participation of polysaccharides. Carbohydrates of HA and FA in terrigenous, diatomaceous and carbonaceous sediments of the surface layer account for 18–64% of the total carbohydrate content in the sediments. Carbohydrates present in the HA are apparently more resistant to decomposition during the diagenesis as compared to the FA carbohydrates. The difference in the monosaccharide composition of HA and Fa are particularly well-expressed in the upper parts of the sediment cores (Yamaoka 1983).

The fact that amino acids and carbohydrates are present in HA of the sediments means that separate blocks which were originally parts of the proteins and carbohydrates of the organisms continue to exist in decomposition chains in geopolymers such as, HA, FA and ROM. However, our knowledge of their polymeric composition, the structure-forming laws or stability under changing environmental conditions is insufficient.

Table 70. The amino-acid composition of bottom sediments (0 – 20 cm layer) and of humic acids isolated from them (% of total amino acids)

Amino acids	Kuril-Kamchatka trough[a]		California borderland[b]	
	in sediments	in HA	in sediments	in HA
Neutral				
glycine	13.6	6.1	12.3	8.5
α-alanine	14.0	6.2	10.7	7.9
β-alanine	–	–	1.2	3.0
valine	5.9	4.8	9.4	5.4
leucine and isoleucine	10.0	9.3	10.2	9.4
serine	10.4	4.0	6.8	6.0
threonine	5.2	4.1	6.3	5.4
proline	not found	4.6	9.7	9.4
Basic				
histidine	15.7	6.6	2.0	2.4
lysin	–	8.0	13.3	5.1
arginine	traces	8.9	5.4	2.4
ornithine	–	–	2.4	1.5
Acid				
aspartic acid	9.5	17.6	0.1	12.1
glutamic acid	traces	11.2	0.8	12.1
Aromatic				
phenylalanine	5.3	5.0	3.7	5.7
tyrosine	7.8	3.5	2.4	3.6
Sulphur containing				
cysteine	not found	–	–	–
methionine	2.6	–	–	–
Amino sugars[c]			0.4[c]	0.3[c]
Amino acid content in sediments μg g^{-1} of sediment (dry weight)	533	–	3281	–
% of OM	2.1	–	~13	–
Amino acids contained in humic acids, calculated with reference to				
dry sediment, μg g^{-1}	–	121	–	331
total amino acids, %	–	22.7	–	10.1
humic acids, %	–	8.6[d]	–	3.3[d]
Sedimentary OM, %	–	0.5	–	1.3

[a] The present work. The total C_{org} content in sediments 1.20%; the content of HA in OM 6.42%
[b] Degens et al. (1964). The OM content in sediments 2.5%; the content of HA in OM 40%
[c] The concentration of amino sugars in OM
[d] With reference to HA ash-free substance

A difference between HA from typically marine and oceanic sediments and those of a continental origin becomes particularly clear in the comparison of their functional composition. Thus, if HA from soils, peats, and coals are polyphenol compounds containing large quantities of free and methylated phenol

Table 71. Oxygen-containing functional groups in humic acids hydrolyzed with 1 – 6 N HCl, mg-eq g^{-1}

Sampling site and bottom depth	COOH	$OH_{phen.}$	OCH_3	C=O (quinoid)
Amur Gulf (1 – 1.5 m)	3.35	1.45	1.45	3.22
East China Sea (276 m)	4.61	0.75	0.95	1.65
Sea of Japan (3000 m)	6.01	0.41	0.41	2.13
East of Honshu i. (6000 m)	5.51	0.09	0.42	1.75
Peruvian shelf (70 m)	1.47	2.55	0.43	3.02
Peruvian slope (550 m)	1.46	1.28	0.33	4.90
Peruvian slope (1060 m)	1.65	3.07	0.29	4.43
Peruvian slope (2890 m)	1.72	1.97	0.30	4.96
Bottom of the Peruvian trench (6250 m)	1.87	1.62	0.28	4.86

hydroxyls, HA from the sediments with the originally planktogenic OM are very poor in them (Table 71; Vasilevskaya et al. 1977, Vasilevskaya and Maksimov 1982). The presence of a large quantity of quinoid groups in HA from recent sediments is possibly due to transformations of aromatic amino acids (through compounds of the dopa-quinone type), as takes place in the synthesis of melanines. The fact that the content of the quinoid groups obtained by two methods (Fe^{2+} reduction in triethanolamine and by alkaline solution of $SuCl_2$) was found to be nearly the same enables us to presume that polycyclic quinones were unavailable in the samples studied. HA from the sediments in the Amur Gulf receiving larger quantities of OM from land differ from those contained in planktonogenic OM by a higher content of methoxyls, amino acids and hydrolyzable nitrogen, as well as higher $E_{465}^{0.02}/E_{665}^{0.02}$ ratio (Vasilevskaya et al. 1977).

Infra-red spectra of HA of recent sediments from such areas as South Caspian, the continental slope of Peru are characterized by strong absorption bands of CH_2-, CH_3-groups at 2920 cm^{-1} and of oxygen aliphatic structures at 1035 – 1040 cm^{-1}. The spectra of these HA are observed to comprise distinct absorption bands of amide groups at 1650 – 1640 and 1540 cm^{-1}, thus indicating that some protein fragments are preserved in their molecules.

9.6 The Composition of Destructive Oxidation Products

The method of destructive oxidation first applied to HA by Ishiwatari has now become very popular in studying HA of river, lake and marine sediments (Ishiwatari 1973, Vasilevskaya et al. 1977, Vasilevskaya and Maksimov 1982). Table 72 shows the group composition of destructive oxidation products obtained by the treatment of HA from the bottom sediments of various Pacific areas with alkaline permanganate. To protect phenol hydroxyls, HA had been first methylated with diazomethane. Identification of methyl ethers was carried out by means of gas-liquid chromatography and chromato-mass-spectrometry.

Table 72. The group composition of the oxidation products of HA in marine sediments

Methyl ethers of humic acids	Yield of methyl ethers of oxidation products, mg g^{-1} of HA				
	Amur Gulf	East China Sea	Sea of Japan	East of Honshu i.	Peruvian upwelling area
Aliphatic monocarboxylic	2.18	–	1.67	0.54	31.2–66.1
Aliphatic dicarboxylic	26.24	161.15	44.88	21.93	59.7–181.9
Benzenepolycarboxylic	113.44	128.11	248.06	117.31	11.3–100.8
Oxy(methoxy)-benzoic	6.62	–	–	–	2.3–6.0
Oxy(methoxy)-benzene-polycarboxylic	49.09	16.86	44.35	45.40	7.8–35.6
Altogether identified	197.57	306.12	338.94	185.21	119.3–311.5
Methyl ethers of oxydation products, volatile under gas-liquid chromatography conditions	238.90	330.74	347.68	363.50	225.0–531.0

Quantitative determinations of aliphatic mono-, dicarboxylic, benzenecarboxylic and oxygenbenzene-carboxylic acids in HA were made for all the investigated sediments. Specific structural lignin units: n-methoxybenzoic, veratric, 3,4,5-trimethoxybenzoic and 4,5-dimethoxy-1,3-benzene-dicarboxylic acids were discovered only in the off-shore silts of the Amur Gulf where much OM is supplied from the land. A characteristic feature of HA in planktonogenic OM was a low content of aromatic oxyacids, and HA of the East China Sea sediments a predominance of aliphatic dicarboxylic acids was observed. The basic structural parts of HA from deep-water muds of the Sea of Japan and of the northwestern basin were complex multiple-substituted and condensed aromatic systems almost devoid of hydroxyl groups.

9.7 The Origin of Humic Acids in Bottom Sediments

Almost completely opposite viewpoints have been expressed on the origin of HA in marine and oceanic sediments. Degens believed that greater part of humic components of the sediments were decomposition products of lignin which is not present in lower algae (Degens et al. 1964).

Relying mainly on the magnitude of $\delta^{13}C$ in HA of marine sediments (between −17.2 and −27.4‰, usually −21 to −23‰) and comparing it with the corresponding values for HA of littoral sediments (between −19.1 and −27.3‰, mostly −21 to −26‰) and those from continental environment (between −14.8 and −29.1‰, mostly from −25 to −27‰), Nissenbaum and Kaplan have arrived at the conclusion that under marine conditions HA can be formed in situ from the products of plankton decomposition (Nissenbaum and Kaplan 1972). Data on the deuterium content in marine sedimental OM do not contradict this conclusion (Nissenbaum 1974).

Karavayev, Vener and Koroleva (1964), while studying fresh-water sapropels, did not find the condensed aromatic structures typical of HA of soils and peats. Not a single fraction of the alkaline extract following nitrobenzene oxidation was found to contain the remnants of lignin: syringaldehyde, vaniline or n-oxybenzaldehyde.

It might be well to emphasize the great importance of lithologic-geochemical proofs of the genesis of HA in oceanic bottom sediments. They are generally assumed to be not as significant as the chemical ones. A summary of certain materials related to this problem is given below.

The distribution of HA in marine and oceanic sediments seems to be directly correlated with plankton production and the content of C_{org} in sediments. The bulk of HA and FA is formed in the two most active zones of biochemical transformation of planktonogenic OM, that is in the photic layer and in the uppermost layer of bottom sediments.

Every year about 28.0 million tons of HA carbon deposited in the world ocean's bottom sediments. Of this total about 26.1 million tons are accumulated in the region of submarine margin, about 1.7 million tons in the marginal area of the ocean floor and about 0.2 million tons in the central parts of the ocean floor. The initial data for these calculations were the absolute masses of C_{org} deposited in oceanic sediments (about 100 million tons of C_{org} year^{-1}), their distribution in the structural zones mentioned above (87%, 10% and 3%, respectively) and the content of HA in OM. Of the total quantity of HA in the sediments, a significant proportion is accumulated in arid zones with a high biological productivity, where the supply of HA and FA from land is insignificant. These sediments are characterized by the maximal humification of OM. Significant humification is also typical of the sediments on the shelf and the upper slope of the Antarctic (see Table 62) where practically no HA or FA are delivered from land.

HA and FA in marine sediments are characterized by a high nitrogen content, compared to those in littoral deposits and in many types of soils and peats. Presumably, this is inherited from marine plankton which contains more nitrogen than land plants, kelp or marine angiospermae.

It was found that the infrared spectra of HA contained in marine sediments manifest distinct absorption bands of amide groups and the bands due to C−O bonds in aliphatic alcohols, ethers and carbohydrates, as well as the absence, in some samples, of absorption of aromatic cycles (3030, 1600, 750−900 cm^{-1}). This, as well as a significant proportion of carbohydrates in HA and FA, indicates the proteinic-carbohydrate genesis of a large part of HA and FA from marine and oceanic sediments (e.g., Huc and Durand 1974).

The supply of a great amount of OM from land is revealed through the presence of plant remnants, charred wood pieces (Kuril-Kamchatka, Japanese, Bougainville troughs), an increased humification of the sediments, and a very light isotopic composition of HA in certain estuarian sediments, for example those of the Amazon (Nissenbaum and Kaplan 1972).

According to average data, the carbon of HA contained in marine sediments is lighter by some 1 – 2‰ than that of plankton, which may be the result of diagenetic transformations and the admixture of OM of continental origin. For example, significant admixture of OM of continental genesis was recorded in near-coastal sediments of the Caspian Sea and the Amur Gulf. This becomes manifest in a decrease of the nitrogen content and the H/C ratio in HA, in an elevated content of acidic functional groups, phenol hydroxides, methoxide and quinoid groups, in the declining intensity of absorption bands of the CH_2^-- and CH_3^--groups at 2920 and 2860 cm^{-1}, in the appearance of specific lignin fragments (Pentina 1971, Vasilevskaya and Maksimov 1982). In all other investigated sediments with largely planktonogenic OM no such structural lignin units were identified.

Between all high-molecular aromatic compounds of oceanic sediments identified in the process of destructive oxidation of HA and those of the continental origin there exist basic differences. The appearance of aromatic structures in the OM of oceanic sediments may be due to the dehydration of cyclic polymers in the highly-saturated fatty acids of marine organisms, to the accumulation and transformation of aromatic bacterial metabolites, to the melanine-like substances arising from aromatic amino acids through the dopaquinone type compounds as well as after the Mayard reaction (Vasilevskaya and Maksimov 1982).

On analysing this material the author has arrived at the conclusion that, within the limits of the submarine margin of the continents, both HA and FA are mostly formed in situ during the decomposition of OM of lower algae and must be referred to the category of amicaginic acids (Vassoyevich 1982). It therefore appears to be incorrect to extend the terminology elaborated for the continental OM to the majority of specific acidic organic substances of marine genesis (see Conclusion). In the near-continental humid areas, HA, FA and ROM are apparently partly of phytobenthonic and allochthonous origin. The latter, particularly in the areas notable for large supplies of OM with river the discharge, can play quite an essential role.

Thus these studies throw light on the origin of the bulk of HA and ROM in marine and oceanic sediments and contribute to the development of the general theory of humification. It is already clear that humification is a very important process of OM transformation on dry land and in the world ocean, providing optimal conditions for the evolution of future generations of organisms and of the cycle of OM on the Earth.

Conclusion

At their present stage of development, biogeochemistry and organic geochemistry of marine and ocean environment have certain special aspects. These include, firstly, a transition to the systematic analysis of all principal forms of OM in the ocean (living, dissolved, suspended and that contained in bottom sediments) and of their interrelations and transformations under varying environmental conditions; secondly, the elaboration, based on quantitative data, of the pattern of the general cycling, evolution and balance of OM in the ocean; thirdly, a detailed study of the morphology and composition of OM on the molecular and atomic level; fourthly, a significant improvement of sampling methods and techniques and, finally, the supplementation of the formerly dominant chemico-analytical methods with the new and/or improved biogeochemical ones.

An important result of recent research is the establishment of the magnitude of masses of dissolved, suspended and living OM which exist in the ocean in dynamic equilibrium. Their relationship is 100 : 1.7 : 0.16 or (in absolute figures, 10^9 tons C_{org}) 1800 : 30 : 2.8. This pyramid of masses is a reflection of an equilibrium maintained between the processes of synthesis, supply and decomposition of OM in the modern ocean and can, therefore, be regarded as a planetary constant.

The bulk of OM in the ocean is present either in actual solution or in a colloidal state. The latest data corroborate the estimate of the average C_{org} content in dissolved organic matter ($\sim 1.3\ \mathrm{mg\,l^{-1}}$) and of its total quantity ($\sim 1.8 \times 10^{12}$ t C_{org}) which has been given by Soviet research workers (Skopintsev 1950, Romankevich 1977, Romankevich et al. 1982).

It is not possible today to give an answer to the long-debated problem of the relation of the OM of deep waters with the living matter in the upper water layers. The residence time of the main portion of DOM in the ocean is believed to be about 1500 years. Decomposition of OM at great depths takes place slowly due to low temperatures, high hydrostatic pressure and the nature of the OM itself.

The characteristic features in the distribution of POM are a direct correlation with biological productivity, a higher content in inland and marginal seas than in the open ocean; a declining concentration in the shelf-pelagial direction, the presence, within the water mass, of layers enriched with living and non-living OM which play an important role in the formation of the biogeochemical structure of the waters and, finally, a deceleration of POM decomposition rate in the mass of water. The residence time of its main portion in the ocean is 10 – 25 years. The turnover time and correspondingly the lability of the main body of POM is about 100 times higher than that of DOM.

A synthesis of available data on the composition of bottom sediments has revealed that the distribution of C_{org} is controlled by the complex of facial sedimentation conditions and reveals circumcontinental, latitudinal and vertical (bathymetric) zonality inherited from suspended matter. In the near-shore zone and, in most cases, within the limits of submarine continental margins and marginal parts of the ocean floor, the content of OM in the sediments is largely affected by the following factors:

a) biological productivity and the input of autochthonous POM; in certain areas a similar factor is the bioproductivity of dry land and the supply of allochthonous OM (direct correlation);
b) the granulometric composition of the sediments and the conditions of OM deposition which are dependent on the hydrodynamics of waters, bottom relief, the aggregation of sedimentary material by benthos, the supply rate and size of fluvial suspended matter and abrasion products (complex correlation);
c) the living conditions of the benthos which are determined by the quantity of assimilated OM, the gaseous regime of the habitation environment and the physiological potentialities of the heterotrophs;
d) the diluting effect of mineral material (inverse dependence).

A direct correlation between the content of C_{org} and the <0.01 mm fraction, which is generally known to be true for inland and marginal seas as well as for water basins of the past, is preserved in the ocean solely within the confines of submarine margins. It has not been observed in pelagic deposits. This, as well as certain other features characterizing the composition of OM in pelagic sediments, such as, for example, its content of nitrogen, lipids, carbohydrate-like substances, humic acids (all other conditions being equal), should be made use of in the identification of pelagic deep-water deposits in the Earth's history.

The principal and most marked feature of the distribution of percentage concentrations and absolute masses of OM in the world ocean proves to be the circumcontinental zonality which finds expression in some hundred-fold decrease of the absolute masses of OM in transition from terrigenous sediments of submarine margins to red deep-water clays of ocean floor basins. As a result, the peripheral ocean areas comprising shelves, continental slopes and near-continental deep-water trenches account for the deposition of as much as 87% of total organic matter, marginal ocean floor for 10%, and central ocean areas for 3%. Thus pelagic deep-water sediments fail to accumulate any significant masses of OM. The most favourable facies for the accumulation of OM appear to be terrigenous, siliceous and carbonaceous sediments on submarine margins of continents and in some deep-water trenches, which facies are well developed in highly productive oceanic zones.

The study of OM at the sedimentogenesis stage has shown that a broad spectrum of facies is favourable to the accumulation of large masses of OM. These comprise different types of terrigenous, siliceous and carbonaceous reduced sediments of submarine margins and deep-water trenches located in the zones of high biological productivity where thick masses of silts and sands with planktonogenic OM accumulate. This OM is initially characterized by a high degree of bitumenization (β_{chl} up to 15 – 19%) and, at the same time, of humification (up to 85% of OM). The position of these sediments is not very dependent on the distribution of latitudinal climatic zones, but is in many respects controlled by the biological productivity of the photic zone, by the overall degree of balance between the processes of biosynthesis and decay of OM and by the ratio of the rate of supply of organic matter to that of mineral material to the seabed.

It has now for the first time become possible to undertake a system analysis of all principal forms of OM and thus to elaborate the scheme of the overall biogeochemical cycle of OM in the world ocean and to furnish a quantitative estimate of the principal elements of its balance at the sedimentogenesis and early diagenesis stages (Figs. 39 and 67). The chief constituents of the OM cycle in the ocean are: primary OM production (unit of OM synthesis); supply of allochthonous OM from land (unit of OM input from land); decomposition, transformation and transfer of suspended and dissolved OM in the mass of water (unit of OM sedimentational transformations); diagenetic, mainly biogenic, transformations in the process of deposition of OM (unit of OM diagenetic transformations).

The above block-scheme takes account mainly of summary flows of substances at the sedimentogenesis and diagenesis stages and reflects the ultimate result of the processes of transformation of OM in the water mass and in the upper layer of bottom sediments. It visually demonstrates that the biogeochemical processes of OM transformations and accumulation in the sediments are intimately associated with the activity of the organisms inhabiting the ocean, with the cycle of the suspended and dissolved organic matter.

At present the balance of OM in the world ocean appears to be as follows.[7] The primary phytoplankton production (gross) is estimated to be $30-70 \times 10^9$ tons C_{org} year^{-1}, the net production being $(25-40) \times 10^9$ tons C_{org} year^{-1}. The most representative data indicate that the average value for the ocean is close to the given lower limit and can be assumed to be 25×10^{-9} tons C_{org} year. It is the net production (the difference between the gross production and phytoplankton spending for respiration) that characterizes the quantity of OM taking part in the formation and the cycle of dissolved and suspended OM in the seas and oceans.

The primary production of phytobenthonic OM in ocean, according to the latest data, amounts to 0.6×10^9 tons C_{org} year^{-1}, whilst the supply of OM from dry land is 1.0×10^9 tons C_{org} year^{-1}. It has been found that among allochthonous OM sources the most important are river discharge + subterranean runoff and eolian transfer, the latter occasionally consisting almost entirely of OM. The contribution of other sources of allochthonous OM is about 20×10^6 tons C_{org} year^{-1}. Among them even now the main portion of OM is accounted for by anthropogenic pollution.

Altogether in the ocean as much as 26.6×10^9 tons C_{org} year^{-1} is being reproduced annually by phototrophs and supplied from dry land, phytoplankton, phytobenthos and allochthonous OM accouting, correspondingly, for 94%, 2.2% and 3.8% of the total. Thus by taking full account of OM supply with river and subterranean discharges, eolian transfer, ice and wave abrasion material, it has become possible to demonstrate that, although the primary source of OM

7 It is difficult to estimate how accurate are the balance figures given below. Beyond doubt, further research will bring changes of absolute figures; it is, however, unlikely that variations would be greater than 1.5 fold

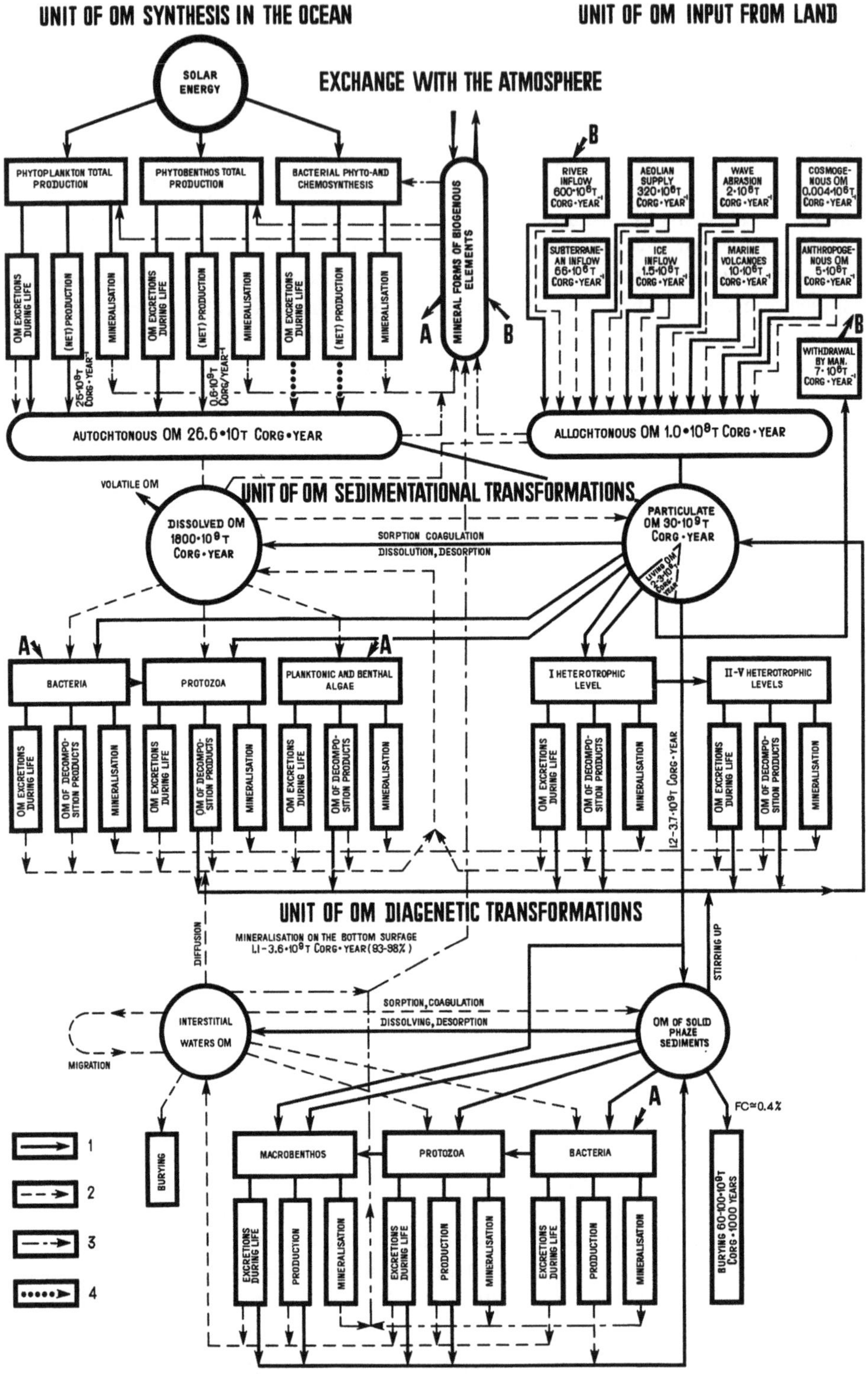

Fig. 67. The general cycle of organic matter in the world ocean. *1* cycling of suspended and precipitated OM; *2* cycling of dissolved OM; *3* cycling of inorganic forms of biogenic elements; *4* partly considered flows; *FC* organic matter fossilization coefficients

in the ocean is phytoplankton, the quantity of allochthonous OM, as well as its effect on the chemical and biological processes taking place in the ocean, had formerly been underestimated. The proportion of OM deriving its origin from phytobenthos in the general OM balance in the ocean is also much larger than was formerly assumed. In the future it seems expedient that we should look for tracers that would enable us to determine directly, in the water mass and in bottom sediments, the relationship between allochthonous and autochthonous OM, including that of phytoplanktonic, phytobenthonic, animal and bacterial genesis. These tracers may be the decomposition products of certain substances, such as lipids, chlorophylls, lignin, and certain saccharides. When using lignin as an indicator of allochthonous OM one must know its quantity and chemical nature in lower algae, as well as in pollen and spore cases.

The analysis of the data obtained has revealed that as much as 86–95% of OM production (net) due to photosynthesis and OM supplied from land disintegrate in the mass of water to mineral compounds. The amount reaching the bottom in particulate form attains $(1.2-3.7)\times10^9$ tons C_{org} or 5–14% of the primary production (net) plus the supply of OM from dry land. On the surface of the sea bottom and in the upper sedimentary layer as much as 92–97% of suspended OM, having reached the bottom, is combusted in chemical reactions, partly dissolves and goes over into the reserve of deep-water OM.

Thus although the absolute masses of OM mineralizing in the mass of water (chiefly within the photic layer) and in the upper sedimentary layer differ by one order, the overall balancing of the two largest systems of OM decomposition is approximately the same.

This enables us to assume that benthonic organisms principally consume the relatively little humified OM arriving at the seabed as part of the suspended material. At the same time, some of them are capable of utilizing the deeply transformed (humified) OM contained in bottom sediments.

The quantitative estimate OM masses mineralizing in the upper sedimentary layer indicates that geothermal heat flow at the ocean floor ($\sim10^{-4}$ cal cm^{-2} min^{-1}) is due, by some 5–15%, to the energy of organic compounds.

From the above data on the quantitative distribution of sedimentary material, including OM, it follows that about 100×10^6 tons C_{org} is buried annually in marine and ocean sediments (the total for the Holocene being about 1×10^{12} tons C_{org}). Of this quantity about 30% are deposited in the marginal and intracontinental seas. The fossilization factors of OM (the proportion of buried OM and the summary reproduction of OM in the seas and oceans plus the supplies from dry land) amount, for the ocean as a whole, to ~0.4%, for submarine margins to ~0.7% and for the ocean floor to ~0.02%.

These data thus reflect the colossal scale of formation, transfer and cycling of OM in the seas and oceans, of which only an insignificant part of the initially phototrophic organic matter and the solar energy bound in it reaches the bottom sediments annually.

The unified character of variability, from the shelves to the central parts of the ocean floor, of phyto- and zooplankton biomasses and production, of the content of POM, of percentage concentrations of C_{org} and its absolute masses in the sediments emphasizes the unity between the physico-geographical environ-

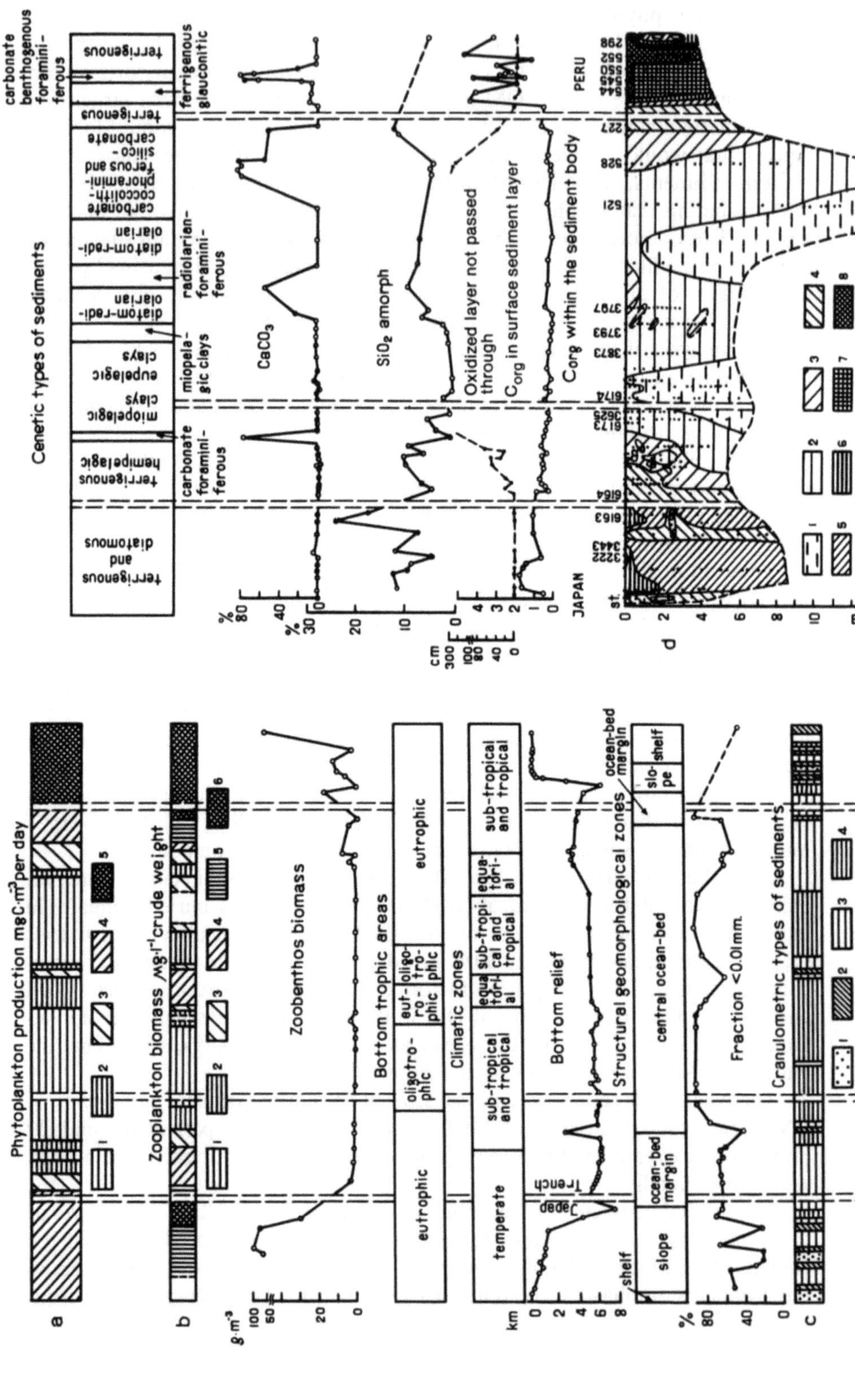
Phytoplankton production mgC·m⁻³per day
Zooplankton biomass mg·l⁻¹crude weight
Zoobenthos biomass
Bottom trophic areas
eutrophic
oligotrophic
Climatic zones
temperate
sub-tropical and tropical
Bottom relief
Japan Trench
Structural geomorphological zones
shelf
slope
ocean-bed margin
central ocean-bed
Fraction <0.01mm
Cranulometric types of sediments
Cenetic types of sediments
terrigenous and diatomous
terrigenous hemipelagic
carbonate foraminiferous
miopelagic clays
eupelagic clays
diatom-radiolarian
radiolarian-foraminiferous
carbonate coccolith-phoraminiferous and silico-carbonate
terrigenous
ferrigenous glauconitic
carbonate benthogenous foraminiferous
CaCO3
SiO2 amorph
Oxidized layer not passed through
Corg in surface sediment layer
Corg within the sediment body
JAPAN
PERU

Fig. 68 a – d

ment and the accumulation of OM in the sediments, which serves as the carrier of transformed space energy. The dependable and, for certain parameters, conjugate character of the variability of some indices is quite apparent from the profile running across the Pacific from the coast of Japan to Peru (Fig. 68). The available data provide corroborating evidence of the presence of a genetically conditioned relationship between the distributions of phytoplankton, its production, POC in the photic zone and in mass of water, heterotrophic microorganisms, benthos, percentage concentrations of OM in the sediments and its absolute masses.

Closely dependent on the quantity of OM arriving at the sea bottom is the quantity of saprophytic bacteria, the distribution of benthonic biomass and of trophic benthos groupings.

The thickness of the upper oxidized layer, the spatial localization and the intensity of oxidation-reduction processes are related to the distribution of the absolute masses of OM (to a lesser extent to its composition). Their characteristic manifestation is the increasing thickness of the oxidized layer in transition from the submarine margin sediments to the hemipelagic ones and further to pelagic sediments proper. Consequently, the environment influences the intensity of diagenetic processes through the absolute masses of OM and the energy confined therein.

Thus, analysis of the data on the quantitative distribution of all principal forms of OM has indicated that in the world ocean there is no complete equilibrium between the production in the sea plus supply from dry land and the decay of OM. Because of the non-closed character of all main metabolic systems (grazing, suspended matter cycle, ecological metabolism), a small part of OM produced in the ocean or arriving from dry land avoids complete decay and becomes buried in bottom sediments.

Owing to spatial irregularity of the primary production and the supply of OM from dry land, to the manner of its utilization by heterotrophic organisms, to differences in the chemical nature and the transport of OM onto the seabed, to the complex effect of the water reservoir morphometry and to the lithologic and chemical properties of the bottom sediments, the degree of balancing between production (and supply from dry land) and destruction differs in different ocean areas. As a result, different absolute masses of OM of a dissimilar chemical composition accumulate on the seabed.

Following the deposition, the processes of biogenic destruction of OM do not cease but continue at a different rate and embrace varying thicknesses of the sediments according to the combined effect of environmental factors, ecological conditions and physiological potentialities of the microorganisms. Consequently,

Fig. 68a – d. The effect of environmental factors on the biogeochemical composition of bottom sediments and the accumulation of organic matter in them along the profile running from Japan to Peru.
a *1* = <2; *2* = 2 – 5; *3* = 5 – 10; *4* = 10 – 100; *5* = >100.
b *1* = <25; *2* = 25 – 50; *3* = 50 – 100; *4* = 100 – 200; *5* = 200 – 500; *6* = >500.
c *1* = sands; *2* = aleurolites; *3* = aleurolitic-pelitic muds; *4* = pelitic muds.
d (% on dry mass basis): *1* = <0.10; *2* = 0.10 – 0.30; *3* = 0.31 – 0.50; *4* = 0.51 – 1.00; *5* = 1.01 – 1.50; *6* = 1.51 – 2.0; *7* = 2.0 – 3.0; *8* = >3.0. For the position of the profile see Fig. 35

the ocean is at present by no means a "waste-free production enterprise" as concerns OM.

Nor was it the case in past epochs. This is testified to by about 150×10^{14} tons of C_{org} confined in the sedimentary rocks of the oceanic and continental sectors of the Earth (Vassoyevich et al. 1976).

In the seas and oceans the composition of dissolved, suspended and precipitated organic material was found to comprise all principal types and many classes of OM. Calculations of correlation links have demonstrated that there is a close direct relationship between the content of C_{org} and of different groups and fractions of OM in the suspended material and in the bottom sediments. Their distribution therefore generally follows the patterns established for suspended and precipitated OM.

In studying the group composition and, in some cases, the molecular and atomic compositions of OM it was found that at the sedimentogenesis and early diagenesis stages its transformation displays a common trend, and that typical process of change in the chemical nature of OM is humification, that is, the predominantly enzymatic depolymerization of biomolecules and the formation of new geopolymers highly stable outside the organisms. In the series plankton – suspended material – surface layer sediments the content of high-molecular protein-like compounds in OM declines 100–200 fold, of amino acids 10–20 fold, of carbohydrate-like compounds 1.2–2 fold, of lipids 2–4 fold, while the content of specific heteropolycondensed structures increases as much as 10–20 times.

In the ocean there are two most pronounced zones where the main masses of OM are transformed, some compounds destroyed and other synthesized, and where the larger part of geopolymers originates. These are the photic layer and the surface sedimentary layer.

The transformation of the elementary composition of POM at the sedimentogenesis stage is expressed in the disintegration of nitrogen- and phosphorus-containing compounds which proceeds at a faster rate than that of OM as a whole. This process also takes place inside the mass of marine and ocean sediments and has been followed up to and including the Miocene. Thus sedimentogenesis and diagenesis are accompanied by a large-scale process involving the differentiation of the organic forms of nitrogen and phosphorus: living OM becomes enriched with them, while POM and OM of bottom sediments are impoverished. The hydrogen content in the OM varies but little in transition from plankton to suspended material and bottom sediments.

An analysis of the abundant material on the distribution and composition of carbohydrate-like compounds in the seas and oceans has shown that in transition from phytoplankton to suspended material and bottom sediments the proportion of water-insoluble polysaccharides genetically interrelated with the highly stable structural carbohydrates of the organisms rises in the composition of the carbohydrate complex. At the sedimentogenesis stage the monomeric composition of carbohydrate-like compounds is subject to restructuring. The carbohydrate complex of suspended material is represented by glucuronomannans similar to the polysaccharides of diatom algae.

The carbohydrate complex of bottom sediments is more diversified than that of seawater and differs from suspended material in having a larger content of ramnose and xylose polymers and occasionally by the absence of glucose. In the sediments the majority of carbohydrate-like compounds enter the composition of geopolymers and are therefore resistant to disintegration.

An important fact, which becomes quite apparent in considering the fate of lipids in the ocean, is that at the sedimentogenesis and early diagenesis stages, the transformation of dissolved, suspended and precipitated lipids in any environment (oxidizing or reducing) tends towards a selective preservation of low-polar compounds including hydrocarbons, fatty acids, occasionally sterols. In the seas and oceans, natural lipids form the following series of increasing stability: bound fatty acids < polar lipids < free fatty acids < fatty alcohols + sterols < hydrocarbons. The content of hydrocarbons in OM of recent marine sediments varies from $<0.1\%$ to 4%. In their composition the content of cyclanes amounts to $\gtrsim 50\%$, alkanes $\lesssim 30\%$, isoalkanes $\sim 2\%$, arenes $\sim 15\%$.

The isotopic composition of carbon in lipids varies in accordance with the magnitudes of thermodynamic isotopic effects and may serve as an indicator of the rate of biogenic transformation of OM. Hydrocarbons and fatty acids are isotopically lighter than polar lipids. In the direction shelf sediments → pelagic sediments, with the growing depth of OM transformation the lipids become more deficient in ^{13}C.

The comparison between the composition of HA from marine and ocean sediments and those from various natural dry land formations has revealed fundamental differences. HA deriving their origin from plankton transformations are distinguished by a higher content of nitrogen, amino acids, carbohydrates, a larger value of the H/C ratio, a lower content of acidic functional groups, of free and methylated hydroxyls, of methoxylic and quinoid groups, as well as by a lower concentration of paramagnetic centres and the presence of distinctly expressed absorption bands of the amide grouping, CH_2-, CH_3-groups, and oxygen aliphatic structures. HA of marine and oceanic sediments with the typically planktonogenic OM lack the specific phenyl-propane structural units of lignin.

These data, along with the analysis of the the distribution of absolute masses of HA and ROM in the ocean, bring us to the conclusion that within the confines of submarine continental margins most of the geopolymers such as HA and FA, as well as ROM formed from them, are formed in situ in the course of transformation of lower algal OM and belong to amicahynic acids. Their formation probably occurs after the melanoidin (amicahyne) reaction. The structure of amicahyne substances is based on the derivatives of the carbohydrate-proteinic biocomplex.

It is clear now, that humic and fulvic acids of bottom sediments and sea water (called "water humus" by B. A. Skopintsev) on the one hand and organic substances of soils and other continental formations on the other, are quite different from chemical and genetic points of view. Geopolymers of varying composition of the first type (of marine genesis) are formed by phytoplankton, phytobenthos and bacteria and are transformed by zooplankton and zoobenthos. Geopolymers of the second type (of substantially continental genesis) are formed by terrestrial plants and are transformed by animals and microflora. The structures of the first type of geopolymers are mostly aliphatic and hydroaromatic, and those of the second condensed aromatic. These two types of OM differ essentially in element composition (H/C, N) as well. Thus it is now necessary to discontinue the use of the terms "humic acids" (general notion) "humic", "fulvic acids" (specific notions) and to use instead "water humus" to indicate marine geopolymers which represent the major part of OM of bottom sediments and sea water (Pocklington 1977, Dawson and Duursma 1981, Vassoyevich 1982). Further support for this point of view is the fact that these terms have lost their clarity of definition because of differing usage, leading to misunderstanding and causing unnecessary difficulties in communications at the interdisciplinary level.

Our increased knowledge of the composition and genesis of geopolymers of bottom sediments and sea water (the latter in lower degree) allows us, following N. B. Vassoyevich, to apply the term "amicahynes" in this case, formed as an acronyme (ami + ca + hy + ne) from two words – amino acids and carbohydrates (Vassoyevich 1982).

Since in the stratisphere (sedimentary cover of the Earth) scattered OM predominates, it is not reasonable to use the classification, presented by an outstanding German geologist and palaeobotanicist H. Potonie. A prominent scientist-encyclopedist, lithologist and petroleum geologist N. B. Vassoyevich has suggested a new classification and terminology of organic substances in recent and fossil sediments, which is worthy of international consideration. The great chemist Antoine Lavoisier noted: science consists of three things – a fact series, that construct science, of ideas, that they create, and of the words, that express them ... and, later ... How reliable the facts are, how correct are the ideas, based on them, they may express only wrong understanding if they are not presented in adequate expressions.

The formation of the principal compositional features of OM does not terminate at the sedimentogenesis stage. At the early diagenesis stage there takes place a major restructuring of its composition and there arise OM geopolymers firmly bound with the mineral basis and more resistant to enzymatic disintegration. It has now been firmly established that the strengthening of the linkup between the OM geopolymers and the mineral basis and the change-over of amicahyne and sapropelic acids into ROM (into "insoluble OM") is one of the principal paths for the preservation of OM in its transition to the fossil state.

Thus, the studies of the alkaline-soluble fraction of OM from marine and ocean sediments have made a significant contribution to the development of the general theory of humification. They have made clear that humification is a highly important process of OM transformation not only on land but also in the world ocean, securing adequate conditions for the growth of the next generations of organisms and for the cycle of OM on the Earth.

The further process of biogeochemistry, organic marine and ocean geochemistry should be based on the following principles reflecting the general development trend of the Earth's sciences. Firstly, it is the application of a more profound system analysis to the cycles of natural and technogenic substances in the seas and oceans, particularly organic compounds of carbon, hydrogen, oxygen, nitrogen, and phosphorus; secondly, undertaking researches on the molecular and atomic (isotopic) level; thirdly, considering the problem of the cycles of elements and their compounds in the world ocean as part of a more universal global system of cycles of substances in the Earth's geospheres. It is to be emphasized that only the studies on the molecular and atomic level lead to a multiple increase of information, to the penetration into the mechanism of transformation of substances, while the consideration of the cycles of substances in the world ocean as an integral part of a larger system enables us to solve problems in the geochemistry of the biosphere and of the Earth's other geospheres closely associated with it.

Practical tasks facing marine and ocean lithology and geochemistry, particularly the protection of seas and oceans against technogenic pollution, call for a speedy solution of the problems in hand.

References

Ackman RG, Eaton CA, Sipos JC, Hooper SN, Castell JD (1970) Lipids and fatty acids of two species of North Atlantic krill (Meganyctiphanes norvegica and Thysanoessa inermis) and their role in the aquatic food web. J Fish Res Bd Can 27:513 – 533

Ackman RG, Hopper SN (1970) Analyses of fatty acids from Newfoundland copepoda and sea water with remarks on the occurrence of arachidic acid. Lipids 5:417 – 421

Agatova AI, Bogdanov YuA (1972) Biochemical composition of particulate organic matter in the Tropical Pacific ocean (in Russian). Okeanologiya 12:267 – 276

Aizatullin TA, Lebedev VL, Khailov KM (1979) The Ocean: the active surfaces and life (in Russian). Gidrometeoizdat, Leningrad

Akulin VN, Karedin EP, Pervuninskaya TA (1975) Lipid fatty acid composition of large Pacific zooplankters (in Russian). Gidrobiol Zh 11:45 – 51

Alam M, Chakravarti A, Ikawa M (1971) Lipid composition of the brown alga Fucus vesiculosus. J Phycol 7:267

Aller RC, Benninger LK (1981) Spatial and temporal patterns of dissolved ammonium, manganese, and silica fluxes from bottom sediments of Long Island Sound, USA. J Mar Res 39:295 – 315

Andersen NR (1977) (ed) Concepts in Marine Organic Chemistry. Marine Chemistry 5 (special Issue)

Anderson DQ (1939) Distribution of organic matter in marine sediments and its availability to further decomposition. J Mar Res 2:225 – 235

Anderson WW, Gehringer JW (1958) Physical oceanographic, biological and chemical data – South Atlantic Coast of the United States. US Fish Wildl Serv Spec Sci Rep Fish 265:1 – 99

Andreyev PF (1956) Bearing on the mechanism of organic matter transformation in anoxic environments of bottom sediments (in Russian). Tr VNIGRI 95:298 – 316

Andreyeva NM, Agatova AI (1981) Organic matter in the bottom sediments of the Posiet Bay (in Russian). Biol morya 2:40 – 49

Andreyeva NM, Agatova AI (1982) Biogeochemical characteristic of the organic matter from the surface active layer of bottom sediments of the Caspian Sea and the Sea of Japan. In: Gershanovich DE (ed) Environmental conditions and the bioproductivity of the sea (in Russian). Legkaya i pishchevaya promyshlennost, Moscow, p 25

Andrusov NI (1892) Some results of the "Chernomorets" expedition (in Russian). Izv Russk Geogr ob-va 28:370 – 397

Andrusov NI (1894) Problems related to the further studies of the Black Sea and surrounding countries (in Russian). Zap Akad Nauk, Ser 8, 1:1 – 10

Andrusov NI (1908) Bearing on the Origin and Occurrence of Petroleum. Tr Bakinsk ORTO, 1 – 2:1 – 26

Antia NJ, Lee CY (1963) Studies on the determination and differential analysis of dissolved carbohydrates in sea-water. Ms Rep Ser Fish Res Bd Can 168

Arkhangelsky AD (1927) The conditions of petroleum formation in the North Caucasus (in Russian). Neftianoye khoz, Moscow Leningrad

Arkhangelsky AD (1927b) The Black Sea sediments and their significance for understanding sedimentary rocks (in Russian). Bull. Mosk. ob-va ispytatelei prirody, otd. geol. 5:199 – 289

Artemiev VE (1973) Particulate carbohydrates in the Pacific Ocean (in Russian). Okeanologiya 13:809 – 813

Artemiev VE (1974) A comparative study of the carbohydrate composition of phytoplankton, particulate matter and bottom sediments in the Ocean (in Russian). Okeanologiya 14:1012 – 1015

Artemiev VE (1976) Carbohydrates in ocean waters and bottom sediments. In: II Volkov (ed) Biogeochemistry of the diagenesis of oceanic bottom sediments (in Russian). Nauka, Moscow, p 20

Artemiev VE (1981) Organic carbon and carbohydrates in river estuaries of the south seas of the USSR (in Russian). Litologiya i poleznyie iskopayemyie 3:142 – 148

Artemiev VE (1983) The biogeochemistry of suspended matter in the profile Limpopo-river mouth-the Mozambique strait. In: Geodekian AA, Trotsiuk VYa, Romankevich EA (eds) Hydrocarbons-genetic research in the Indian Ocean (in Russian). PP Shirshov Inst Oceanol
Artemiev VE, Melnikov IA (1974) Organic matter in the subtropical and antarctic waters of the South Atlantic. In: Filatova ZA (ed) Biological research in the Atlantic sector of the Antarctic (in Russian). Tr Inst Okeanol Akad Nauk SSSR, vol 98. Nauka, Moscow, p 254
Artemiev VE, Romankevich EA (1973) Carbohydrates in bottom sediments and interstitial waters of the Ocean (in Russian). Geokhimiya 12:1871 – 1879
Artemiev VE, Scherbinin AD (1975) On the relationship between dissolved organic carbon distribution and hydrodynamics (in Russian). Okeanologiya 15:272 – 275
Artemiev VE, Krayushkina LN, Romankevich EA (1971) Determination of the total carbohydrate content in oceanic sediments (in Russian). Okeanologiya 11:1125 – 1128
Artemiev VE, Bobyleva NV, Liutsarev SV (1981) Dissolved and particulate organic matter in the sea/river intermixing zone of the Riga bay (in Russian). Okeanologiya 21:286 – 292
Avilov IK (1965) Topography and bottom sediments of the shelf and continental slope of the north-western Atlantic. Tr VNIRO 57:173 – 234
Bada JL, Luyendyk BP, Maynard JB (1970) Marine sediments. Dating by the racemization of amino acids. Science 170:730 – 732
Bader RG (1955) Carbon and nitrogen relations in surface and subsurface marine sediments. Geochim Cosmochim Acta 7:205 – 211
Bader RG, Hood DW, Smith JB (1960) Recovery of dissolved organic matter in sea-water and organic sorption by particulate material. Geochim Cosmochim Acta 19:236 – 243
Balch N (1972) ATP content of Calanus finmarchicus. Limnol Oceanogr 17:606 – 608
Barashkov GK (1972) Comparative biochemistry of algae (in Russian). Pishchevaya prom, Moscow
Barber RT (1968) Dissolved organic carbon from deep water resists microbial oxidation. Nature 220:274 – 275
Barbier M, Joly D, Saliot A, Tourres D (1973) Hydrocarbons from sea water. Deep-Sea Res 20:305
Barlow RG, Swart MJ (1981) Analysis of protein and carbohydrate in particulate matter. Fish Bull S Afr 15:89 – 93
Barrick RC, Hedges JI, Peterson ML (1980) Hydrocarbon geochemistry of Puget Sound region. I. Sedimentary acyclic hydrocarbons. Geochim Cosmochim Acta 44:1349 – 1362
Barskov IS (1975) Biochemical and microstructural methods in application to paleontology. In: Stratigraphy, paleontology (scientific accomplishments), vol 6 (in Russian). Izv VINITI, Moscow, pp 5 – 59
Baturin GN (1978) Phosphorites on the ocean bottom (in Russian). Nauka, Moscow
Baturin GN (1982) Phosphorites on the sea floor. Origin, composition, distribution. Elsevier, Amsterdam London New York
Baturin GN (1983) Biosedimentation of phosphorus in the world ocean. In: Lisitzin AP (ed) Biosedimentation in the seas and oceans (in Russian). Inst Oceanol Akad Nauk SSSR, Moscow, p 83
Baturina MV (1980) Ultrafiltration of sea water. In: Romankevich EA (ed) Methods for studying the organic matter in the ocean (in Russian). Nauka, Moscow, p 21
Baturina MV, Mishustina IE (1976) Electron-microscopic investigations of the organic matter fraction in sea water (in Russian). Izv Akad Nauk SSSR Ser Biol 2:281 – 288
Baturina MV, Liutsarev SV, Romankevich EA (1975) Fractionation of marine aquatic organic matter by ultrafiltration (in Russian). Tr Inst Okeanol Akad Nauk SSSR, vol 102. Nauka, Moscow, p 76
Baturina MV, Mishustina IE, Romankevich EA (1979) Fractionation and electron-microscopic study of dissolved organic matter of sea water. In: Interaction between water and living matter, vol 1 (in Russian). Proceedings of the International Symposium, Odessa 6 – 10 Oct 1975. Nauka, Moscow, p 51
Beardsley GF, Pak Jr H, Carder K, Lingren B (1970) Light scattering and suspended particles in the Eastern Equatorial Pacific Ocean. J Geophys Res 75:2837 – 2845
Beers JR (1966) Studies on the chemical composition of the major zooplankton groups in the Sargasso sea off Bermuda. Limnol Oceangr 11:520 – 528
Beklemishev KV (1954) The nutrition of certain mass planktonic copepods in the Far East seas. Zool Zh 33:1210 – 1230
Beliayev GM (1974) On the age of the deep-sea fauna of the ocean and the ultra-abyssal fauna of ocean trenches (in Russian). Bull Mos Obshchestva Ispyt Prirody, Otd Biol 79:94 – 112

Beliayeva AN (1974) Free lipid composition in bottom sediments of the Western Tropical Pacific and South Atlantic Oceans (in Russian). Okeanologiya 14:77–81

Beliayeva AN (1980a) Method of determination of the group composition of lipids in the sea water, suspended matter and bottom sediments. In: Romankevich EA (ed) Methods for studying the organic matter in the ocean (in Russian). Nauka, Moscow, p 143

Beliayeva AN (1980b) Extraction of lipids from marine and interstitial waters and bottom sediments. In: Romankevich EA (ed) Methods for studying the organic matter in the ocean (in Russian). Nauka, Moscow, p 129

Beliayeva AN (1980c) A method for establishing the concentration of lipids in sea water, suspended matter and bottom sediments. In: Romankevich EA (ed) Methods for studying the organic matter in the ocean (in Russian). Nauka, Moscow, p 139

Beliayeva AN, Romankevich EA (1976) Chemical transformations of lipids in the process of sedimentogenesis. In: Vassoyevich NB, Timofeev PP (eds) Investigations of organic matter in recent and fossil sediments (in Russian). Nauka, Moscow, p 81

Beliayeva AN, Romankevich EA (1983) Organic matter in the Barents sea water column (in Russian). Okeanologiya 23:255–263

Beliayeva NV (1962) The distribution of plankton Foraminifera in the water masses of the Indian ocean. Biul Mosk ob-va ispytatelei prirody. otd geol 37:98–101

Belov NA, Lapina NN (1961) Bottom sediments of the Arctic Basin (in Russian). Morskoi transport, Leningrad

Benson AA, Lee RF (1975) The role of wax in oceanic food chains. Sci Am 3:77–83

Bezrukov PL (1960) Bottom sediments of the Okhotsk Sea (in Russian). Tr Inst Okeanol Akad Nauk SSSR, 32:15

Bezrukov PL (1964a) Zonality and irregularity of oceanic sedimentation. In: Modern Problems of Geography (in Russian). Nauka, Moscow, p 249

Bezrukov PL (1964b) Sedimentation in the Northern and Central Indian Ocean. – Ocean and sea bottom geology. Papers by Soviet geologists at the XXII session of the International Geological Congress (in Russian). Nauka, Moscow, p 41

Bezrukov PL (1971) The problems of the geological exploration of the ocean's mineral resources (in Russian). Okeanologiya 11:969–981

Bezrukov PL (1973) Geological testing grounds and their significance for the study of ocean floor geology (in Russian). Okeanologiya 13:280–283

Bezrukov PL, Lisitzin AP (1975) Organic carbon distribution in bottom sediments. – Geological-geophysical atlas of the Indian Ocean (in Russian). Izv Glavn Upr Geodezii i Kartografii, Moscow

Bezrukov PL, Romankevich EA (1970) Rate of sedimentation in the Pacific Ocean. In: Pacific Ocean, vol 2, Pacific Sedimentation. Nauka, Moscow, p 288

Bezrukov PL, Murdmaa IO, Saidova KM, Filatova ZA (1958) On the bottom sediments and benthic fauna of the North-Eastern China Sea (in Russian). Oceanol Limnol Sin 1:269–315

Bezrukov PL, Lisitzin AP, Romankevich EA, Skornyakova NS (1961) Contemporary sedimentation in the North Pacific. In: Strakhov NM (ed) Recent sediments in seas and oceans (in Russian). Izv Akad Nauk SSSR, Moscow, p 98

Bezrukov PL, Petelina VP, Skorniakova NS (1970) The mineral resources of the ocean. In: Bezrukov PL (ed) Sedimentation in the Pacific Ocean, vol 6 (in Russian). Nauka, Moscow, book 2 p 322

Biogéochimie de la matière organique à l'interface eau-sédiment marine (1980) In: Colloques Internationaux du CNRS N 293. Ed CNRS, Paris

Blackhurn TH (1980) Seasonal variations in the rate of organic-N mineralization in anoxic marine sediments. In: Colloques, Internationaux du CNRS, N 293 – Biogéochimie de la matière organique à l'interface eau-sédiment marin. Publ CNRS, Paris, p 173

Blazhchishin AI (1972) Bottom sediments of the Bay of Biscay (in Russian). Okeanologiya 12:277–290

Blazhchishin AI (1976) Major chemical constituents of bottom sediments. In: Gudelis V, Emelyanov E (eds) The Geology of the Baltic Sea (in Russian). Mokslas, Vilnius, p 255

Blazhchishin AI, Emelianov EM (1981) Absolute masses of Holocene sediments in the Atlantic ocean (in Russian). Okeanologiya 21:84–88

Blinova EI, Vozzhinskaya VB (1971) Marine macrophytes and plant resources of the ocean. In: Beklemischev KV (ed) Fundamentals of biological productivity of the ocean and its utilization (in Russian). Nauka, Moscow, p 137

Blumer M, Guillard RRL, Chase T (1971) Hydrocarbons of marine phytoplankton. Mar Biol 8:183–189
Bobyleva NB, Beliayeva AN (1982) The Genesis and composition of lipids in suspended matter and bottom sediments of the Baltic sea (in Russian). Okeanologiya 22:757–763
Bogdanov YuA, Levitan MA (1980) The history of accumulation of organic carbon. In: Monin AS, Lisitsin AP (eds) Oceanology – Geology of the Ocean – Geological history (in Russian). Nauka, Moscow, p 252
Bogdanov YuA, Lisitzin AP (1968) Distribution and composition of particulate organic matter in the Pacific Ocean waters (in Russian). Okeanol Issl, No 18, p 75–155
Bogdanov YuA, Ovchinnikova LI (1965) To the analytical determination of bituminous substances in the particulate fraction. Okeanologiya 5:366–372
Bogdanov YuA, Romankevich EA (1976) The evolution of the accumulation of organic carbon in the Pacific during the Late Cretaceous and the Cainozoic. In: Vassoevich NB (ed) Abstracts of papers for the 5th USSR seminar on organic matter in recent and fossil sediments (in Russian). Moscow University Press, p 43–44
Bogdanov YuA, Shaposhnikova MG (1970) Particulate organic matter in the suspended material of the Kuril-Kamchatka Basin (in Russian). Tr Inst Okeanol Akad Nauk SSSR, 86:46–58
Bogdanov YuA, Lisitzin AP, Romankevich EA (1971) Organic matter in the suspended matter and bottom sediments of seas and oceans. In: Vassoevich NB (ed) Organic matter in recent and fossil sediments (in Russian). Nauka, Moscow, p 35
Bogdanov YuA, Lisitzin AP, Gurvich EG (1979) A model of accumulation of organic carbon in the Pacific sediments (in Russian). Geokhimiya 6:918–927
Bogorov VG (1960) Geographical variations in the plankton fatty composition in the ocean. Dokl Akad Nauk SSSR, 134:1441–1442
Bogorov VG (1967) Biological transformation, energy and substance exchange in the ocean (in Russian). Okeanologiya 7:839–859
Bogorov VG (1971) On the quantity of materials in living organisms of the world ocean. In: Vassoyevich NB (ed) Organic Matter in recent and fossil sediments (in Russian). Nauka, Moscow, p 12
Bogorov VG, Bordovsky OK, Vinogradov ME (1966) Biogeochemistry of oceanic plankton. Distribution of certain chemical constituents of plankton in the Indian Ocean (in Russian). Okeanologiya 6:314–325
Bohling H (1970) Untersuchungen über freie gelöste Aminosäuren in Meereswasser. Mar Biol 6:213–225
Bojanowski R, Pempkowiak J (1980) Preliminary results of the study of humic substances in the bottom deposits of the southern Baltic. In: Okolotowicz G (ed) Ecosystems of the Baltic, part 1 (in Russian). Inst Mar Fish, Gdynia, p 128
Bolin B, Degens ET, Duvigneaud P, Kempe S (1977) The Global Biogeochemical Carbon Cycle. In: Bolin B, Degens ET, Kempe S, Ketner P (eds) The global carbon cycle. John Wiley, Chichester New York Brisbane Toronto, p 1
Bordovsky OK (1964) Organic matter accumulation and transformation in marine sediments (in Russian). Nedra, Moscow
Bordovsky OK (1968) Organic matter in the East Antarctic iceberg sediments (in Russian). Okeanologiya 8:69–77
Bordovsky OK (1969) Organic matter in recent sediments of the Caspian Sea (in Russian). Okeanologiya 9:996–1007
Bordovsky OK (1974) Organic matter in sediments of seas and oceans at the early diagenesis stage (in Russian). Nauka, Moscow
Bordovsky OK, Skolova MN, Smirnov BA, Akhmetieva EA, Zezina ON (1974) Bearing on the role of benthic populations in the transformation of organic matter in sediments (with special reference to the deep water detritus-feeders of the Kuril-Kamchatka Trench) (in Russian). Okeanologiya 14:161–166
Bordovsky OK, Akhmetieva EA, Nemirovskaya IA (1981) Molecular composition of the dissolved organic matter of sea-water and bottom sediments solutions. In: Bordovsky OK (ed) Biohydrochemistry of the Northwestern Indian Ocean (in Russian). Nauka, Moscow, p 102
Bottino NR (1972) Fatty acid exchange among trophic levels of the Ross sea: phytoplankton, copepods and euphausiids. Antarct JUS 7:178

Boussuge C, Gout M, Saliot A, Tissier MJ (1979) Acides gras et hydrocarbures aux interfaces eau de mer-sédiment et eau interstitielle – sédiment en Atlantique tropical est. In: Géochimie organique des sédiments marins profonds. Ed CNRS, Paris, p 304

Boussuge C, Gout M, Tissier MJ, Tusseau D, Saliot A (1981) Sédimentation organique en mer d'Arabie: biologie et transfer de lipides entre zone euphotique et interface océan – sédiment. In: Géochimie organique des sédiments marins profonds. Ed CNRS, Paris, p 415

Bray EE, Evans ED (1965) Hydrocarbons in non-reservoir rock source beds. Bull Am Assoc Petr Geol 49:248 – 257

Breck WG (1978) Biomonitors of aquatic pollutants in time and space. Thalassia Jugosl 14:157 – 170

Brooks J, Shaw G (eds) (1971) Sporopollenin. Academic, London New York

Brown M (1975) High-molecular weight materials in Baltic seawater. Mar Chem 3:253 – 258

Bruyevich SV (1953) Chemistry and biological productivity of the Black Sea (in Russian). Tr Inst Okeanol Akad Nauk SSSR, vol 7, p 12 – 56

Bruyevich SV, Gudkov MP (1954) Atmospheric dust over the Caspian Sea (in Russian). Izv Akad Nauk SSSR, Ser Geogr, 4:18 – 28

Burkov VA (1980) General circulation of the world ocean (in Russian). Gidrometeoizdat, Leningrad

Burney CM, Sieburth JMcN (1977) Dissolved carbohydrates in sea-water II. A spectrophotometric procedure for total carbohydrate analysis and polysaccharide estimation. Mar Chem 5:15 – 28

Burney CM, Johnson KM, Sieburth JMcN (1981) Diel flux of dissolved carbohydrate in a salt marsh and a simulated estuarine ecosystem. Mar Biol 63:175 – 187

Burns KA, Vileneve J-P (1983) Biogeochemical processes affecting the distribution and vertical transport of hydrocarbon residues in the coastal Mediterranean. Geoch Cosm Acta 47:995 – 1006

Bursa AS (1968) Starch in the oceans. J Fish Res Bd Can 25:1269 – 1284

Calvert SE, Price NB (1971) Recent sediments of the South West African shelf. In: Delany FM (ed) The geology of the East Atlantic continental margin. Inst Geol Sci Rep 70/16:171

Caratini C, Bellet J, Tissot C (1978) Etude microscopique de la matière organique: palynologie et palynofaciès. In: Géochimie organique des sédiments marins profonds. Orgon II. Atlantique – NE Brésil. Ed CNRS, Paris, p 157

Caratini C, Bellet J, Tissot C (1979) Étude microscopique de la matière organique: palynologie et palynofaciès. In: Géochimie organique des sédiments marine profonds. Orgon III Mauritanie-Sénégal-iles du Cap-Vert. Ed CNRS, Paris, p 215

Caratini C, Bellet J, Tissot C (1981) Etude microscopique de la matière organique: palynologie et palynofaciès. In: Géochimie organique des sédiments marins profonds. Orgon IV. Golfe d'Aden, d'Oman. Ed CNRS, Paris, p 265

Cauwet G (1981) Non-living particulate matter. In: Duursma EK, Dawson R (eds) Marine Organic Chemistry. Elsevier, Amsterdam Oxford New York, p 71

Chesselet R (1980) Modes of settling and organic input to the sediment-seawater interface: a review. In: Colloques Internationaux du CNRS, No 293 – Biogéochimie de la matière organique à l'interface eau-sédiment marin. Ed CNRS, Paris, p 27

Chester R, Stoner JH (1974) The distribution of particulate organic carbon and nitrogen in some surface waters of the world ocean. Mar Chem 2:263 – 276

Chetverikova OP, Pentina TYu (1982) On the generation and composition of the major part of scattered organic matter. In: Sidorenko AV, Vassoyevich NB (eds) Geochemistry of recent and fossil sediments (in Russian). Nauka, Moscow, p 96

Chetverikova OP, Rodionova KF, Viktorova NS, Chutkerashvili SE (1972) Residual organic matter (in Russian). Tr VNIGNI 85:25 – 45

Chetverikova OP, Pentina TYu, Dubovik VI (1974) On the structure of the major part of scattered organic matter in sedimentary rocks. In: Vassoyevich NB (ed) Organic Matter in recent and fossil sediments and methods of its investigation (in Russian). Nauka, Moscow, p 268

Childress JJ, Nygaard (1974) Chemical composition and buoyancy of midwater crustaceans as function of depth of occurrence off Southern California. Mar Biol 27:225 – 238

Chudayeva VA (1980) The organic matter of river suspended material. In: Mikhailov MA (ed) Geochemistry and petrochemistry of sedimentary complexes of the Far East (in Russian). Far East Science Centre Academy of Sciences of the USSR, Vladivostok, p 128

Chuecas L, Riley IP (1969) Component fatty acids of total lipids of some marine phytoplankton. J Mar Biol UK 49:97 – 116

Clarke FW (1924) The data of geochemistry. 5th ed Washington Government printing office

Clarke HT, Mazur A (1941) The lipids of diatoms. J Biol Chem 141:283
Coastal Upwelling: its sediment record (1981), Symposium. Abstr Pap Adv Res Inst, Vilamoura/ Algarve, Portugal
Collier A (1958a) Some biochemical aspects of red tides and related oceanographic problems. Limnol Oceanogr 3:33–39
Collier A (1958b) Gulf of Mexico physical and chemical data from Alaska cruises. US Fish Wildl Serv Spec Sci Rep Fish, p 249
Collier A, Ray SM, Magnitzky AW (1950) A preliminary note on naturally occurring substances in seawater affecting the feeding of oysters. Science 111:151–152
Collier A, Ray SM, Magnitzky AW, Bell JO (1953) Effect of dissolved organic substances on oysters. US Fish Wildl Serv Bull 84:167–185
Collins KJ, Williams PJ (1977) An automated photochemical method for the determination of dissolved organic carbon in sea and estuarine water. Mar Chem 5:123–141
Collyer DM, Fogg GE (1955) Studies on fat accumulation by algae. J Exp Bot 6:256–275
Combaut G, Codomier L, Teste J (1981) Seasonal chemical evolution of the alga Cystoseira elegans. Phytochemistry 20:2036–2037
Combaz A, Bellet J, Poulain D, Caratini C, Tissot C (1977) Etude microscopique de la matière organique de sédiments quaternaires de mer de Norvège. In: Géochimie organique des sédiments marins profonds. Orgon I. Mer de Norvège. Ed CNRS, Paris, p 139
Conover RJ (1961) The turnover of phosphorus by Calanus finmarchicus. J Mar Biol Assoc UK 41:484–488
Conover RJ (1978) Transformation of organic matter. In: Kinne O (ed) Marine Ecology, vol IV Dynamics. Wiley, Chichester New York Brisbane Toronto, p 221
Copin-Montegut, C, Copin-Montegut G (1973) Comparison between two processes of determination of particulate organic carbon in sea water. Mar Chem 1:151–156
Copin-Montegut C, Copin-Montegut G (1978) The chemistry of particulate matter from the south Indian and Antarctic oceans. Deep-Sea Res 25:911–931
Coughenower DD, Herbert CC (1975) An automated technique for total dissolved free amino acids in seawater. Limnol Oceanogr 20:128–131
Cowell EB (1976) Oil pollution of the sea. In: Johnston R (ed) Marine Pollution. Academic, London New York San Francisco
Craig H (1971) The deep metabolism: oxygen consumption in abyssal ocean water. J Geophys Res 76:5078–5086
Culkin F, Morris RJ (1969) The fatty acids of some marine crustaceans. Deep-Sea Res 16: 109–116
Culkin F, Morris RJ (1970) The fatty acids of some cephalopods. Deep-Sea Res 17:171–174
Dal Pont G, Newell B (1963) Suspended organic matter in the Tasman sea. Aust J Mar Freshwater Res 14:155–163
Daniushevskaya AI, Matveyeva VA (1979) Humic acids and amino acids as the reflection of the genesis and transformation level of the scattered organic matter. In: Vassoyevich NB, Timofeyev PP (eds) Accumulation and transformation of *sedicahytes* (in Russian). Nauka, Moscow, p 65
Daniushevskaya AI, Romankevich EA (1982) Organic matter in suspensions and bottom deposits of the Baltic Sea. In: Vassoyevich NB, Timofeyev PP, Nazarevich IA (ed) *Sedicahytes* at various lithogenesis stages (in Russian). Nauka, Moscow, p 19
Daniushevskaya AI, Dibner VD, Shkatov EP, Ragozina EA, Glebovskaya EA, Kulikova EM, Norenkova IK, Shchelovanov YuS (1978) Complex geochemical research of bottom sediments of the arctic shelf (in Russian). Okeanologiya 18:70–78
Daniushevskaya AI, Yashin DS, Kirillov OV (1980) Geochemical stipulations of the distribution of organic carbon in the bottom sediments of the Arctic Seas (in Russian). Okeanologiya 20:281–289
Dastillung M, Corbet B (1978) Hydrocarbures saturés et insaturés des sédiments. In: Géochimie organique des sédiments marins profonds. Orgon II. Atlantique NE Brésil. Ed CNRS, Paris, p 293
Datsko VG (1959) Organic matter in the waters of the USSR Southern Seas (in Russian). Izv Akad Nauk USSR, Moscow
Datsko VG, Datsko VE (1950) Microquantitative determination of organic carbon in seas and fresh waters. Tr AzCher NIRO 14:263–270
Daumas RA (1976) Variations of particulate proteins and dissolved amino acids in coastal seawater Mar Chem 4:225–242

Daumas R, Laborde P, Paul R, Romano JC et Sautriot D (1978) Les mécanismes de transformation de la matière organique en Atlantique intertropical. Etude de la minéralisation et de la diagenèse dans les sédiments superficiels. In: Géochimie organique des sédiments marins profonds. Orgon II. Atlantique – NE Brésil. Ed CNRS, Paris, p 43

Dawson R, Duursma EK (1981) State of the art. In: Marine organic chemistry. Elsevier, Amsterdam Oxford New York, p 497

Degens ET (1965) Geochemistry of sediments. A brief summary. Prentice-Hall, Englewood Cliffs, New Jersey

Degens ET (1967) Geochemistry of sediments. MIR, Moscow (in Russian, transl from English)

Degens ET (1970) Molecular nature of nitrogenous compounds in sea water and recent marine sediments. In: Organic matter in natural waters. Inst Mar Sci Univ Alaska, p 77

Degens ET (1971) Early and late diagenesis of organic matter. In: Larsen G, Chilingar CV (eds) Diagenesis in Sediments (in Russian, transl from English), MIR, Moscow, p 307

Degens ET (1976) Molecular mechanisms of carbonate, phosphate, and silica deposition in the living cell. In: Topics in Current Chemistry, vol 64. Springer, Berlin Heidelberg New York, p 1

Degens ET (1979) Carbon in the sea. Nature 279:191 – 192

Degens ET, Mopper K (1976) Factors controlling the distribution and early diagenesis of organic material in marine sediments. In: Riley JP, Chester R (eds) Chemical Oceanography, 2nd ed, vol 6. Academic, London New York San Francisco, p 59

Degens ET, Ittekkot V (1983) The carbon cycle – tracking the path of organic particles from sea to sediments. In: Brooks J, Flett A (eds) Proceeding of the special Conference on Marine Petroleum Source Rocks (May 17th – 18th 1983) at the Royal Society, London. Geol Soc London Blackwell Sci Publ p 2

Degens ET, Reuter JH, Shau KNF (1964) Biochemical compounds in off-shore California sediments and sea waters. Geochim Cosmochim Acta 28:45 – 66

Derenbach JB (1969) Partikuläre Substanz und Plankton anhand chemischer und biologischer Daten gemessen in den oberen Wasserschichten des Gotlandes-Tief im Mai 1968. Kiel Meeresforsch 25:279 – 289

De Vooys CGN (1977) Primary production in aquatic environments. In: Bolin B, Degens ET, Kempe S, Ketner P (eds) The Global Carbon Cycle. SCOPE 13. Wiley, Chichester New York Brisbane Toronto, p 259

Didyk BM, Simoneit BTR, Brassel SC, Eglinton G (1978) Organic geochemical indicators of paleoenvironmental conditions of sedimentation. Nature 272:216 – 222

Diester-Haass L, Müller PJ (1979) Processes influencing sand fraction composition and organic matter content in surface sediments off W Africa (12 – 19 °N). Meteor Forschungsergebnisse, Reihe C 31:21 – 47

Dietrich R, Hohnk W, Manzel WD (1966) Studien zur Chemie ozeanischer Bodenproben, IV Veröff Inst Meeresforsch Bremerhaven 10:121 – 136

Djamalov RG, Zektser IS, Meskheteli AB (1977) Subterranean run-off into the seas and the world ocean (in Russian). Nauka, Moscow

Drozdova TV (1977) The geochemistry of the amino acids (in Russian). Nauka, Moscow

Duursma EK (1961) Dissolved organic carbon, nitrogen and phosphorus in the sea. Neth J Sea Res 1:1 – 148

Duursma EK (1965) The dissolved organic constituents of sea water. In: Riley JP, Skirrow G (eds) Chemical Oceanography, vol 1. Academic, New York, p 433

Duursma EK, Dawson R (eds) (1981) Marine Organic Chemistry. Evolution, Composition, Interactions and Chemistry of Organic Matter in Seawater. Elsevier, Amsterdam Oxford New York

Ehrhardt M (1969) A new method for the automatic measurement of dissolved organic carbon in sea water. Deep-Sea Res 16:393 – 396

El Wakeel SK, Riley JP (1965) Chemical and mineralogical investigations of deep water sediments. In: Vdovenko VM Problems of geochemistry and geochronology of the ocean (in Russian, transl from English). MIR, Moscow

Emelianov EM (1973 a) Major types of recent bottom sediments in the Mediterranean, their mineralogy and geochemistry (in Russian). Litol I Polezn Iskop 1:29 – 46

Emelianov EM (1973 b) Distribution and composition of muds of the Southwestern African shelf (in Russian). Tr Inst Okeanol Akad Nauk SSSR 95:211 – 238

Emelianov EM (1975) Organic carbon in the Atlantic Ocean Sediments (in Russian). Dokl Akad Nauk SSSR 220:1186–1189
Emelianov EM, Romankevich EA (1977) Atlantic Ocean geochemistry. Organic matter and phosphorus (in Russian). Nauka, Moscow
Emelianov EM, Romankevich EA (1979) Geochemistry of the Atlantic ocean. The organic matter and phosphorus (in Russian). Nauka, Moscow
Emelianov EM, Senin YuM (1969) Composition specificities of bottom sediments of the Southwestern African Shelf (in Russian). Litol Polezn Iskop 2:10–25
Emery KO (1956) Sediments and water of Persian Gulf. Bull Am Assoc Petr Geol 40:2354–2383
Emery KO (1960) The sea off Southern California. Wiley, New York London
Emery KO (1964) Sediments of Gulf of Aqaba (Eilat). In: Miller RL (ed) Papers in Marine Geology: Shepard Commemorative Volume, New York London, p 257
Emery KO (1970) Continental margins of the World. I. General and Economic Papers, ICSU/SCOR Working Party 31, Sympos Cambridge, Rep N 70/13
Emery KO, Niino H (1963) Sediments of the Gulf of Thailand and adjacent continental shelf. Bull Geol Soc Am 74:541–554
Emergy KO, Rittenberg SC (1952) Early diagenesis of California basin sediments in relations to origin of oil. Bull Am Assoc Petr Geol 35:2174
Emery KO, Ross DA (1968) Topography and sediments of small area of the continental slope south of Marth's Vineyard. Deep-Sea Res 15:415–422
Engler K (1888) On the origin of petroleum (in Russian). J Gornyi 3:306–365
Eppley RW, Carlucci AF, Holm-Hansen O, Kiefer D, McCarthy JJ, Venrick E, Williams PM (1971) Phytoplankton growth and composition in shipboard cultures supplied with nitrate, ammonium, or urea as the nitrogen source. Limnol Oceanogr 16:741–751
Erdman JG, Marlett EM, Hanson WE (1956) Survival of amino acids in marine sediments. Science 124:1026
Filatova ZA (1977) Deep-sea bottom fauna. In: Vinogradov ME (ed) Oceanology. Biology of the ocean, vol 1 (in Russian). Nauka, Moscow, p 88
Finenko ZZ (1965) Organic matter content of sestone in the Black and Azov Seas. In: Vodianitsky VA (ed) Plankton investigations in the Black and Azov seas (in Russian). Nauk Dumka, Kiev, p 12
Finenko ZZ, Ostapenia AP (1971) Vertical distribution of organic matter in the tropical waters of the Southern Pacific. In: Vinogradov ME (ed) Functions of pelagial communities in tropical areas of the ocean (in Russian). Nauka, Moscow, p 241
Finenko ZZ, Zaïka VE (1970) Particulate organic matter and its role in the productivity of the sea. In: Steele JH (ed) Marine food chains. Oliver Boyd, Edinburgh, p 32
Florovskaya VN (1957) The luminescence-bituminological method in the oil geology (in Russian). Moscow University, Moscow
Florovskaya VN, Gursky YuN (1970) Specificities of organic matter composition in deep water sediments of the Black Sea. In: Chemical resources of seas and oceans (in Russian). Nauka, Moscow, p 226
Fogg GE (1956) Photosynthesis and formation of fats in a diatom. Ann Bot NS, 20:265–285
Fournier RO (1971) Light and electron microscopy of some aphotic oceanic microorganisms. Proc Joint Oceanol Assoc, The Ocean World, Tokyo
Fox DF, Oppenheimer CH, Kitterge JS (1953) Microfiltration in oceanographic research. J Mar Res 12:233–243
Fox DL, Updegraff DM, Novelli GD (1944) Carotenoid pigments in the ocean floor. Arch Biochem 5:1–23
Gadel F (1980) Géochimie des composés humiques dans les dépots superficiels. In: Colloques Internationaux du CNRS N 293 – Biogéochimie de la matière organique à l'interface eau-sédiment marine. Ed CNRS, Paris, p 45
Gagosian RB (1975) Sterols in the western Atlantic Ocean. Geochim Cosmochim Acta 39:1443–1454
Gagosian RB (1980) Transformation reactions of biogenic organic compounds at the sediment-water interface. Colloques Internationaux du CNRS N 293 – Biogéochimie de la matière organique à l'interface eau-sédiment marin. Ed CNRS, Paris, p 211
Gagosian RB, Lee C (1981) Processes controlling the distribution of biogenic organic compounds in seawater. In: Duursma KE, Dawson R (eds) Marine organic chemistry. Elsevier, Amsterdam Oxford New York, p 91

Gagosian RB, Stuermer DH (1977) The cycling of biogenic compounds and their diagenetically transformed products in seawater. Mar Chem 5:605 – 632
Gagosian RB, Nigrelli GE, Volkman JK (1983) Vertical transport and transformation of biogenic organic compounds from a sediment trap experiment off the coast of Peru. In: Suess E, Thiede J (eds) Coastal Upwelling: Its Sediment Record Part A, Plenum Press, New York London, p 241
Galimov EM (1968) Geochemistry of stable carbon isotopes (in Russian). Nedra, Moscow
Galimov EM (1981) The nature of the biological fractioning of the isotopes (in Russian). Nauka, Moscow
Galwey AK (1972) The rate of hydrocarbon desorption from mineral surfaces and the contribution of heterogeneous catalytic-type processes to petroleum genesis. Geochim Cosmochim Acta 36:1115
Garret WD (1967) The organic chemical composition of the ocean surface. Deep-Sea Res 14:221 – 228
Garret WD (1970) Organic chemistry of natural sea surface films. In: Hood DW (ed) Organic matter in natural waters. Inst Mar Sci Univ Alaska, p 469
Gastaud JM (1973) Comparaisons biochimiques des fractions unsaponifiables de zooplankton des mers chaudes et froides. Rapp et proc-verb reuna Commiss internat explor sci Mer Mediterr, Monaco 22:17 – 18
Geinrikh AK (1971) Productivity cycles in marine pelagial areas. In: Beklemishev KV (ed) Fundamentals of the biological productivity in the ocean and its utilization (in Russian). Nauka, Moscow, p 43
The geochemical activity of microorganisms in the bottom sediments of the Pacific ocean (1981) Ivanov MV (ed). Izv nauchnogo tsentra biol issl, Pushchino (in Russian)
Géochimie organique des sédiments marins profonds. Orgon I, Mer de Norvège (1977). Ed CNRS, Paris
Géochimie organique des sédiments marins profonds. Orgon II, Atlantique – NE Brésil (1978) Combaz A, Pelet R (eds) Ed CNRS, Paris
Geodekyan AA (1968) Geological-geochemical specificity of petroleum formation in the South Caspian basin (in Russian). Nedra, Moscow
Germanov AI, Borzenkov IA, Yusupov IF (1981) The biogenic processes of sulphate reduction and methane formation in the system subterranean waters – rocks (in Russian). Sov Geol 8:116 – 121
Gershanovich DE (1956) Silica acids, calcium carbonate and organic carbon in deep-sea sediments of the Sea of Japan (in Russian). Tr GOIN 31(43):72 – 79
Gershanovich DE (1965) Major results of marine geological investigations in the Bering Sea, in the Gulf of Alaska. In: Belousov IM (ed) IGY Oceanological Investigations, N 13 (in Russian). Nauka, Moscow
Gershanovich DE (1981) Zonalities in the distribution of coastal upwellings sediments in the Ocean. In: Lisitzin AP, Gershanovich DE (eds) The climatic zonality and sedimentation (in Russian). Nauka, Moscow, p 73
Gershanovich DE, Koniukhov AI (1980) Recent sedimentation in the Peruvian current area. In: Timofeyev PP (ed) Current problems in the geology of seas and oceans (in Russian). Nauka, Moscow, p 48
Gershanovich DE, Muromtsev AM (1982) Oceanographic foundations of the biological productivity of the Ocean (in Russian). Gidrometeoizdat, Leningrad
Gershanovich DE, Neiman AA (1964) Bottom sediments and benthic faunas of the East China Sea (in Russian). Okeanologiya 4:1089 – 1095
Gershanovich DE, Avilov IK, Zarikhin IP (1972) Bottom sediments of the South Atlantic continental margins (in Russian). Estimated productivity of the world ocean, N 1. Pischevaya promishlennost, Moscow
Gershanovich DE, Gorshkova TI, Koniukhov AI (1974) Organic matter in recent sediments of continental margins. In: Organic matter in recent and fossil sediments and methods of its investigation (in Russian). Nauka, Moscow, p 63
Ginzburg-Karagicheva TL, Prianishnikov ND, Rodionova KF (1934) Some data on the microbiological and chemical analysis of Black Sea deep-bottom slimes (in Russian). Mikrobiologiya 3:513 – 522
Gjiessing ET (1970) Ultrafiltration of aquatic humus. Environ Sci Technol 4:437 – 438
Glagoleva MA (1959) Forms of element migration in river waters. In: Strakhov NM (ed) To elucidation of sediment diagenesis (in Russian). Izv Akad Nauk SSSR, Moscow, p 5 – 28
Glagoleva MA (1961) Contribution to the geochemistry of the Black Sea sediments. In: Strakhov NM, Bezrukov PL, Jablokov VS (eds) Recent sediments in seas and oceans (in Russian). Izv Akad Nauk SSSR, Moscow, p 448

The Global Carbon Cycle (1977) Bolin B, Degens ET, Kempe S, Ketner P (eds). Wiley, Chichester New York Brisbane Toronto

Goldberg ED (1975) Marine pollution. In: Riley JP, Skirrow G (eds) Chemical oceanography, 2nd ed, vol 3. Academic, London New York San Francisco, p 39

Goldberg ED, Walker TJ, Whisenand A (1951) Phosphate utilization by diatoms. Biol Bull 101:274–284

Golterman HL (1972) The role of phytoplankton in detritus formation. Mem Ist Ital Idrobiol, "Dott M Marchi" 29:89–103

Goncharova TO, Kolosov IV (1980) The role of acidic-basic properties of clayey minerals and fulvic acids in the sorption balance fulvic acids-clayey minerals'. Gidrokhim. Materialy 68: 55–65

Gorbunova ZN, Lisitzin AP (1979) Clayey minerals. In: Bezrukov PL (ed) Geology of the ocean. Sedimentation and magmatism of the ocean (in Russian). Ser Okeanologiya, Nauka, Moscow, p 214

Gordeyev EI (1968) Chemical composition of the surface water particulate fraction in the Northern and Central Indian Ocean. In: Lisitzin AP (ed) Oceanological investigations, N 18 (in Russian). Nauka, Moscow, p 175

Gordeyev VV (1983) River input in the ocean and some features of its geochemistry (in Russian). Nauka, Moscow

Gordon DC (1970a) Some studies on the distribution and composition of particulate organic carbon in the North Atlantic Ocean. Deep-Sea Res 17:233–243

Gordon DC (1970b) A microscopic study of organic particles in the North Atlantic Ocean. Deep-Sea Res 17:175–185

Gordon DC (1971) Distribution of particulate carbon and nitrogen at an oceanic station in the Central Pacific. Deep-Sea Res 18:1127–1134

Gordon DC, Sutcliffe WH (1973) A new dry combustion method for the simultaneous determination of total organic carbon and nitrogen in sea water. Mar Chem 1:231–244

Gordon DC, Wangersky PJ, Sheldon RW (1979) Detailed observations on the distribution and composition of particulate organic material at two stations in the Sargasso Sea. Deep-Sea Res 26:1083–1092

Goriunova SV, Rzhanova GN, Orleanskii VK (1969) Blue-green algae, their physiology, biochemistry and practical role (in Russian). Nauka, Moscow

Gorshkova OM (1983) The study of the distribution of dissolved colloidal organic substances by molecular size in marine and interstitial waters. In: Geodekian AA, Trotsiuk VYa, Romankevich EA (eds) Hydrocarbons-genetic research in the Indian ocean (in Russian). PP Shirshov Inst of oceanology, Moscow

Gorshkova TI (1936) Some data on the chemical composition of Greenland Sea sediments. In: Fersman AE (ed) Papers devoted to the 50th anniversary of Academician VI Vernadsky's scientific and pedagogic activities (in Russian). Akad Nauk SSSR, Moscow, p 923

Gorshkova TI (1962) Organic matter in sediments of the Norwegian Sea and conditions of its accumulation (in Russian). Tr VNIRO, vol 46, p 38

Gorshkova TI (1970) Biogeochemistry of recent sediments in the USSR Seas. Abstract of a thesis (in Russian). Inst Okeanol Akad Nauk SSSR, Moscow

Gorshkova TI (1973) Organic matter accumulation in recent sediments of the USSR Seas. In: Vassoyevich NB (ed) Nature of organic matter in recent and fossil sediments (in Russian). Nauka, Moscow, p 60

Gorshkova TI (1975) Organic matter of recent shelf sediments in the northern seas of the USSR. In: Nevessky EN (ed) Problems of shelf geology (in Russian). Nauka, Moscow, p 66–72

Gorskaya AI, Glebovskaya EA (1966) On the chemical features of oceanic particulate organic matter. In: Bruyevich SV (ed) Chemical processes in seas and oceans. Nauka, Moscow, p 49

Grill EV, Richards FA (1964) Nutrient regeneration from phytoplankton decomposing in sea water. J Mar Res 22:51–69

Gross MG (1967) Organic carbon in surface sediment from the northeast Pacific Ocean. J Oceanol Limnol 1:46–54

Gubkin IM (1975) Naphtology. 3rd ed (in Russian). Nauka, Moscow, 384 p

Guilcher A, Pruleau M (1962) Morphologie et sédimentologie sous-marines de la rade de Brest. Com Trav Hist Sci Bull Sect Geogr 74:1–11

Guseva AN, Leifman IE (1979) On the processes of transformation of biomolecules during the fossilisation of organic matter. In: Vassoyevich NB, Timofeyev PP Accumulation and transformation of sedicahytes (in Russian). Nauka, Moscow, p 95 – 104
Hamilton RD, Holm-Hansen O, Strickland JDH (1968) Notes on the occurrence of living microscopic organisms in deep water. Deep-Sea Res 15:651 – 656
Hamilton RJ, Rail MG, Weatherston J, Brooks CI, Borthwick JH (1975) Crustacean surface waxes, part I. The hydrocarbons from the surface of Ligia oceanica. J Chem Soc Perkin Trans 4:354 – 357
Handa N (1966) Distribution of dissolved carbohydrates in the Ocean. J Oceanogr Soc Jpn 22:16 – 20
Handa N (1967) The distribution of the dissolved and particulate carbohydrates in the Kuroshio and adjacent areas. J Oceanogr Jpn 23:115 – 123
Handa N (1969) Carbohydrate metabolism in the marine diatom Skeletonema costatum. Mar Biol 4:208 – 214
Handa N (1970) Dissolved and particulate carbohydrates. In: Hood DW (ed) Organic Matter in Natural Waters. Inst of Mar Sci Univ Alaska, p 129
Handa N, Tominaga H (1969) A detailed analysis of carbohydrates in marine particulate matter. Mar Biol 2:228 – 235
Handa N, Yanagi K (1969) Studies on water-extractable carbohydrates of the particulate matter from the Northwest Pacific Ocean. Mar Biol 4:197 – 207
Handa N, Yanagi K, Matsunaga K (1972) Distribution of detrital materials in the Western Pacific Ocean and their biochemical nature. Mem Ist Ital Idrobiol 29:53 – 71
Handbook on the analyses of bitumens and dispersed organic matter in rock deposits (1966) Uspensky VA, Rodionova KF, Gorskaya AI, Shishkova AP (eds) (in Russian). Nedra, Leningrad
Harding GCH (1974) The food of deep sea copepods. J Mar Biol Assoc UK 54:141 – 155
Harrington GW, Beach DH, Dunhan JE, Hols GG (1970) The polyunsaturated acids of marine dinoflagellates. J Protozool 17:213 – 219
Harris JE (1972) Characterization of suspended matter in the Gulf of Mexico. I. Spatial distribution of suspended matter. Deep-Sea Res 19:719 – 726
Haug A, Myklestad S (1976) Polysaccharides of marine diatoms with special reference to Chaetoceros species. Mar Biol 34:217 – 222
Hayashi K, Kida S, Kato K, Yamada M (1974) The composition of fatty acids of acetone-soluble lipids of 17 marine algal species making part of benthos. Bull Jpn Soc Sic Fish 40:609
Hecky RE, Mopper K, Kilham P, Degens ET (1973) The amino acid and sugar composition of diatom cellwalls. Mar Biol 19:323 – 331
Hedges JI (1980) Interaction between humic substances and inorganic matter in surface marine sediments. In: Colloques Internationaux du CNRS N 293 – Biogéochimie de la matière organique à l'interface eau-sédiment marin. Ed CNRS, Paris, p 61
Hellebust JA (1965) Excretion of some organic compounds by marine phytoplankton. Limnol Oceanogr 10:192 – 206
Herring PJ (1972) Depth distribution of the carotenoid pigments and lipids of some oceanic animals. I. Mixed zooplankton, copepods and euphausiids. J Mar Biol Assoc UK 52:179 – 190
Hirayama H (1974) Fluorimetric determination of carbohydrates in sea water. Anal Chim Acta 70:141 – 148
Hirsch J, Ahrens EH (1958) Séparation of complex lipid mixtures by the use of silic acid chromatography. J Biol Chem 233:311
Hitchcock GL (1977) The concentration of particulate carbohydrate in a region of the West Africa upwelling zone during march, 1974. Deep-Sea Res 24:83 – 93
Høјerslev NK (1980) On the origin of yellow substance in the marine environment. Rep Inst fys oceanogr Københavns univ 42:39 – 56
Hobbie JE, Holm-Hansen O, Packard TT, Pomeroy LR, Sheldon RW, Thomas JP, Wiebe WJ (1972) A study of the distribution and activity of microorganisms in ocean water. Limnol Oceanogr 17:544 – 555
Hobson LA (1967) The seasonal and vertical distribution of suspended particulate matter in an area of the Northeast Pacific Ocean. Limnol Oceanogr 12:642 – 649
Hobson LA (1971) Relationships between particulate organic carbon and microorganisms in upwelling areas off Southwest Africa. Invest Pesq 35:195 – 208
Hobson LA, Menzel DW (1969) The distribution and chemical composition of organic particulate matter in the sea and sediments off the east coast of South America. Limnol Oceanogr 14:159 – 162

Hobson LA, Menzel DW, Barber RT (1973) Primary productivity and sizes of pools of organic carbon in the mixed layer of the ocean. Mar Biol 19:298–306
Holm-Hansen O (1968) Ecology, physiology and biochemistry of bluegreen algae. Annu Rev Microbiol 22:47–70
Holm-Hansen O (1969) Determination of microbial biomass in ocean profiles. Limnol Oceanogr 14:740–747
Holm-Hansen O (1970) Determination of microbial biomass in deep ocean water. In: Hood DW (ed) Organic matters in natural waters. Inst Mar Sci Univ Alaska, p 287
Holm-Hansen O, Booth CR (1966) The measurement of adenosine triphosphate in the ocean and its ecological significance. Limnol Oceanogr 11:510–518
Holm-Hansen O, Strickland JDH, Williams PM (1966) A detailed analysis of biologically important substances in a profile off Southern California. Limnol Oceanogr 11:548–561
Huc AY, Durand B (1974) Etudes des acides humiques et de l'humine de sédiments récents considérés comme précurseurs des kérogènes. In: Advances in Organic Geochemistry, 1973. Technip, Paris, p 53
Hunter KA, Liss PS (1981) Organic sea surface films. In: Duursma EK, Dawson R (eds) Marine organic chemistry. Elsevier, Amsterdam, p 259
Ichikawa T (1982) Particulate organic carbon and nitrogen in the adjacent seas of the Pacific ocean. Mar Biol 68:49–60 (N 1)
Iliffe TM, Calder JA (1974) Dissolved hydrocarbons in the eastern Gulf of Mexico loop current and the Caribbean Sea. Deep-Sea Res 21:481–487
Iseki K (1981) Vertical transport of particulate organic matter in the deep Bering Sea and Gulf of Alaska. J Oceanogr Soc Jpn 37:101–110
Ishiwatari B (1973) Chemical characterization of fractionated humic acids from lake and marine sediments. Chem Geol 12:113–126
Ittekkot V, Brockmann U, Michaelis W, Degens ET (1981) Dissolved free and combined carbohydrates during a phytoplankton bloom in the Northern North Sea. Mar Ecol Progr Ser 4:299–305
Ittekkot V, Degens ET, Brockmann U (1982) Monosaccharide composition of acid-hydrolyzable carbohydrates in particulate matter during a plankton bloom. Limnol Oceanogr 27:770–776
Ivanenkov VN (1979) General data on the nitrogen, phosphorus and silica. In: Bordovsky OK, Ivanenkov VN (eds) Chemistry of the ocean, vol 1 Chemistry of ocean waters (in Russian). Nauka, Moscow, p 176–184
Ivanov MV (1979) The distribution and geochemical activity of bacteria in the oceanic sediments. In: Volkov II (ed) Oceanology. Chemistry of the ocean, vol 2 Geochemistry of bottom sediments (in Russian). Nauka, Moscow, p 312
Iwata I (1964) Lipids of algae. Fatty acids of Chorella and Scenedesmus. Agric Biol Chem 28:610
Iwata I, Sakurai J (1963) Lipids of algae. Agric Biol Chem 27:253
Jannasch HW (1970) Threshold concentrations of carbon sources limiting bacterial growth in sea water. In: Hood DW (ed) Organic Matter in Natural Waters. Inst Mar Sci Univ Alaska, p 321
Jannasch HW, Wirsen CO (1973) Deep sea microorganisms: in situ response to nutrient enrichment. Science 180:641–643
Jannasch HW, Eimhjellen K, Wirsen CO, Farmanfarmain A (1971) Microbial degradation of organic matter in deep sea. Science 171:627–675
Jannasch HW, Truper HG, Tuttle JH (1974) Microbial sulfur cycle in Black Sea. In: Degens ET, Ross DA (eds) Black Sea – Geology, Chemistry and Biology. Tulsa, Oklahoma, p 419
Jeffrey LM (1970) Lipids of marine waters. In: Hood DW (ed) Symposium on organic matter in natural waters. Inst Mar Sci Univ Alaska, p 55
Jeffrey LM, Pasby BF, Stevenson B, Hood DW (1964) Lipids of ocean water. In: Ingerson E (ed) Advances in organic geochemistry. Pergamon, Oxford, p 175–198
Jeffrey LM, Pasby BF, Stewenson, Hood DW (1966) Lipids in sea water. J Am Oil Chem Soc 43:211–214
Jocteur-Monrozier L (1977) Etude des formes de l'azote dans les échantillons de surface de sédiments récents (mer de Norvège). In: Géochimie organique des sédiments marins profonds. Orgon I Mer de Norvège. Ed CNRS, Paris, p 269
Joctour-Monrozier L (1980) Azote organique: nature et évolution dans les sédiments récents, relations avec l'origine du matériel organique. In: Colloques Intern du CNRS N 293 –

Biogéochimie de la matière organique à l'interface eau-sédiment marin. Ed CNRS, Paris, p 167
Jocteur-Monrozier L, Jeanson P (1979) L'azote organique et ammoniacal-combinaisons et stabilité chimique. In: Arnould M, Pelet R (eds) Géochimie organique des sédiments marins profonds. Orgon III Mauritanie-Sénégal-Iles du Cap-Vert. Ed CNRS, Paris, p 405
Jocteur-Monrozier L, Jeanson P (1981) Caractères généraux et évolution de l'azote organique dans les sédiments superficiels. In: Géochimie organique des sédiments marins profonds. Orgon IV Golfe d'Aden, Mer d'Oman. Ed CNRS, Paris, p 503
Johannes RE (1972) Some roles of detritus in the flux of energy and nutrients in coral reef community. Mem Ist Ital Idrobiol "Dott M Marchi" 29:525 – 527
Johnson RG (1974) Particulate matter at the sediment-water interface in coastal environments. J Mar Res 32:313 – 333
Jurkovskis AK, Fominykh TA (1980) The biohydrochemical aspects of the dynamics of the forms of phosphorus in the Baltic Sea. In: Okolotowicz G (ed) Ecosystems of the Baltic, part 1 (in Russian). Inst Mar Fish, Gdynia, p 40
Kalle K (1966) The problem of the Gelbstoff in the sea. Oceanogr Mar Biol Ann Rev 4:91 – 104
Kanazawa A, Teshima S (1971) Sterols of the suspended matters in sea water. J Oceanogr Soc Jpn 27:207 – 219
Kanazawa A, Yoshioka M, Teshima S (1972) Sterols in some red algae. Mem Fac Fish Kagoshima Univ 21:103 – 108
Karavaev NM, Vener RA, Koroleva KI (1964) Composition and chemical nature of sapropelic acids. Dokl Akad Nauk SSSR, 156:877 – 879
Kato K (1956) Chemical investigations on marine humus in bottom sediments. Mem Fac Fish Hokkaido Univ 4:91 – 209
Kato K (1961) Some aspects of biochemical characteristics of sea water and sediments in Mochima Bay. Bol Inst Oceanogr Univ Oriente Cumana-Venezuela 1:343 – 358
Kawahara H, Ishii J, Maita Y (1966) Studies of clay minerals and organic matter of the recent marine sediments from Japan Trench. Mar Geochem Invest Recent Sediments 72:75 – 84
Kawahara H, Maita G, Ishii J (1967) Studies on the organic matter in sea water. I. The distribution of carbohydrate in water of southern Hokkaido. Bull Jpn Soc Sci Fish 33:825 – 833
Kay H (1954) Eine Mikromethode zur chemischen Bestimmung des organisch gebundenen Kohlenstoffes im Meerwasser. Kiel Meeresforsch Bd 10:26 – 36
Kemp S (1979) Carbon in the freshwater cycle. In: Bolin B, Degens ET, Kemp S, Ketner P (eds) The global carbon cycle. Wiley, Chichester Toronto, p 317 – 342
Kennicutt MC, Jeffrey LM (1981 a) Chemical and GC-MS characterisation of marine dissolved lipids. Mar Chem 10:367 – 387
Kennicutt MC, Jeffrey LM (1981 b) Chemical and GC-MS characterisation of marine particulate lipids. Mar Chem 10:389 – 407
Kenyon CN, Stanier RY (1970) Possible evolutionary significance of polyunsatured fatty acids in blue-green algae. Nature 227:1164 – 1166
Ketchum BH, Redfield AC (1949) Some physical and chemical characteristics of algae growth in mass culture. J Cell Comp Physiol 33:281 – 299
Ketseridis G, Hahn J, Jaenicke R, Junge C (1976) The organic constituents of atmospheric particulate matter. Atmos Environ 10:603 – 610
Khailov KM (1968) On organic macromolecules dissolved in the sea-water (in Russian). Geokhimiya 5:595 – 603
Khailov KM (1971) The ecological metabolism in the sea (in Russian). Naukova Dumka, Kiev
Khailov KM, Finenko GA, Burlakova LP, Smirnov VA (1973) On the relation between stationary concentrations of organic matter basic forms in sea water and unit rates of their trophic consumption by organisms of coastal associations (in Russian). Dokl Akad Nauk SSSR 209: 1210 – 1212
Khan DV (1951) On the relationship between organic matter and soil minerals (in Russian). Dokl Akad Nauk SSSR 81:461 – 464
Khrustalev YuP, Reznikov SA, Turovsky DS (1977) The lithology and geochemistry of bottom sediments of the Aral sea (in Russian). Rostov University Press, Rostov-on-Don
Kinney PJ, Loder TC, Croves J (1971) Particulate and dissolved organic matter in the Amerasian Bassin of the Arctic Ocean. Limnol Oceanogr 16:132 – 136

Kizevetter IV (1954) On the nutrient value of plankton in the Okhotsk and Japan Seas (in Russian). Izv TINRO 39:97
Kizevetter IV (1973) The biogeochemistry of water-originated raw materials (in Russian). Pishchevaya prom., Moscow
Kizevetter IV, Sukhoveyeva MV, Shmelkova LP (1981) Commercial marine plants and weeds of the far-eastern seas (in Russian). Legkaya u pishchevaya prom., Moscow
Kleimenova GI (1983) The role of the the spore-pollen biomass in marine sedimentation in the Late- and Post-Glacial Time (as examplified by the Baltic Sea). In: Lisitzin AP (ed) Biosedimentation in the seas and oceans (in Russian). Inst Oceanol Akad Nauk SSSR, Moscow, p 141
Klem A (1932) Contribution to the study of the oils of marine Crustacea. Hvalradets Skr 6:24–27
Klubova TT (1971) Mechanism of inter-reaction between clay minerals and organic matter in sediments. In: Organic matter in recent and fossil sediments (in Russian). Nauka, Moscow
Knauer GA, Martin JH (1980) Studies on the biological transport of materials from surface to deep ocean waters. I Fluxes of carbon, nitrogen and phosphorus. II Fluxes of trace elements. In: Fedorov VD (ed) Man and Biosphere, vol 5. Moscow University Press, Moscow, p 63
Koblents-Mishke OI (1977) Primary production. In: Vinogradov ME (ed) Oceanology. Biology of the ocean, vol 1 (in Russian). Nauka, Moscow, p 62
Koblents-Mishke OI, Vedernikov VI (1977) Primary production. In: Vinogradov ME (ed) Oceanology. Biology of the ocean, vol 2 (in Russian). Nauka, Moscow, p 183
Koblents-Mishke OI, Volkovinsky VV, Kabanova YuG (1970) Primary productivity of plankton in the world ocean. In: Zenkevich LA (ed) Program and methods of investigations of aquatic biogeocoenoses (in Russian). Nauka, Moscow, p 66
Kochenov AV, Baturin GN, Kovalyova SA, Emelianov EM, Shimkus KM (1965) Uranium and organic matter in sediments of the Black and Mediterranean Seas. Geokhimia 3:302–313
Kondratieva TM, Finenko ZZ (1971) Primary productivity and some physiological aspects of planktonic algae. In: Vodianitsky VA (ed) Problems of marine biology (in Russian). Naukova Dumka, Kiev, p 179–184
Koniukhov AI (1975) Facies characteristics of sediments off the Southern Primorye continental margin. In: Nevessky EN (ed) Problems of Shelf Geology (in Russian). Nauka, Moscow
Koniukhov AI, Kuprin PN, Gershanovich DE, Potapova LI, Sharno NM (1978) Pecularities of the content of organic matter in the sapropel sediment layers in the seas and oceans. In: Vassoyevich NB (ed) Accumulation and transformation of organic matter of recent and fossil sediments (in Russian). Nauka, Moscow, p 86
Korzhikova LI, Smirnov BA, Bordovsky OK (1979) On the characteristic of chemical composition of organic matter of oceanic sediments from the massive accumulation areas (in Russian). Okeanologiya 19:421–426
Koshimizu N (1961) On the distribution of organic matter in the sea water and bottom samples of Maizuru Bay. Bull Maizuru Mar Observ 7:1–11
Koyama T (1962) Organic compounds in sea water. J Oceanogr Soc Jpn 5:563–576
Koyama M, Tsuchiya Y, Mead JE (1963) A model experiment of aquatic food chain with special significance in fatty acid conversion. Bull Jpn Soc Sci Fish 19:452–458
Koziar LA, Skolova MN, Zezina ON (1974) Spores and pollens of highly-organized plants in the nutrition of marine invertebrates (in Russian). Okeanologiya 14:522–525
Krey J (1958) Chemical determination of net plankton, with special reference to equivalent albumin content. J Mar Res 17:312–324
Krey J (1961a) Beobachtungen über den Gehalt an Mikrobiomasse und Detritus in der Kieler Bucht 1958–1960. Kiel Meeresforsch 17:163–175
Krey J (1961b) Der Detritus im Meere. J Conceil Intern Explor Mer 26:263–280
Krey J (1967a) The vertical distribution of microbiomass in Northern and in tropical seas. In: Studies in tropical oceanography N 5. Univ Miami Inst Mar Sci, p 247
Krey J (1967b) Detritus in the Ocean and adjacent seas. In: Lauff GH (ed) Estuaries, N 83. Publ Am Assoc Adv Sci, Washington, p 389
Kriss AE (1959) Marine (deep-water) microbiology (in Russian). Izv Akad Nauk SSSR, Moscow
Kriss AE (1973) Metabolic processes and hydrostatic pressure (in Russian). Nauka, Moscow
Kriss AE (1976) Microbiological oceanography (in Russian). Nauka, Moscow
Kriss AE, Stupakova TP, Tzyban AV (1972) Ecologo-geographical aspects of microbial (heterotrophic) distribution in the central part of the Pacific Ocean. Microbiologija 41:542–549

Krogh A (1931) Dissolved substances as food of aquatic organisms. Rep at Cambridge Philosophical Society. Biol Rev 6:412–442
Krogh A, Keys E (1934) Methods for the determination of dissolved organic carbon and nitrogen in sea water. Biol Bull (Woods Hole) 67:132–144
Krogh A, Lange E, Smith W (1930) On the organic matter given off by algae. Biochem J 24:1666–1670
Kuprin PN, Bagirov VI (1971) To the elucidation of the depositional conditions of the central and southern Caspian sediments. In: Leontiev OK, Mayev EG (eds) Complex investigations of the Caspian Sea, N 2 (in Russian). Moscow Univ Press, p 91
Kuprin PN, Potapova LI (1973) Some features of organic matter diagenetic modification in sediments of the Southern Seas. In: Vassoyevich NB (ed) Nature of organic matter in recent and fossil sediments (in Russian). Nauka, Moscow, p 186
Kuprin PN, Potapova LI (1978) Early diagenetic transformations of organic matter in sediments. In: Nalivkin VD, Aliyev MM (eds) The principal conditions for the generation and accumulation of hydrocarbons (in Russian). Nauka, Moscow, p 17
Kuprin PN, Shatov AS, Shlykov VG, Potapova LI, Poliakov AS (1976) Zonality in the changing properties of bottom sediments of the Caspian Sea along a profile. In: Mayev EG (ed) Complex research of the Caspian Sea, vol 5 (in Russian). Moscow University Press, p 107
Kutner AS (1964) Conteudo organico dos sedimentos de fundo de Cananeia, Sao Paulo. Oceanogr Fisica 7:1–14
Kuznetsov SI (1970) Microflora in lakes and its geochemical activities (in Russian). Nauka, Leningrad
Laflamme RE, Hites RA (1978) The global distribution of polycyclic aromatic hydrocarbons in recent sediments. Geochim Cosmochim Acta 42:289–303
Laur MH, Liem PQ (1970) Sur les lipides neutres de trois fucasécs des côtes françaises. CR Acad Sci Paris, D 271:1752–1755
Lebedeva LP, Vinogradov ME, Shushkina EA, Sazhin AF (1982) An estimate of the intensity of the detritus-formation process in marine planktonic communities (in Russian). Okeanologiya 22:652–659
Ledet EJ, Laseter JL (1974) Alkanes at the air/sea interface from offshore Lousiana and Florida. Science 186:261–263
Lee C, Bada JL (1975) Amino acids in Equatorial Pacific waters. Earth Planet Sci Lett 26:61–68
Lee C, Wakeham SG, Farrington JW (1983) Variations in the composition of particulate organic matter in a time-series sediment trap. Mar Chem 13:181–194
Lee RF (1974) Lipids of zooplankton from Bute Inlet, British Columbia. J Fish Res Bd Can 31:1577–1582
Lee RF, Hirota J (1973) Wax esters in tropical zooplankton and nekton and geographical distribution of wax esters in marine copepods. Limnol Oceanogr 18:227–239
Lee RF, Nevenzel JC, Paffenhöfer GA, Benson AA (1970) The metabolism of wax esters and other lipids by the marine copepod Calanus helgolandicus. J Lipid Res 11:237–240
Lee RF, Nevenzel JC, Paffenhöfer GA (1971) Importance of wax esters and other lipids in marine food chain-phytoplankton and copepods. Mar Biol 9:99–108
Lein AYu, Namsaraev BB, Trotsiuk VYa, Ivanov MV (1981) Bacterial Methanogenesis in Holocene Sediments of the Baltic Sea. J Geomicrobiol 2:299–315
Leo RF, Parker PL (1966) Branched chain fatty acids in sediments. Science 152:649–650
Leontiev OK, Lukianova SA, Kalinina LI (1975) Areas of the major morphological-structural elements of the ocean floor (in Russian). Dokl Akad Nauk SSSR 222:1441–1444
Lerman A (1979) Geochemical Processes. Water and Sediment Environments. Wiley, New York
Lewin RA (1956) Extracellular polysaccharides of green algae. Can J Microbiol 2:665–672
Lewis GJ, Rakestraw NW (1955) Carbohydrates in sea water. J Mar Res 14:253–258
Lewis RW (1962) Temperature and pressure effects on the fatty acids of some marine ectotherms. Compar Biochem Physiol 6:75–89
Liebezeit G, Bölter M, Brown IF, Dawson R (1980) Dissolved free amino acids and carbohydrates at pycnocline boundaries in the Sargasso Sea and related microbial activity. Oceanol Acta 3:357–362
Lisitzin AP (1955) Atmospheric and aquatic suspended matter as the initial source of sediment formation (in Russian). Inst Okeanol Akad Nauk SSSR 13:16
Lisitzin AP (1961) Distribution and composition of suspended matter in seas and oceans. In: Strakhov NM (ed) Recent sediments in seas and oceans (in Russian). Nauka, Moscow, p 124

Lisitzin AP (1964) Distribution and chemical composition of the particulate fraction in the Indian Ocean waters. In: IGY Oceanological Investigations, N 10 (in Russian). Nauka, Moscow, p 5
Lisitzin AP (1966) Contemporary sedimentation in the Bering Sea (in Russian). Nauka, Moscow
Lisitzin AP (1974) Sedimentation in the oceans (in Russian). Nauka, Moscow
Lisitzin AP (1978) The processes of the oceanic sedimentation (in Russian). Nauka, Moscow
Lisitzin AP (1980) General regularities in the structure of the sedimentary mass. In: Monin AS, Lisitzin AP (eds) Oceanology. Geology of the ocean. The geologic history of the ocean (in Russian). Nauka, Moscow, p 36
Lisitzin AP (1983) Biodifferenciation of matter in the ocean. In: Lisitzin AP (ed) Biosedimentation in the seas and oceans (in Russian). Inst Oceanol Akad Nauk SSSR, Moscow, p 3
Lisitzin AP, Vinogradov ME (1982) The global laws of the distribution of life in the ocean and their reflection in the composition of bottom sediments. Generation and distribution of biogenic sediments (in Russian). Izv Akad Nauk SSSR, Ser Geol 4:5 – 24
Lisitzin AP, Bogdanov YuA, Potapova LI, Maksimov AN (1971) Bituminous substances in the oceanic suspended matter (in Russian). Okeanologiya 11:53 – 66
Lisitzin AP, Bogdanov YuA, Emelianov EM, Maximov AN, Pustelnikov OS (1975) Suspended substances in the Atlantic waters. In: Lisitzin AP (ed) Sedimentation in the Atlantic ocean (in Russian). Izd Kalinigradskaya Pravda, Kaliningrad, p 5
Lisitzin AP, Bogdanov YuA, Emilianov EM, Maksimov AN, Pustelnikov OS, Serova VV (1975) Suspended matter in the Atlantic Ocean waters. In: Lisitzin AP (ed) Sedimentation in the Atlantic Ocean (in Russian). Izd Kaliningradskaya Pravda, Kaliningrad, p 5
Litchfield CD, Munro ALS, Massie LC, Floodgate GD (1974) Biochemistry and microbiology of some Irish Sea sediments. I. Amino acid analyses. Mar Biol 26:249 – 260
Littlepage JL (1964) Seasonal variation in lipid content of two Antarctic marine Crustacea. Actual Sci Ind 1312:463 – 470
Liutsarev SV (1973) Analytical techniques for low-background organic carbon measurements in sea waters. In: Modern techniques in fish industry marine hydrochemical investigations (in Russian). Pishchevaya prom, Moscow, p 114 – 126
Liutsarev SV, Pashkova EA (1983) Colloid fraction of the organic matter of river waters and Baltic waters. In: Lisitzin AP, Emelianov EM (eds) Geological history and geochemistry of the Baltic Sea (in Russian). Nauka, Moscow (in print)
Liutsarev SV, Romankevich EA (1975) Organic carbon, phosphorus and nitrogen in Japan Sea waters (in Russian). Okeanologiya 15:261 – 265
Liutsarev SV, Smetankin AV (1980a) The determination of carbon of the dissolved organic matter in terrestrial and marine waters. In: Romankevich EA (ed) Methods for studying the organic matter in the ocean (in Russian). Nauka, Moscow, p 32
Liutsarev SV, Smetankin AV (1980b) The determination of carbon in aquatic suspensions. In: Romankevich EA (ed) Methods for studying the organic matter in the ocean (in Russian). Nauka, Moscow, p 46
Liutsarev SV, Romankevich EA, Semenov AD, Finenko ZZ (1972) Preliminary results of organic matter studies in sea waters (the 50th cruise of r/v "Vityaz") (in Russian). Eksp inf TsNII Rybn Khoz Ser 9 3:14 – 20
Liutsarev SV, Mirkina SD, Romankevich EA, Smetankin AV (1975) Dissolved organic carbon and phosphorus in the Eastern Equatorial Pacific Waters (in Russian). Tr Inst Okeanol Akad Nauk SSSR 102:70 – 75
Liutsarev SV, Gorshkova OM, Chubarov VV (1984) The study of the dissolved colloidal organic matter of marine and interstitial waters by means of laser fluorometry (in Russian). Okeanologiya 24:95 – 101
Livingstone DA (1963) Chemical composition of rivers and lakes. In: Data of Geochemistry, 6th ed. US Geol Surv, Prof Paper 440G:1 – 64
Logvinenko NV, Romankevich EA (1973) Recent sediments of the Southeastern Pacific off the coast of Peru and Chile (in Russian). Litol Polezn Iskop 1:3 – 16
Logvinenko NV, Nikolayeva IV, Romankevich EA (1973) Authigenous minerals in recent bottom sediments of the Southeastern Pacific (in Russian). Litol Polezn Iskop 4:15 – 27
Lohmann H (1908) Untersuchungen zur Feststellung des vollständigen Gehaltes des Meeres an Plankton. Wiss Meeresuntersuch, Kiel 10:129 – 370

Lorenzen CJ, Welschmeyer NA, Copping AE, Vernet M (1983a) Sinking rate of organic particles. Limnol Oceanogr 28:766–769

Lorenzen CJ, Welschmeyer NA, Copping AE (1983b) Particulate organic carbon flux in the subarctic Pacific. Deep-Sea Res 30:639–643

Low EM (1955) Studies on some chemical studies on diatoms. J Mar Res 14:199–204

MacKinnon MD (1978) A dry oxidation method for the analysis of TOC in sea water. Mar Chem 7:17–37

MacKinnon MD (1981) The measurement of organic carbon in sea water. In: Duursma EK, Dawson R (eds) Marine Organic Chemistry. Elsevier, Amsterdam Oxford New York, p 415

Maksimova MP (1973) Specificities of the organic matter balance in the southern and northern inland seas of the USSR (in Russian). Tr VNIRO 80:7–17

Maksimova MP (1977) Organic phosphorus in Indian Ocean waters. In: Skopintzev BA, Ivanenkov VN (eds) Chemical-oceanological investigations (in Russian). Nauka, Moscow, p 97

Maltseva AV (1980) Mean perennial discharge of organic substances from the USSR territory and its temporal variations (in Russian). Gidrokhim materialy, vol 68. Gidrometeoizdat, Leningrad, p 14

Manheim FT, Hathaway JC, Uchupi E (1972) Suspended matter in surface waters of the northern Gulf of Mexico. Limnol Oceanogr 17:17–27

Manskaya SM, Drozdova TV (1964) Geochemistry of organic matter (in Russian). Nauka, Moscow

Manskaya SM, Kodina LA (1975) Lignin Geochemistry (in Russian). Nauka, Moscow

Marti YuYu (1971) Biological productivity of the Peruvian current area. In: Beklemishev KV (ed) Fundamentals of oceanic biological productivity and its utilization (in Russian). Nauka, Moscow, p 73

Matthews WS, Smith LL (1968) Sterol metabolism. III. Sterols of marine waters. Lipids 3:239–246

Maurer LG (1976) Organic polimers in sea water: changes with depth in the Gulf of Mexico. Deep-Sea Res 23:1059–1064

McLachland J, McInnes AG, Falk M (1965) Studies on the chitan (chitin: poli-N-acetylglucosamine) fibers of the diatom Thalassiosira fluviatilis Hustedt. I. Can J Bot 43:707–715

Melnikov IA (1975a) Microplankton and organic detritus in waters of the Southeastern Pacific (in Russian). Okeanologiya 15:146–156

Melnikov IA (1975b) The fine-dispered fraction of suspended organic matter in the Eastern Pacific waters (in Russian). Okeanologiya 15:266–271

Melnikov IA (1980) The ecosystem of the Arctic drifting ice. In: Vinogradov ME, Melnikov IA (eds) The Biology of the Central Arctic basin (in Russian). Nauka, Moscow, p 61

Melnikov IA, Pavlov GL (1978) The characteristic features of the distribution of organic carbon in the water and ice of the Arctic basin (in Russian). Okeanologiya 18:248–254

Menzel DW (1964) The distribution of dissolved organic carbon in the Western Indian Ocean. Deep-Sea Res 11:757–766

Menzel DW (1967) Particulate organic carbon in the deep sea. Deep-Sea Res 14:229–238

Menzel DW (1974) Primary productivity of dissolved and particulate organic matter, and the sites of oxidation of organic matter. In: Goldberg ED (ed) The Sea, vol 5. Marine Chemistry. Wiley, New York London Sydney Toronto, p 659

Menzel DW, Goering JJ (1966) The distribution of organic detritus in the ocean. Limnol Oceanogr 11:333–337

Menzel DW, Ryther JH (1964) The composition of particulate organic matter in the Western North Atlantic. Limnol Oceanogr 9:179–186

Menzel DW, Ryther JH (1970) Distribution and cycling of organic matter in the ocean. In: Hood DW (ed) Symposium on organic matter in natural waters. Inst Mar Sci Univ Alaska 1: 31–54

Menzel DW, Vaccaro RF (1964) The measurement of dissolved organic and particulate carbon in sea water. Limnol Oceanogr 9:138–142

Meyers PA, Quinn JG (1971) Interaction between fatty acids and calcite in sea water. Limnol Oceanogr 16:992–997

Meyers PA, Quinn JG (1973) Association of hydrocarbons and mineral particles in saline solution. Nature 224:23–24

Michaelis W, Ittekkot V, Degens ET (1980) Amino acids and carbohydrates as possible indicators of sedimentation processes. In: Colloques Internationaux du CNRS N 293 – Biogéochimie de la matière organique à l'interface eau-sédiment marin, p 129

Mitskevich IN (1975) Microbiological investigations in the 11th cruise of R/V "Akademic Kurchatov". In: Filatova ZA (ed) Biological investigations in the South Atlantic Ocean and in the region of Peru-Chile trench (in Russian). Tr Inst Okeanol Akad Nauk SSSR 103: 48–59

Mobius K (1880) Die Auster und die Austernwirtschaft (Berlin transl 1880). The oyster and oyster culture. Rep US Fish Commiss

Mohamed AF (1949) The distribution of organic matter in sediments of the Northern Red Sea. Am J Sci 247:116

Mohsen AF, Khaleafa AF, Hashem MA, Metwalli A (1974) Effect of different nitrogen sources on growth, reproduction, amino acid, fat and sugar contents in Ulva fasciata Delile, part III. Bot Mar 17:218–222

Moïseyev PA (1969) Biological resources of the world ocean (in Russian). Pishchevaya prom, Moscow

Monin AS, Nelepo BA (eds) (1982) 2nd USSR congress of oceanographers Abstracts of papers, vol 4, part 1. Symp 14–15 Dynamics and forecasts, ways and means of abating the pollution of the Ocean, biogeochemical destruction of pollutants (in Russian), Morskoi gydrofiz inst Akad Nauk SSSR, Sebastopol, p 84

Monin AS, Romankevich EA (1979) The problems in the biogeochemistry of the world ocean. In: Kovalsky VV, Letunova SV (eds) Current tasks and problems of the biogeochemistry, vol 17 (in Russian). Tr Biogeokhim Lab, Nauka, Moscow, p 74

Moore HB (1933) Faecal pellets from marine deposits. Discovery Rep 7:17–26

Moore RM, Burton JD, Williams PL, Young MR (1979) The behavior of dissolved organic material, iron and manganese in estuarine mixing. Geochim Cosm Acta 43:919–926

Mopper K (1980) Carbohydrates in the marine environment: recent developments. In: Colloques Internationaux du CNRS, N 293 – Biogéochimie de la matière organique à l'interface eau-sédiment marin. Ed CNRS, Paris, p 71

Mopper K, Lindroth P (1982) Diel and depth variations in dissolved free amino acids and ammonium in the Baltic Sea determined by ship-board HPLC analysis. Limnol Oceanogr 27:336–347

Mopper K, Dawson R, Liebezeit G, Ittekkot V (1980) The monosaccharide spectra of natural waters. Mar Chem 10:55–66

Morris RJ (1972) The occurrence of wax esters in crustaceans from the Northeast Atlantic Ocean. Mar Biol 16:102–107

Müller PJ, Mangini A (1980) Organic carbon decomposition rates in sediments of the Pacific manganese nodule belt dated by ^{230}Th and ^{231}Pa. Earth Planet Sci Lett 51:94–114

Mullin MM (1965) Size fractionation of particulate organic carbon in the surface waters of the western Indian Ocean. Limnol Oceanogr 10:459–462

Myklestad S (1974) Production of carbohydrates by marine planktonic diatoms. I Comparison of nine different species in culture. J Exp Mar Biol Ecol 15:261–274

Nace RL (1970) World Hydrology: status and prospects. Proc. IASH-UNESCO-WHO Sympos on World water balance, Publ 92, 1

Nakai Z (1952) The chemical composition, volume weight and size of the important marine plankton. Nippon Kaiyo Gakkaishi 1:45

Nakajima K (1973) Suspended particulate matter in the Western North Pacific Ocean. Mem Fac Fish Hokkaido Univ 20:1–106

Naletova IA (1979a) Dissolved amino acids in Atlantic waters (in Russian). Okeanologiya 19:258–264

Naletova IA (1979b) The role of amino acids in the syntrophy of a marine ecosystem (in Russian). Tr VNIRO 136:87–94

Naletova IA (1982) Dissolved amino acids in the waters of the Bellingshausen Sea and SE Pacific. In: Gershanovich DE (ed) Environmental conditions and the bioproductivity of the sea (in Russian). Legkaya i pishchevaya prom, Moscow, p 20

Nazarkin LA (1979) The influence of sedimentation rate and weathering upon oil and gas potentials of sedimentary basins (in Russian). Saratov State Univ Publ House Saratov

Nemirovskaya IA, Nesterova MP (1982) Oil pollution of the waters of the Atlantic and Indian Oceans (in Russian). Okeanologiya 22:955–961

Nemirovskaya IA, Romankevich EA, Bystrova MG (1983) Lipids of the suspended matter from the Atlantic and Indian Oceans. In: Geodekian AA, Trotsiuk VA, Romankevich EA (eds) Hydro-

carbons-genetic research in the Indian Ocean (in Russian). PP Shirshov Inst Okeanol Akad Nauk SSSR, Moscow, p 59
Nemoto T, Ishikawa K (1969) Organic particulate and aggregate matters stained by histological reagents in East China Sea. J Oceanogr Soc Jpn 25:281 – 290
Nesterova IL (1960) Chemical composition of particulate and dissolved matter in the Ob river (in Russian). Geokhimia 4:355 – 361
Nesterova MP, Simonov AI (1979) The chemical pollution of the ocean and the means for abating it. In: Bordovsky OK, Ivanenkov VN (eds) Chemistry of the ocean. Vol 1 Chemistry of oceanic waters (in Russian). Nauka, Moscow, p 436
Newell BS (1969) Seasonal variations in the Indian Ocean along 110°. II. Particulate carbon. Aust J Mar Freshwater Res 20:51 – 54
Newell BS, Kerr JD (1968) Suspended organic matter in the South-Eastern Indian Ocean. Aust J Mar Freshwater Res 19:129 – 138
Nienhuis PH (1981) Distribution of organic matter in living marine organisms. In: Duursma EK, Dawson R (eds) Marine organic chemistry. Elsevier, Amsterdam Oxford New York, p 31
Niino H, Emery KO (1961) Sediments of shallow portions of East China Sea and South China Sea. Bull Geol Soc Am 72:731 – 762
Niino H, Emery KO (1966) Continental shelf sediments off North-Eastern Asia. J Sed Petrol 36:152 – 161
Nishizawa S (1971) Concentration of organic and inorganic material in the sea surface skin at the equator 155°W. Bull Plankton Soc Jpn 18:42 – 44
Nishizawa S, Nakajima K (1971) Concentration of particulate organic material in the sea surface layer. Bull Plankton Soc Jpn 18:12 – 19
Nissenbaum A (1974) The organic geochemistry of marine and terrestrial humic substances: Implications of carbon and hydrogen isotope studies. In: Tissot B, Bienner F (eds) Advances in organic geochemistry. Technip, Paris, p 39
Nissenbaum A, Kaplan IR (1972) Chemical and isotopic evidence for the in situ origin of marine humic substances. Limnol Oceanogr 17:570 – 582
Odum EP, Cruz AA (1967) Particulate organic detritus in a Georgia salt marsh-estuarine ecosystem. In: Lauff GH (ed) Estuaries. Am Ass Adv Sci Publ Washington 83:383 – 388
Ogura N (1970) The relation between dissolved organic carbon and apparent oxygen utilization in the Western North Pacific. Deep-Sea Res 17:221 – 232
Ogura N (1974) Molecular weight fractionation of dissolved organic matter in coastal seawater by ultrafiltration. Mar Biol 24:305 – 312
Ogura N (1977) High molecular weight organic matter in seawater. Mar Chem 5:535 – 549
Ohta N (1953) Chemical investigation of deep-sea deposits. J Chem Soc Jpn, Pure Chem Sec 74:873 – 876
Ohwada M (1960) Vertical distribution of living and dead diatoms down to one thousand meters off Sauriki, Northern Japan. Mem Kobe Observ Jpn XIV:69 – 73
Okuda T, Gomez JR (1964) Distribucion del carbono y nitrogeno organicos de los sedimentos en la region Nororiental de Venezuela. Bol Inst Oceanogr, Univ Oriente 3:91 – 106
Olausson E (1960) Description of sediment cores from Central and Western Pacific and the adjacent Indonesian region. Rep Swed Deep-Sea Exped Rep Göteborg, 1947 – 1948, 9:53 – 88
Omori M (1969) Weight and chemical composition of some important oceanic zooplankton in the North Pacific Ocean. Mar Biol 3:4 – 10
Oparin AI, Bezinger EN (1960) On the carbohydrates in Zostera, Cladophora and Lyngbya and their breakdown after the death of the plants. In: Senukov VM (ed) Contemporary analogues of oil-bearing facies (in Russian). Gostoptekhizdat, Moscow, p 290
Oppenheimer CH (1960) Organic matter in shallow marine bays. Bull Geol Soc Am 71:1938
Orcutt DM, Patterson GW (1975) Sterol, fatty acid and elemental composition of diatoms grown in chemically defined media. Com Biochem Physiol 50:579 – 584
Orlov DS (1974) Humic acids in soils (in Russian). Moscow University Press
Ostapenia AP, Kovalevskaya RZ (1965) On the particulate organic matter content in the surface layer of sea waters (in Russian). Okeanologiya 5:649 – 653
Östlund HG (1974) Expedition "Odysseus 65": radiocarbon age of Black Sea deep water. In: Degens ET, Ross DA (eds) Black Sea – Geology, Chemistry and Biology. Tulsa, Oklahoma, p 127

Ostroumov EA, Volkov II (1957) On the interconnection of phosphorus, vanadium and organic matter in the Black sea sediments (in Russian). Geokhimia 6:518–528

Packard TT, Healy ML, Richards FA (1971) Vertical distribution of the activity of the respiratory electron transport system in marine plankton. Limnol Oceanogr 16:60–70

Pakhomova AS (1959) On the chemical composition of the particulate matter in the Volga delta bottom sediments and Northern Caspian Sea (in Russian). Tr Gos Okeanogr Inst 45:117–143

Pakhomova AS (1961) Organic matter in bottom sediments of the Caspian Sea (in Russian). Tr Gos Okeanogr Inst 59:58–84

Pakhomova MV (1969) Biochemistry of blue-green algae. In: Fedorov VD, Telitchenko MM (eds) Biology of blue-green algae, N 2 (in Russian). Moscow University Press, p 84

Parker FL (1973) Living planktonic foraminifera from the Gulf of California. J Foraminiferal Res 3:70–77

Parker PL, Calder JA (1970) Stable carbon isotope ratio variations in biological systems. In: Hood DW (ed) Organic matter in natural waters. Inst Mar Sci Univ Alaska, p 107

Parsons TR (1963) Suspended organic matter in sea water. In: Sears M (ed) Progress in oceanography, vol 1. Pergamon, Oxford London New York Paris, p 205

Parsons TR (1975) Particulate organic carbon in the sea. In: Riley JP, Skirrow G (eds) Chemical oceanography, vol 2, 2nd ed. Academic, London New York, p 365

Parsons TR, Strickland JDH (1962a) Oceanic detritus. Science 136:313–315

Parsons TR, Strickland JDH (1962b) On the production of particulate organic carbon by heterotrophic processes in sea water. Deep-Sea Res 8:211–222

Parsons TR, Takahashi M (1973) Biological oceanographic processes. Pergamon, Oxford

Parsons TR, Stephens K, Strickland JDH (1961) On the chemical composition of eleven species of marine phytoplankters. J Fish Res Bd Can 18:1001–1016

Parsons TR, Takahashi M, Hargrave B (1977) Biological oceanographic processes. Pergamon, Oxford

Patterson GW (1970) Effect of culture temperature on fatty acid composition of Chlorella sorokiniana. Lipids 5:597–600

Peake E, Casagrande DJ, Hodgson GW (1974) Fatty acids, chlorins, hydrocarbons, sterols and carotenoids from a Black Sea core. In: Ross DA, Degens ET (eds) Black Sea – Geology, Chemistry and Biology. Tulsa, Oklahoma Mem 20:505

Pelet R (1981) Géochimie organique des sédiments marins profonds du Golfe d'Aden et de la mer d'Oman: vue d'ensemble. In: Géochimie organique des sédiments marins profonds, Orgon IV Golfe d'Aden, mer d'Oman. Ed CNRS, Paris, p 529

Pentina TYa (1973) Humic Acids in Recent and Late Quarternary deposits of the Caspian Sea (in Russian). Tr VNIGNI 138:201–213

Petersen CCJ (1918) The sea bottom and its production of fish food. Dan Biol Sta Rep 25, Copenhagen

Petipa TS (1964) Fat metabolism in Calanus helgolandicus (Claus) under experimental conditions (in Russian). Dokl Akad Nauk SSSR 155:470–473

Petipa TS (1968) Nutrition of planktonic crustaceans in the Black Sea. In: Biological investigations of the Black Sea and its commercial resources (in Russian). Nauka, Moscow, p 67

Petipa TS (1981) Trophodynamics of the copepods in the marine planktonic communities (in Russian). Naukova Dumka, Kiev

Pierce RW, Van der Veen J, Olscott HS (1969) Proximate and lipid analyses of krill (Euphausia species) and red crab (Pleuroncodes planipes). J Agric Food Chem 17:367–369

Pinet PR, Frey RW (1977) Organic carbon in surface sands seaward of Altamaha and Doboy Sounds, Georgia. Geol Soc Am Bull 88:1731–1739

Platt T, Subba Rao DV (1975) Primary production of marine microphytes. In: Cooper JP (ed) Photosynthesis and productivity in different environments. Cambridge Univ Press, Cambridge, p 249

Plunkett MA (1957) The qualitative determination of some organic compounds in marine sediments. Deep-Sea Res 4:259–262

Pocklington R (1971) Free amino acids dissolved in North Atlantic Ocean waters. Nature 230:374–375

Pocklington R (1977) Chemical processes and interactions involving marine organic matter. Mar Chem 5:479–496

Poliakova LD (1974) Distribution regularities of organic matter components in interstitial waters of the Black Sea in relation to sedimentation and diagenesis. Abstract of a thesis (in Russian). Moscow University Press, Moscow
Pomeroy LR (1960) Residence time of dissolved phosphate in natural waters. Science 131:1731
Postma H (1968) Dissolved organic matter in the oceans. In: Advances in organic geochemistry 1968. Pergamon, New York London, p 47
Potapova LI (1980) The diagenesis of the organic matter of the Caspian and Black Sea sediments. In: Gusev AM, Mayev EG (eds) Complex investigations of the nature of the ocean, vol 7 (in Russian). Moscow University Press, p 66
Potapova LI, Gursky YuN (1972) Process of the Black Sea organic matter transformation according to data of luminescent-microscopical study of sediments. Geokhimia 8:997 – 1005
Potapova LI, Kuprin PN, Rakcheyev AD (1973) Infrared investigations of organic matter in recent sediments. In: Vassoyevich NB (ed) Nature of organic matter in recent and fossil sediments (in Russian). Nauka, Moscow, p 239
Prashnowsky A, Degens ET, Emery KO, Pimenta I (1961) Organic matter in recent and ancient sediments, part I. N. Jahrb Palaeontol Monatsh 8:400
Pugh PR (1971) Changes in the fatty acid composition of Coscinodiscus excentricus with culture-age and salinity. Mar Biol 2:118 – 124
Pustelnikov OS (1973) The distribution and composition of suspended matter of the SE Atlantic. The formation of bioproductivity and bottom sediments in relation to water circulation patterns in the SE Atlantic (in Russian). Tr Inst okeanol Akad Nauk SSSR, vol 95 Kaliningrad, p 193 – 210
Pustelnikov OS (1975) Organic matter in suspensions and its supply to the bottom of the Baltic Sea (in Russian). Okeanologiya 15:1040 – 1045
Pustelnikov OS (1983) The sedimentation patterns of the biogenic substances in the Kursch Bay. In: Lisitzin AP (ed) Biosedimentation in the seas and oceans (in Russian). Inst okeanol Akad Nauk SSSR, Moscow, p 31
Pütter A (1909) Die Ernährung der Wassertiere und der Stoffhaushalt der Gewässer. J Fisher, Jena
Pytkowicz RM (1968) Water masses and their properties at 160°W in the Southern Ocean. J Oceanogr Soc Jpn 24:21 – 31
Radkhakrishna K (1969) Geographical changes in the primary productivity in coastal waters of the oceans. Abstract of a thesis (in Russian). Inst okeanol Akad Nauk SSSR, Moscow
Rakusa-Suszczewski S, Dominas H (1974) Chemical composition of the Antarctic Amphipoda Paramaera walkeri Stebbing and chromatographic analysis of its lipids. Pol Arch Hydrobiol 21:216 – 268
Rao S (1960) Organic matter in marine sediments off the east coast of India. Bull Am Assoc Petrol Geol 44:1705 – 1713
Rashid MA, King LH (1970) Major oxygen-containing functional groups present in humic and fulvic acid fractions isolated from contrasting marine environments. Geochim Cosmochim Acta 34:193 – 201
Rashid MA, Prakash A (1972) Chemical characteristics of humic compounds isolated from some decomposed marine algae. J Fish Res Bd Can 29:55 – 60
Rateyev MA, Gorbunova ZN, Lisitzin AP, Nosov GM (1968) Clay minerals in world ocean sediments. In: Kossovskaya AG (ed) Physical and chemical processes and facies (in Russian). Nauka, Moscow, p 213
Raymont JEG, Austin J, Linford E (1967) The biochemical composition of certain oceanic zooplanktonic decapods. Deep-Sea Res 14:113 – 116
Raymont JEG, Srinivasagam RT, Raymont JKB (1969) Biochemical studies on marine zooplankton. IV. Deep-Sea Res 16:141 – 156
Raymont JEG, Srinivasagam RT, Raymont JKB (1971) Biochemical studies on marine zooplankton. VIII. Deep-Sea Res 18:1166 – 1178
Redfield AC, Ketchum BH, Richards FW (1963) The influence of organisms on the composition of sea water. In: The Sea. Vol 2. Wiley, New York, p 26
Reeve MR, Raymont JEG, Raymont JKB (1970) Seasonal biochemical composition and energy sources of Sagitta hispida. Mar Biol 6:357 – 364
Reinheimer G (1973) Bakterien im Stickstoffkreislauf des Meeres. In: Ellenberg H (ed) Ökosystemforschung. Springer, Berlin Heidelberg New York

Richey JE, Brock JT, Neiman RJ, Wissmar RC, Stallard RF (1980) Organic carbon: oxidation and transport in the Amazon River. Science 207:1348 – 1351
Riley GA (1944) The carbon metabolism and photosynthetic efficiency of the Earth as a whole. Am J Sci 32:129 – 134
Riley GA (1970) Particulate organic matter in sea water. Adv Mar Biol 8:1 – 118
Riley GA, Hemert D, Wangersky PJ (1965) Organic aggregates in surface and deep water of the Sargasso Sea. Limnol Oceanogr 10:354 – 363
Riley JP, Segar DA (1970) The seasonal variation of the free and combined dissolved amino acids in the Irish Sea. J Mar Biol Assoc UK 50:713 – 720
Rittenberg SC, Emery KO, Hülsemann J, Degens ET, Fay RC, Reuter JH, Grady JR, Richardson SH, Bray EE (1963) Biogeochemistry of sediments in Experimental Mohole. J Sed Petrol 33:140
Rodina AG (1951) On the role of certain groups of bacteria in the productivity of water bodies (in Russian). Tr Problemn Tematich Sovesch Zool Inst Akad Nauk SSSR 1:23 – 33
Rodionova KF (1951) On the microbial modification of organic matter in marine muds and crude oils, in relation to petroleum formation. In: Commemorative of Academician Gubkin IM (in Russian). Izv Akad Nauk SSSR, Moscow, p 287
Rodionova KF (1967) Organic matter and parental petroleum beds in the Devonian of the Volga-Ural oil-bearing region (in Russian). Nedra, Moscow
Rodionova KF, Pentina TYu (1969) Humic Acids (in Russian). Tr VNIGNI 68:130 – 140
Rodionova KF, Romankevich EA, Diuzhikova TN, Telkova MS (1973) Geochemistry of organic matter in recent marine sediments off the coast of Peru (in Russian). Tr VNIGNI 138: 118 – 129
Romanenko VI (1972) Organic matter destruction in water samples of long-term preservation (in Russian). Biologia vnutrennikh vod. Inf bull N 16, Nauka, Leningrad, p 62
Romankevich EA (1959) Organic matter in sediments of the Kronotsky Bay (in Russian). Tr Inst Okeanol Akad Nauk SSSR 36:32 – 39
Romankevich EA (1962) Organic matter in the surface layer of the West Pacific sediments. In: IGY Oceanological Investigations (in Russian). N 5, Nauka, Moscow, p 67
Romankevich EA (1968) Distribution of organic carbon and nitrogen in Recent and Quarternary sediments of the Pacific Ocean (in Russian). Okeanologiya 8:825 – 838
Romankevich EA (1970) Organic matter in sediments. In: Pacific Ocean. Sedimentation in the Pacific Ocean, vol 2 (in Russian). Nauka, Moscow, p 107
Romankevich EA (1971) The relation of the organic matter of suspended matter, bottom sediments and benthos with the biological productivity (in Russian). Dokl Akad Nauk SSSR, Moscow 198(5):1199 – 1202
Romankevich EA (1974a) Biogeochemical composition of sediments of the Pacific Ocean (in Russian). Litol Polezn Iskop 1:27 – 40
Romankevich EA (1974b) Bioorganic composition of the particulate matter and bottom sediments in the north-western Pacific Ocean. In: Vassoyevich NB (ed) Organic matter in recent and fossil sediments and methods of its investigation (in Russian). Nauka, Moscow, p 33
Romankevich EA (1976) Organic matter of bottom sediments east off Japan and its influence on the oxidation-reduction processes. In: Volkov II (ed) The biogeochemistry of the diagenesis of oceanic sediments (in Russian). Nauka, Moscow, p 5
Romankevich EA (1977) Geochemistry of organic matter in the ocean (in Russian). Nauka, Moscow
Romankevich EA (1979) Carbon of the suspended organic matter. In: Bordovsky OK, Ivanenkov VN (eds) Oceanology, Chemistry of the ocean, vol 1 (in Russian). Nauka, Moscow, p 259
Romankevich EA, Artemiev VE (1969) Organic matter composition of sediments in the Kuril-Kamchatka trench (in Russian). Okeanologiya 9:796 – 806
Romankevich EA, Batrakov SG (1971) Diagenetic changes in the free lipid composition of the sediments in the Western Pacific Ocean (in Russian). Geokhimiya 11:1353 – 1362
Romankevich EA, Baturin GN (1972) Organic matter composition of phosphorites from the shelf off Southwestern Africa (in Russian). Geokhimiya 6:719 – 726
Romankevich EA, Baturin GN (1974) Biogeochemical composition of the sediments from the West-African Shelf (5 – 23 °S) (in Russian). Okeanologiya 14:660 – 664
Romankevich EA, Liutsarev SV (1983) The biogeochemical carbon cycle in the ocean. In: Khain VE, Sokolov BA, Nazarevich IA (eds) Advances in the development of the sedimentary-migrational theory of hydrocarbons formation (in Russian). Nauka, Moscow

Romankevich EA, Urbanovich IM (1971) Organic matter (nitrogen, carbohydrates) in the particulate fraction, bottom sediments and interstitial waters of the Peruvian-Chilean region (in Russian). Tr Inst Okeanol Akad Nauk SSSR 89:106–117
Romankevich EA, Bezrukov PL, Baranov VI, Khristianova LA (1966) Stratigraphy and absolute age of deep-water sediments in the Pacific Ocean. In: IGY oceanological investigations (in Russian). N 14. Nauka, Moscow, p 5
Romankevich EA, Artemiev VE, Beliayeva AN, Liutsarev SV (1982a) The biogeochemistry of the dissolved and particulate organic matter in the ocean. In: Sidorenko AV, Geodekyan AA (eds) Organic geochemistry of waters and prospecting geochemistry (in Russian). Nauka, Moscow, p 7
Romankevich EA, Daniushevskaya AI, Beliayeva AN, Rusanov VP (1982b) The biogeochemistry of the organic matter of the Arctic Seas (in Russian). Nauka, Moscow
Ronov AB, Migdisov AA (1970) Evolutionary changes in the chemical composition of the rocks in shields and the sedimentary cover of the Russian and North American platforms (in Russian). Geokhimiya 4:403–438
Ronov AB, Yaroshevsky AA (1967) Chemical composition of the Earth's crust (in Russian). Geokhimiya 11:1285–1309
Rose AH (1968) Chemical microbiology. 2nd ed. Butterworths, London
Rosenfeld JK (1981) Nitrogen diagenesis in Long Island sound sediments. Am J Sci 281:436–462
Rubenchik LI (1972) Biological indicator microorganisms (in Russian). Naukova Dumka, Kiev
Rutten M (1973) On the origin of life (in natural environments) (in Russian). Mir, Moscow
Ryther JH (1969) Photosynthesis and fish production in the sea. Science 166:72–76
Sackett WM, Poag CW, Eadie BJ (1974) Kerogen recycling in the Ross Sea, Antarctica. Science 185:1045–1047
Saliot A (1981) Natural hydrocarbons in sea water. In: Duursma EK, Dawson R (eds) Marine organic chemistry. Elsevier, Amsterdam Oxford New York, p 327
Saliot A, Barbier M (1974) Acides gras, hydrocarbures et sterols dissous dans l'eau de mer. In: Advances in organic geochemistry (1973). Technip, Paris, 1974, p 616–617
Saliot A, Boussuge C (1978) Inventaire et dynamique des lipides à l'interface eau de sédiment. II. – Acides gras de l'eau de mer et de l'eau interstititielle. In: Géochimie organique des sédiments marins profonds. Ed CNRS, Paris, p 239
Saliot A, Tissier MJ (1977) Interface eau-sédiment: acides gras et hydrocarbures dissous et particulaires dans l'eau de mer. In: Géochimie organique des sédiments marins profonds. Orgon I. Mer de Norvège. Ed CNRS, Paris, p 197
Saliot A, Tissier MJ (1978) Inventaire et dynamique des lipides à l'interface eau de mer/sédiment. IV. Hydrocarbures aliphatiques et alicycliques de l'eau de mer et de l'eau interstitielle. In: Géoçhimie organique des sédiments marins profonds. Orgon II. Atlantique N-E-Brésil. Ed CNRS, Paris, p 263
Saliot A, Goutx M, Fevrier A, Tusseau D, Andrie C (1982) Organic sedimentation in the water column in the Arabian sea: relationship between the lipid composition of small and large-size, surface and deep particles. Mar Chem 11:257–278
Sargent JR, Lee RF (1975) Biosynthesis of lipids in zooplankton from Saanich Inlet, British Columbia, Canada. Mar Biol 31:15–23
Scherbakov FA, Kuprin PN, Potapova LI, Poliakov AS, Zabelina EK, Sorokin VM (1978) Sediment accumulation on the continental margin of the Black Sea (in Russian). Nauka, Moscow
Schlesinger WH, Melack JM (1981) Transport of organic carbon in the world's rivers. Tellus 33:172–187
Schultz DM, Quinn JG (1972) Fatty acids in surface particulate matter from the North Atlantic. J Fish Res Bd Can 29:1482–1486
Schulz S (1973) Biologische Wirkungen des Wasserauftriebes. Biol Rundsch 11:238–247
Seki H, Nakai T, Otobe H (1972) Regional differences on turnover rate of dissolved materials in the Pacific Ocean at summer of 1971. Arch Hydrobiol 71:79–89
Sellner KG, Hendrikson P, Ochoa N (1983) Relationships between the chemical composition of particulate organic matter and phytoplankton distribution in recently upwelled waters off Peru. In: Suess E, Thiede J (eds) Coastal Upwelling Its Sediment Record Part A. Plenum Press, New York London, p 273
Semenov AD (1971) Organic matter in surface waters of the Soviet Union. Abstract of a thesis (in Russian). Hydrochem Inst, Novocherkassk

Semenov AD, Nemtsova LI, Kishkinova TS, Pashanova AP (1967) On organic matter in atmospheric precipitation (in Russian). Dokl Akad Nauk SSSR 173:1185–1187

Semina GI (1974a) Pacific phytoplankton (in Russian). Nauka, Moscow

Semina GI (1974b) Phytoplankton in the Pacific Ocean (in Russian). Nauka, Moscow

Semina GI (1977) The phytoplankton. In: Vinogradov ME (ed) Oceanology. Biology of the ocean, vol 1 (in Russian). Nauka, Moscow, pp 58, 117

Semina GI (ed) (1982) Oceanic phytoplankton and primary production, vol 114 (in Russian). Trans PP Shirshov Inst Okeanol Akad Nauk SSSR, Moscow

Senin YuM (1970) Distribution specificities of biogenic components in shelf grounds off Northwestern Africa in relation to the biological productivity and upwellings of deep waters (in Russian). Tr Atlant NIRO 27:247–264

Serenkov GP, Pakhomova MV (1959) Biochemical investigations of two diatom algal species (in Russian). Vestnik MGU, Ser Biol Pochv Geol i Geogr 2:39–46

Serenkov GP, Pakhomova MV (1965) A contribution to the biochemistry of blue-green algae. In: Kuzin BS (ed) Ecology and physiology of blue-green algae (in Russian). Nauka, Moscow, p 177

Serenkov GP, Pakhomova MV, Borisova IG (1957) Comparative biochemical studies of two green algal species (in Russian). Vestnik MGU, Ser Biol Pochv Geol i Geogr 3:77–85

Setchell FW (1981) Particulate protein measurement in oceanographic samples by dye binding. Mar Chem 10:301–313

Shabarova NT (1950) Carbohydrate changes in marine plants in the process of decomposition. In: Senukov VM (ed) Contemporary analogues of oil-bearing facies (in Russian). Gostoptekhizdat, Moscow, p 284

Shabarova NT (1955) Biochemical composition of deep-water marine deposits (ocean bottom sediments) (in Russian). Biokhimiya 20:146–151

Shadsky IP, Romankevich EA, Baturin GN (1980) The isotopic composition of carbon in lipids of modern phosphorite concretions and the problem of their formation (in Russian). Dokl Akad Nauk SSSR 254:743–744

Shadsky IP, Romankevich EA, Grinchenko YuI (1982) The isotopic composition of carbon in lipids of the Atlantic sediments (in Russian). Geokhimiya 1:99–105

Sharp JH (1973) Size classes of organic carbon in sea water. Limnol Oceanogr 18:441–447

Sheldon RW, Prakash A, Sutcliffe WH (1972) The size distribution of particles in the ocean. Limnol Oceanogr 17:327–340

Shimkus KM, Trimonis ES (1974) Modern sedimentation in Black Sea. In: Degens ET, Ross DA (eds) The Black Sea – geology, chemistry, and biology. Publ Am Ass Petrol Geol, Tulsa, Oklahoma, USA, p 249

Shirinsky VG, Romankevich EA, Beliayeva AN (1974) Lipid carbon isotope composition of bottom sediments as an indicator of the rate of organic matter transformation. IVth All-Union Seminar Organic matter in recent and fossil sediments. Abstracts of papers (in Russian). Moscow University Press, p 40

Shishenina EP, Popova NV, Chernova TG, Telkova MS, Morozova RM (1974) On the geochemistry of hydrocarbons of the organic matter of recent sediments. Geokhimiya 8:1212–1219

Shvets VM (1973) Organic matter of subterranean waters (in Russian). Nedra, Moscow

Sieburth JM (1971) Distribution and activity of oceanic bacteria. Deep-Sea Res 18:1111–1121

Silen LG (1976) The ocean as a chemical system. Science 156:1189–1196

Simoneit BRT (1977) Organic matter in eolian dusts over the Atlantic ocean. Mar Chem, vol 5. Special issue. Andersen NR (ed) Concepts in marine organic chemistry, p 443

Skopintsev BA (1950) Organic matter in natural waters (aquatic humus) (in Russian). Tr Gos Okeanogr Inst 17

Skopintsev BA (1961) Some results of organic matter studies in sea waters as applied to sedimentation processes. In: Strakhov NM (ed) Recent sediments in seas and oceans (in Russian). Izv Akad Nauk SSSR, Moscow, p 285

Skopintsev BA (1965a) Estimates of the formation and oxidation of organic matter in sea waters. In: IGY Oceanological investigations (in Russian). N 13, Moscow, p 96–106

Skopintsev BA (1965b) Investigations on the oxygen minimum layer in the Northern Atlantic during the autumn of 1959. In: IGY Oceanological investigations (in Russian). N 13, Nauka, Moscow, p 108–114

Skopintsev BA (1966) Certain considerations on the distribution and forms of occurrence of organic matter in ocean waters (in Russian). Okeanologiya 6:441–450

Skopintsev BA (1971) Recent advances in the studies of organic matter in ocean waters (in Russian). Okeanologiya 11:939–956

Skopintsev BA (1975) Formation of the contemporary chemical composition of the Black Sea waters (in Russian). Gidrometeoizdat, Leningrad

Skopintsev BA (1981) Decomposition of organic matter of plankton, humification and hydrolysis. In: Duursma EK, Dawson R (eds) Marine organic chemistry. Elsevier, Amsterdam Oxford New York, p 125

Skopintsev BA, Krylova LP (1955) Organic matter load of largest rivers in the Soviet Union (in Russian). Dokl Akad Nauk SSSR 105:770–773

Skopintsev BA, Timofeyeva SN (1961) The dry combustion method, proposed by LP Krylova, in application to the determination of organic carbon in sea waters (in Russian). Gidrokhim Materialy 32:153–176

Skopintsev BA, Timofeyeva SN (1962) The organic carbon content of the Baltic and North Seas, subtropical and tropical areas of the North Atlantic Ocean (in Russian). Tr Morsk Gidrofiz Inst Akad Nauk SSSR 25:110–117

Skopintsev BA, Timofeyeva SN, Vershinina OA (1966) Organic carbon in the waters of the Equatorial and South Atlantic Ocean and the Mediterranean Sea (in Russian). Okeanologiya 6:251–260

Skopintsev BA, Timofeyeva SN, Danilenko AF, Sokolova MV (1971) Organic matter in the Black Sea waters, and its mineral derivatives. In: Vassoyevich NB (ed) Organic matter of recent and fossil sediments (in Russian). Nauka, Moscow, p 16

Skopintsev BA, Bordovsky OK, Ivanenkov VN (1979) Carbon of the dissolved organic matter. In: Bordovsky OK, Ivanenkov VN (eds) Oceanology. Chemistry of the ocean, vol 1 (in Russian). Nauka, Moscow, p 251

Skorniakova NS (1961 a) Bottom sediments of the Northeastern Pacific Ocean (in Russian). Tr Inst Okeanol Akad Nauk SSSR 45:22–64

Skorniakova NS (1961 b) Bottom sediments of the Sea of Japan. In: Geology and hydrology of the Sea of Japan (in Russian). Izv Akad Nauk SSSR, Moscow, p 23

Skorniakova NS, Petelin VP (1967) Sediments of the central South Pacific region (in Russian). Okeanologiya 7:1005–1019

Slowey JF, Jeffrey LM, Hood DW (1962) The fatty acid content of ocean water. Geochim Cosmochim Acta 26:607–617

Smirnov BA (1976) Hydrocarbons in recent marine sediments and their probable genetic relationship to hydrocarbons of scattered organic matter and crude oils (in Russian). Vth All-Union seminar Organic matter of recent and fossil sediments. Abstracts of papers. Moscow University Press, p 41–42

Smirnov BA (1980) The general pattern of transformation of hydrocarbons in the process of sedimentation of organic matter in the world ocean (in Russian). Okeanologiya 20:856–865

Smith PV (1954) Studies on origin of petroleum: occurrence of hydrocarbons in Recent sediments. Bull Am Ass Petrol Geol 38:377–404

Sokolov VA (1971) Geochemistry of natural gases (in Russian). Nedra, Moscow

Sokolov VS, Sokolova EG (1975) A volumetric method for the determination of carbonates and organic carbon in marine sediments. In: Ostroumov EA (ed) Chemical analysis of marine sediments (in Russian). Nauka, Moscow, p 17

Sorokin YuI (1968) Studies of the bacterial processes in the Black Sea. In: Vodianitsky VA (ed) Biological investigations in the Black Sea and its fishery resources (in Russian). Nauka, Moscow, p 18

Sorokin YuI (1970) Studies of the quantity, production and functional activities of bacteria in the Black Sea. In: Vodianitsky VA (ed) Biology of the sea (in Russian), N 19. Naukova Dumka, Kiev, p 43

Sorokin YuI (1971) The amount and production of bacteria in Central Pacific water column (in Russian). Okeanologiya 11:105–117

Sorokin YuI (1973) Biogeochemical activities and trophical role of bacteria in sea waters (in Russian). J Obsch Biol 34:396–406

Sorokin YuI (1977) The bacterioplankton. In: Vinogradov ME (ed) Oceanology. Biology of the ocean, vol 1 (in Russian). Nauka, Moscow, p 124

Sorokin YuI (1978) Decomposition of organic matter and nutrient regeneration. In: O'Kinne (ed) Marine Ecology. Wiley, Chichester New York Brisbane Toronto, p 501

Sorokin YuI (1982) The Black Sea (in Russian). Nauka, Moscow
Spoehr HA, Milner HW (1959) The chemical composition of chlorella, effect on environment conditions. Plant Physiol 24:120
Stadnikov GL (1937) Origin of coals and petroleum (in Russian). Izv Akad Nauk SSSR, Moscow
Staresinic N (1981) Downward flux of particulate organic matter in the Peru coastal upwelling and its relationship to primary production. In: Coastal upwelling: its sediment record. Symposium. Vilamoura Algarve, Portugal, p 89
Starikova ND (1961) Organic matter in the liquid phase of sediments in seas and oceans (in Russian). Tr Inst Okeanol Akad Nauk SSSR 50:130–169
Starikova ND (1966) Dissolved organic matter. In: Kort VG (ed) The Pacific Ocean. Chemistry of the Pacific Ocean (in Russian). Nauka, Moscow, p 245
Starikova ND (1967) The content and distribution of the organic carbon in the water of the Indian Ocean. In: Bruyevich SV (ed) Chemical processes in the sea water and sediments (in Russian). Tr Inst Okeanol Akad Nauk SSSR 83:38–50
Starikova ND (1970) Regularities in the vertical distribution of dissolved organic carbon in sea waters and ground solutions of marine sediments (in Russian). Okeanologiya 10:988–1000
Starikova ND, Korzhikova LI (1970) Amino acids in the water, suspended matter and bottom sediments of the Kuril-Kamchatka trench area. In: Bogorov VG (ed) The fauna of the Kuril-Kamchatka trench and its habitation conditions, vol 86 (in Russian). Tr Inst Okeanol Akad Nauk SSSR, p 39
Starikova ND, Korzhikova LI (1972) Amino acid content and composition in waters, particulate matter, sediments and ground solutions of the Black Sea (in Russian). Geokhimiya 2:230–240
Starikova ND, Korzhikova LI (1973) Amino acid composition of organic matter in the Northwestern Pacific Ocean. In: Skopintsev BA, Ivanenkov VN (eds) Chemistry of seas and oceans (in Russian). Nauka, Moscow, p 56
Starikova ND, Yablokova OG (1972) Carbohydrates in the Black Sea (in Russian). Okeanologiya 12:431–436
Starikova ND, Yablokova OG (1973) Organic matter in the Northwestern Pacific Waters (along the profile Sangara Strait-Atoll Wake) (in Russian). Okeanologiya 13:1009–1014
Starikova ND, Korzhikova LI, Yablokova OG (1974) Organic carbon, nitrogen, amino acids in the liquid and solid phases of Northwestern Pacific sediments (in Russian). Dokl Akad Nauk SSSR 216:422–425
Steemann-Nielsen E, Jensen A (1957) Primary oceanic production. The autotrophic production of organic matter in the oceans. Galathea Rep 1:47–135
Stepanov VN (1961) The principle dimensions of the world ocean and its major constituents (in Russian). Okeanologiya 1:213–219
Stepanov VN (1982) The nature of the world ocean (in Russian). Prosveshcheniye, Moscow
Stepanov VN, Svetlichnyi LS (1975) In relation to calculations of the velocity of passive vertical migrations of planktonic organisms (in Russian). Okeanologiya 15:321–324
Stepanov VN, Sarkisian AS, Demin YuL, Drozdov VN, Bulatov RP, Galerkin LI, Shcherbinin AD (1977) The diagnostic calculations of the horizontal circulation of ocean waters (in Russian). Morskiye Gidrofiz Issl 1:34–46
Stepanov VN, Galerkin LI, Kutko VP, Oleinikov SA, Panfilova SG, Shcherbinin AD (1982) The climatologo-statistical study of the thermohaline fields in the North Atlantic (in Russian). Gidrometeoizdat, Moscow
Stevenson FJ (1974) Nonbiological transformations of amino acids in soils and sediments. In: Tissot B, Bienner F (eds) Advances in organic geochemistry 1973. Technip, Paris, p 701
Stevenson FJ, Tilo SN 1966 (1970) Nitrogenous constituents of deep-sea sediments. In: Proc 3rd Int Meet Org Geochem, London, 1966, p 227
Stewart RW, Hameed S, Pinto J (1978) The natural and perturbed troposphere. In: Levine JS, Schryer DR (eds) Man's impact on the troposphere. National Aeronautics and Space Administration Ref Publ 1022:27–64
Strakhov NM (1947) Bearing on the principles and mechanism of marine sedimentation. 1. Black Sea (in Russian). Izv Akad Nauk SSSR, Ser Geol N 2:49–90
Strakhov NM (1960) Fundamentals of the theory of lithogenesis, vols 1 and 2 (in Russian). Izv Akad Nauk SSSR, Moscow
Strakhov NM (ed) (1966) Silicon geochemistry (in Russian). Nauka, Moscow

Strakhov NM, Volkov II, Lisitzina NA (1973) On the formation of the element distribution mechanism in the Pacific Ocean (Japan profile) (in Russian). Litol i Polezn Iskop 3:3–28

Strickland JDH (1965) Production of organic matter in the primary stages of the marine food chain. In: Riley IP, Skirrow G (eds) Chemical oceanography, vol 1. Academic Press, London New York, p 477

Strickland JDH, Parsons TR (1968) A practical handbook of sea water analysis. Bull Fish Res Bd Can 167

Strickland JDH, Holm-Hansen O, Eppley RW, Zinn RJ (1969) The use of a deep tank in plankton ecology. I Studies of the growth and composition of phytoplankton crops at low nutrient levels. Limnol Oceanogr 14:23–34

Stuermer DH, Harvey GR (1974) Humic substances from sea water. Nature 250:480–481

Stuermer DH, Harvey GR (1978) Structural studies on marine humus: a new reduction sequence for carbon skeleton determination. Mar Chem 6:55–70

Stupakova TP (1975) The effect of high hydrostatic pressure on the activities of catabolic enzymes of baro-tolerant bacteria (in Russian). Mikrobiol 44:174–175

Suess E (1980) Particulate organic carbon flux in the oceans – surface productivity and oxygen utilization. Nature 288:260–263

Suess E, Müller PJ (1980) Productivity, sedimentation rate and sedimentary organic matter in the oceans II. – Elemental fractionation. In: Colloques Int du CNRS N 293 – Biogéochimie de la matière organique à l'interface eau-sédiment marin. Ed CNRS, Paris, p 17

Suess E, Thiede J (eds) (1983) Coastal upwelling: its sediment record. Part A: Responses of the sedimentary regime to present coastal upwelling. Plenum Press, New York London

Suschenia LM (1961) Some data on the quantity of sestone in the waters of the Aegian, Ionian and Adriatic Seas (in Russian). Okeanologiya 1:664–669

Suschenia LM (1972) The respiration intensity of crustaceans (in Russian). Naukova Dumka, Kiev

Sverdrup HU (1955) The place of physical oceanography in oceanographic research. J Mar Res 14:287–294

Sverdrup HU, Johnson MW, Fleming RH (1942) The oceans, their physics, chemistry and general biology. Prentice Hall, New York

Sviridova IV (1980) The general laws of the distribution, transformation and cycle of nitrogen compounds in the Pacific. Abstract of a thesis (in Russian). PP Shirshov Inst Okeanol Akad Nauk SSSR, Moscow

Szekielda KH (1968a) Le carbon particulaire dans les masses d'eau profondes. Mar Biol 2:71–72

Szekielda KH (1968b) The transport of organic matter by the bottom water of the oceans. Sarsia 34:243–252

Tambiyev AH, Kirikova NN (1981) Release of organic matter by marine algae (in Russian). Usp Sovrem Biol 92:100–114

Tanoue E, Handa N (1979) Distribution of particulate organic carbon and nitrogen in the Bering Sea and northern North Pacific Ocean. J Oceanogr Soc Jpn 35:47–62

Tanoue E, Handa N (1980a) Some characteristic features of the vertical profile of organic matter in recent sediments from the Bering Sea. J Oceanogr Soc Jpn 36:1–14

Tanoue E, Handa N (1980b) Vertical transport of organic materials in northern North Pacific as determined by sediment trap experiment. J Oceanogr Soc Jpn 36:231–245

Tatsumoto M, Williams WT, Prescott JM, Hood DW (1961) Amino acids in samples of surface sea water. J Mar Res 19:89–95

Tissier MJ, Dastillung M (1978) Inventaire et dynamique des lipides à l'interface eau de mer/sédiment. V. Hydrocarbures polyaromatiques des sédiments, de l'eau de mer et de l'eau interstitielle. In: Géochimie organique des sédiments marins profonds. Orgon II. Atlantique N-E – Brésil. Ed CNRS, Paris, p 275

Tissot BP, Welte DH (1978) Petroleum formation and occurrence. Springer, Berlin Heidelberg New York

Towle DW, Pearse JS (1973) Production of the giant kelp, Macrocystis, estimated by in situ incorporation of ^{14}C in polyethylene bags. Limnol Oceanogr 18:155–159

Trask PD (1932) Origin and environment of source sediments of petroleum. Am Petrol Inst Gulf Publ, Houston

Trask PD (1953) The sediments of the Western Gulf of Mexico Part II. Papers Phys Oceanogr and Meteorol. Mass Inst Technol Woods Hole Oceanogr Inst 12:51

Trask PD, Wu CC (1930) Does petroleum form in sediments at time of deposition. Bull Am Assoc Petrol Geol 14:1451
Trequer P, Corre PL, Courtot P (1972) A method for determination of the total dissolved free fatty acid content of sea water. J Mar Biol Assoc UK 52:1045 – 1055
Tseitlin VB (1981) The assessment of the vertical flux of detritus from the surface zone in tropical oceanic areas (in Russian). Okeanologiya 21:713 – 718
Ushakov AN, Vitiuk DM, Vaver VA, Bergelson LD (1966) Fatty acids in the Black Sea waters (in Russian). Okeanologiya 6:891 – 893
Uspensky VA (1956) The carbon balance of the biosphere in relation to carbon distribution in the Earth's crust (in Russian). Gostoptekhizdat, Leningrad
Uspensky VA (1970) An introduction into petroleum geochemistry (in Russian). Nedra, Leningrad
Vallentyne JR (1962) Solubility and the decomposition of organic matter in nature. Arch Hydrobiol 58:423
Vallentyne JR (1963) Geochemistry of carbohydrates. In: Organic Geochemistry. Pergamon, Oxford London New York Paris
Van Andel TH (1964) Recent marine sediments of the Gulf of California. In: Van Andel TjH, Shor GD (eds) Marine geology of the Gulf of California. Am Assoc Petrol Geol Mem 3:216
Van Andel TH, Postma H (1954) Recent sediments of the Gulf Paria. Verh K Ned Akad Wet 20:254
Van Vleet ES, Quinn JG (1979) Diagenesis of marine lipids in ocean sediments. Deep-Sea Res 26:1225 – 1236
Vasilevskaya NA, Maksimov OB (1982) Chemical composition of humic acids of continental and marine sediments from destructive oxidation data. In: Sidorenko AV, Vassoyevich NB, (eds) Geochemistry of recent and fossil sediments (in Russian). Nauka, Moscow, p 34
Vasilevskaya NA, Golyashin VN, Denisenko NM, Maksimov OB (1977) Chemical investigations of humic acids in bottom sediments of the Western Pacific Ocean (in Russian). Okeanologiya 17:459 – 469
Vassoyevich NB (1973) The principal regularities peculiar to organic matter of recent and fossil sediments. In: Vassoyevich NB (ed) The nature of organic matter in recent and fossil sediments (in Russian). Nauka, Moscow, p 11
Vassoyevich NB (1982) On the need of new classifications and new names of carboniferous organic compounds. In: Vassoyevich NB, Timofeyev PP, Nazarevich IA (eds) Sedicahytes at various stages of lithogenesis (in Russian). Nauka, Moscow, p 7
Vassoyevich NB, Kornilova NN, Chernyshov VV (1973) On the content of carboniferous organic matter in the continental sector of the Earth's sedimentary cover (in Russian). Vestnik MGU 2:8 – 23
Vassoyevich NB, Konyukhov AI, Lopatin NV (1976) Common features and specificities in the formation of coals, petroleum and gaseous hydrocarbons. In: Fossil fuel (in Russian). Nauka, Moscow, p 7
Vedernikov VI, Starodubtsev EG (1971) Primary productivity and chlorophyll in the Southeastern Pacific (in Russian). Tr Inst Okeanol Akad Nauk SSSR 89:33 – 43
Verigo AA (1881) Investigations of Odessa medicinal estuaries and muds. Rep on the activities of Odessa balneological soc (in Russian). N 1. Odessa, p 1 – 108
Verigo AA (1888) Investigations of the medicinal muds from the Balaklava Bay shallows. Rep on the activities of Odessa balneological Soc 1883 – 1887 (in Russian), N 3. Odessa, p 21 – 26
Vernadsky VI (1921) Concerning the problems of geochemical investigations in the Azov Sea and its basin. Selected works, vol 5 (in Russian). Izv Akad Nauk SSSR, Moscow
Vernadsky VI (1934) Studies on geochemistry (in Russian). Gorgeonefteizdat, Moscow
Vernadsky VI (1960) Mineral history of the Earth's crust. Selected works, vol 4 (in Russian). Izv Akad Nauk SSSR, Moscow
Vernadsky VI (1965) Chemical composition of the Earth's biosphere and its surroundings (in Russian). Nauka, Moscow
Vernadsky VI (1967) Biosphere (selected works on biogeochemistry) (in Russian). Mysl, Moscow
Vernadsky VI (1980) The problems of biogeochemistry (in Russian). Nauka, Moscow
Victor RG, Anantha PS (1973) Distribution of phosphorus and organic carbon in the near-shore sediments of Goa. Indian J Mar Sci 2:84 – 89
Vigneaux M, Dumon JC, Faugeres JC, Grousset F, Jouanneau JM, Latouche C, Poutiers J, Pujol C (1980) Matière organique et sédimentation en milieu marin. In: Biogéochimie de la matière organique à l'interface eau-sédiment marin. Ed CNRS, Paris, p 113

Vikherenko NM (1964) Organic carbon in bottom sediments of the North Atlantic Ocean (in Russian). Okeanologiya 4:437–446
Vikhrenko NM (1966) Distribution and composition of organic matter in the surface layer of Atlantic Ocean sediments. In: Bruevich SV (ed) Chemical processes in seas and oceans (in Russian). Nauka, Moscow, p 59
Vinberg GG (1960) Primary productivity of water bodies (in Russian). Izv Akad Nauk SSSR, Minsk
Vinogradov AP (1953) The elementary chemical composition of marine organisms. Mem 2, Sears Found Mar Res, New Haven
Vinogradov AP (1962) Average content of chemical elements in the major igneous rock types of the Earth's crust (in Russian). Geokhimiya 7:555–571
Vinogradov AP (1964) The organic matter in the chemistry of the Earth. In: Manskaya SM, Drozdova TV (eds) Geochemistry of organic matter (in Russian). Nauka, Moscow, p 4
Vinogradov AP (1967) An introduction into ocean geochemistry (in Russian). Nauka, Moscow
Vinogradov AP (1973) Technical progress and environmental control of the biosphere (in Russian). Vestn Akad Nauk SSSR 9:7–15
Vinogradov ME (1968) Vertical distribution of oceanic zooplankton (in Russian). Nauka, Moscow
Vinogradov ME, Lisitzin AP (1981) The global laws of the distribution of life in the ocean and their reflection in the composition of bottom sediments. Distribution patterns of plankton and benthos in the ocean (in Russian). Izv Akad Nauk SSSR 3:5–25
Vinogradov ME, Bordovsky OK, Akhmetieva EA (1970) Biochemistry of oceanic plankton. Chemical composition of plankton at various depths in the Northwestern Pacific (in Russian). Okeanologiya 10:871–878
Vinogradova ZA (1967a) Biochemical aspects of marine plankton investigations. In: Vodianitsky VA (ed) Problems of biooceanography (in Russian). Naukova Dumka, Kiev, p 52
Vinogradova ZA (1967b) Some aspects of zooplankton biochemistry. In: Vinogradov KA (ed) Biology of the Northwestern Black Sea (in Russian). Naukova Dumka, Kiev, p 93
Vinogradova ZA (1967c) A contribution to the plankton biochemistry of the Equatorial Atlantic. In: Vinogradova ZA (ed) Biochemistry of marine organisms (in Russian). Naukova Dumka, Kiev, p 27
Vinogradova ZA (1967d) Biochemical composition of the antarctic plankton. In: Vinogradova ZA (ed) Biochemistry of marine organisms (in Russian). Naukova, Dumka, Kiev, p 7
Vitiuk DM (1967) The dissolved fatty acid fraction in the Black Sea. In: Vodianitsky VA (ed) Black Sea hydrodynamics and aspects of hydrochemistry (in Russian). Naukova Dumka, Kiev, p 143
Vitiuk DM (1970) Fatty acid fraction of dissolved organic matter in the Red Sea. In: Bruevich SV (ed) Chemical resources of seas and oceans (in Russian). Nauka, Moscow, p 169
Voyshvillo NA, Akhrem AA, Titov YuA (1972) Algae sterols. Biol Nauk 1:92–95
Vozzhinskaya VB (1974) Seasonal aspects of biology and phytobenthic productivity in the White Sea. In: Golikov AN (ed) Hydrobiology and biogeography of shelf areas in cold and temperate waters of the world ocean (in Russian). Nauka, Leningrad
Vozzhinskaya VB (1977) Sea-bottom vegetation. In: Vinogradov ME (ed) The biology of the ocean, vol 1 (in Russian). Ser Okeanologiya. Nauka, Moscow, p 78
Waksman SA (1933) On the distribution of organic mater in the sea bottom and the chemical nature and origin of marine humus. Soil Sci 36:125–147
Waksman SA, Renn ChE (1936) Decomposition of organic matter in sea water by bacteria. III Factors influencing the rate of decomposition. Biol Bull 70:472–483
Walsh GE (1965) Studies on dissolved carbohydrate in Cape Cod waters. I. General survey. Limnol Oceanogr 10:570–576
Walsh GF, Douglass J (1966) Vertical distribution of dissolved carbohydrate in the Sargasso Sea off Bermuda. Limnol Oceanogr 11:406–408
Wangersky PJ (1959) Dissolved carbohydrates in Long Island Sound, 1956–1958. Bull Bingham Oceanogr Coll 17:87–94
Wangersky PJ (1976) Particulate organic carbon in the Atlantic and Pacific oceans. Deep-Sea Res 23:457–465
Wangersky PJ (1978) Production of dissolved organic matter. In: Kinne O (ed) Marine ecology, vol. IV. Wiley, Chichester New York Brisbane Toronto, p 115
Wangersky PJ, Gordon DC (1965) Particulate carbonate, organic carbon and Mn^{++} in the open ocean. Limnol Oceanogr 10:544–550

Warnke DA, Richter J, Oppenheimer CH (1973) Characteristics of the nearshore environment off the south coast of Anvers Island, Antarctic Peninsula. Limnol Oceanogr 18:131 – 142
Weber VV (1966) Facies, favorable for petroleum formation (in Russian). Nedra, Moscow
Weber VV (1973) Oil-bearing suites and their contemporary analogues (in Russian). Nedra, Moscow
Weber VV, Gorbunova LI (1961) On the role of the mineralogical composition of pelite in the process of bitumenformation (in Russian). Sov Geol 11:73 – 85
Weber VV, Ginsburg-Karagicheva TL, Gelbovskaya EA, Gorskaya AI, Zakharov AA, Manucharova EA, Mekhtiyeva VL, Romm II, Savich VG, Taldykina NN, Fokina NI, Yurkevich IA (1956) Accumulation and transformation of organic matter in recent marine sediments (in Russian). Gostoptekhizdat, Moscow
Weber VV, Gorskaya AI, Glebovskaya EA (1960) Bitumen formation in Quarternary sediments and the origin of petroleum (in Russian). Gostoptekhizdat, Moscow
Wehmiller J, Hare PE (1971) Racemization of amino acids in marine sediments. Science 173:907 – 911
Wells JM, Wells AH, Van Derwalker JG (1973) In situ studies of metabolism in benthic reef communities. Helgoländ Wiss Meeresuntersuch 24:78 – 81
Wheeler J (1972) Some effects of solar levels of ultraviolet radiation on lipids in artificial sea water. J Geophys Res 77:5302 – 5306
Whittle KJ (1977) Marine organisms and their contribution to organic matter in the ocean. Mar Chem 5:381 – 411
Wiebe WJ, Pomeroy LR (1972) Microorganisms and their association with aggregates and detritus in the sea: a microscopic study. Mem Ist Ital Idrobiol 29:325 – 352
Williams PJ (1975) Biological and chemical aspects of dissolved organic material in sea water. In: Riley JP, Skirrow G (eds) Chemical oceanography, vol 2, 2nd ed. Academic Press, London New York San Francisco, p 301
Williams PM (1961) Organic acids in Pacific Ocean waters. Nature 189:219 – 220
Williams PM (1965) Fatty acids derived from lipids of marine origin. J Fish Res Bd Can 22:1107 – 1122
Williams PM (1967) Sea surface chemistry: organic carbon and organic and inorganic nitrogen and phosphorus in surface films and subsurface waters. Deep-Sea Res 14:791 – 800
Williams PM (1968) Stable carbon isotopes in dissolved organic matter of the sea. Nature 219:152 – 153
Williams PM (1969) The association of copper with dissolved organic matter in sea water. Limnol Oceanogr 14:156 – 158
Williams PM, Gordon LI (1970) Carbon 13: carbon-12 ratios in dissolved and particulate organic matter in the sea. Deep-Sea Res 17:19 – 27
Williams PM, Oeschger M, Kinney P (1969) Natural radiocarbon activity of the dissolved organic carbon in the North-east Pacific Ocean. Nature 224:256 – 258
Wilson RD, Monaghan PH, Osanik A, Price LC, Rogers MA (1974) Natural marine oil seepage. Science 184:857 – 865
Wiseman JD, Bennett H (1940) The distribution of organic carbon and nitrogen in sediments from the Arabian Sea. Sci Rept John Murray Exped 1933 – 1934, vol 3, N 4. Br Mus, London
Wolff WJ (1980) Biotic aspects of the chemistry of estuaries. In: Olausson E, Cato J (eds) Chemistry and biogeochemistry of estuaries. Wiley, Chichester New York Brisbane Toronto, p 263
Wollast R (1980) Degradation mechanism of organic nitrogen in surfacial sediments and its mathematical modelling. In: Colloques Int du CNRS N 293 – Biogéochemie de la matière organique à l'interface eau-sediment marin. Ed CNRS, Paris p 155
Woodwell GM, Whittaker RH, Reiners WA, Likens GE, Delwiche CC, Botkin DB (1978) The biota and the world carbon budget. Science 199:141 – 146
The world water balance and aquatic resources of the earth (1974) Korzoun VI, Sokolov AA, Budyko MI, Voskresensky KP, Kalinin GP (ed board) (in Russian). Gidrometeoizdat, Leningrad
Yamada M (1972) New observations on the lipids of aquatic origin. Fac Fish Hokkaido Univ 19: p 35 – 52
Yamaoka Y (1983) Carbohydrates in humic and fulvic acids from Hiroshima bay sediments. Mar Chem 13:227 – 237
Yerokhin VE (1972) Dissolved carbohydrates of some biotopes from the coastal zone of the sea. Okeanologiya 12:291 – 298

Zadkova II, Maliushko LD, Sarochan VF (1975) Geochemistry of the Sakhalin Bousset Lagoon (in Russian) Nauka, Novosibirsk

Zakharov LA (1974) A contribution to organic matter geochemistry of bottom sediments in the southwestern Atlantic (in Russian). Tr VNIRO 98:145 – 152

Zalmanzon ES (1957) Determination of certain forms of elements and analysis of the colloidal fraction in clays. In: Strakhov NM (ed) Methods of investigation of sediments, vol 2 (in Russian). Gosgeolizdat, Moscow, p 53

Zaslavsky EM (1981) The content and biochemical composition of organic detritus in marine sediments (in Russian). Okeanologiya 21:1039 – 1046

Zeitschel B (1970) The quantity composition and distribution of suspended particulate matter in the Gulf of California. Mar Biol 7:305 – 318

Zenkevich LA (1967) Studies of sea and ocean faunas. In: Byhovsky BE (ed) Advances in biology in the USSR (in Russian). Nauka, Moscow, p 323

Zenkevich LA (1970) General features of oceanic biogeocoenoses and their comparison with the land biogeocoenoses. In: Zenkevich LA (ed) Program and methods of studies of aquatic biogeocoenoses (in Russian). Nauka, Moscow, p 7

Zenkevich LA (1971) On the age of the ocean and of its fauna. In: Zenkevich LA (ed) World ocean history (in Russian). Nauka, Moscow, p 77

Zhivago VN (1980) Collection of aquatic suspensates by centrifuge separators. In: Romankevich EA (ed) Methods for the study of organic matter in the ocean (in Russian). Nauka, Moscow, p 17

Zlobin VS, Perliuk MF, Orlova TA (1975) Amino acid composition of dissolved organic matter in the Atlantic waters (in Russian). Okeanologiya 15:643 – 648

Zobell CE (1945) The role of bacteria in the formation and transformation of petroleum hydrocarbons. Science 102:364 – 369

Zobell CE (1972) Microbial oxygen biogeochemistry (in Russian). Izv Akad Nauk SSSR, Ser Biol 1:23 – 42

Zsolnay A (1971) Preliminary study of the dissolved hydrocarbons and hydrocarbons of particulate material in the Gotland Deep of the Baltic. Kiel Meeresforsch 27:120 – 134

MIX
Papier aus verantwortungsvollen Quellen
Paper from responsible sources
FSC® C105338

If you have any concerns about our products,
you can contact us on
ProductSafety@springernature.com

In case Publisher is established outside the EU,
the EU authorized representative is:
Springer Nature Customer Service Center GmbH
Europaplatz 3, 69115 Heidelberg, Germany

Printed by Libri Plureos GmbH
in Hamburg, Germany